15.09.
SJ
2|04

Mechanical Eng

Mechanical Engineering

BTEC National Option Units

Alan Darbyshire

Newnes

AMSTERDAM BOSTON HEIDELBERG LONDON NEW YORK OXFORD
PARIS SAN DIEGO SAN FRANCISCO SINGAPORE SYDNEY TOKYO

Butterworth-Heinemann
An imprint of Elsevier
Linacre House, Jordan Hill, Oxford OX2 8DP
200 Wheeler Road, Burlington MA 01803

First published 2003

British Library Cataloguing in Publication Data
A catalogue record for this book is available from the British Library

Library of Congress Cataloguing in Publication Data
A catalogue record for this book is available from the Library of Congress

ISBN 0 7506 5761 8

For information on all Butterworth-Heinemann publications
visit our website at www.bh.com

Typeset by Integra Software Services Pvt. Ltd, Pondicherry, India
www.integra-india.com
Printed and bound in Italy

Contents

Introduction

Welcome to the challenges of mechanical engineering! This book has been written to help you get through the core unit Mechanical Principles and a further five optional units which may form part of your BTEC National Certificate or Diploma award in mechanical engineering. It accompanies the book BTEC National Engineering by Mike Tooley and Lloyd Dingle and provides the essential under-pinning knowledge required by a student who wishes to pursue a career in mechanical engineering.

The book has been written by a highly experienced lecturer with over thirty years experience of industry and higher education. It has been designed to cover the requirements of the revised and updated BTEC Engineering programme. The author has adopted a common format and approach throughout the book with numerous student activities, examples, problems and key points.

About the BTEC National Certificate and Diploma

The BTEC National Certificate and National Diploma qualifications have long been accepted by industry as appropriate qualifications for those who are about to enter industry or who are receiving training at the early stages of employment in industry. At the same time, these qualifications have become increasingly acceptable as a means of gaining entry into higher education.

BTEC National programmes in Engineering attract a very large number of registrations per annum such that there is in excess of 35 000 students currently studying for these qualifications in the UK by both part-time and full-time modes of study.

The BTEC National syllabus was recently reviewed and exten-sively updated and new programmes have been launched with effect from September 2002. The new scheme is likely to be adopted by *all* institutions that currently offer the programme as well as a number of others who will now be able to mix and match parts of the BTEC qualification with vocational GCSE and AVCE awards.

Many organizations have contributed to the design of the new BTEC National Engineering programme including the Qualifica-tions and Cirriculum Authority (QCA), the Engineering Council, and several National Training Organizations (NTO).

The Engineering Council continues to view the BTEC National Certificate/Diploma as a key qualification for the sector. They also recognise that BTEC National qualifications are frequently used as a means of entry to higher education courses, such as HNC/HND programmes and Foundation Degree courses.

In revising and updating the Engineering BTEC programme, Edexcel has taken into consideration a number of issues, including:

- Occupational standards and NTO requirements.
- Professional requirements (particularly with regard to the engineering technician).
- Progression into employment.
- Progression to Higher National qualifications (with particularly close match in the core units of Engineering Science and Mathematics).
- A flexible course structure that is commensurate with the broad aims of curriculum 2000.
- Relevant QCA criteria.
- External assessments as required.
- Key skills signposting.
- Several other issues, including the way in which the study of the area can contribute to an understanding of spiritual, moral, ethical, social and cultural issues.

How to use this book

Chapter 1 of the book covers the core unit Mechanical Principles, which is essential for a BTEC National Certificate/Diploma award in Mechanical Engineering. Succeeding chapters cover the optional units Further Mechanical Principles, Mechanical Technology, Engineering Materials, Fluid Mechanics and Thermodynamics. Each chapter contains Text, 'Key points', 'Test your knowledge' questions, Examples, Activities and Problems.

The 'Test your knowledge' questions are interspersed with the text throughout the book. These questions allow you to check your understanding of the preceding text. They also provide you with an opportunity to reflect on what you have learned and consolidate this in manageable chunks.

Most 'Test your knowledge' questions can be answered in only a few minutes and the necessary information, formulae, etc., can be gleaned from the preceding text. Activities, on the other hand, make excellent vehicles for gathering the necessary evidence to demonstrate that you are competent in Key Skills. Consequently, they normally require a significantly greater amount of time to complete. They may also require additional library or resource area research time coupled with access to computing and other information technology resources.

Many tutors will use 'Test your knowledge' questions as a means of reinforcing work done in class whilst Activities are more likely to be 'set work' for students to do outside the classroom. Whether or not this approach is taken, it is important to be aware that this student-centred work is designed to complement a programme of lectures and tutorials based on the BTEC syllabus. Independent learners (i.e. those not taking a formal course) will find complete syllabus coverage in the text.

The units Mechanical Principles, Further Mechanical Principles, Fluid Mechanics and Thermodynamics involve a considerable amount of mathematical calculation. The worked examples will show you this should be done and in order to successfully tackle the work you will need to have a good scientific calculator (and know how to use it). The units Mechanical Technology and Engineering Materials are essentially descriptive and investigative. Access to a computer with word processing and drawing software will be of advantage when carrying out activities and other set-work in these units.

Finally, here are some general points to help you with your studies:

- Allow regular time for reading – get into the habit of setting aside an hour, or two, at the weekend. Use this time to take a second look at the topics that you have covered during the week or that you may not have completely understood.
- Make notes and file these away for future reference – lists of facts, definitions and formulae are particularly useful for revision.
- Look out for inter-relationships between subjects and units – you will find many ideas and a number of themes that crop up in different places and in different units. These can often help to reinforce your understanding.
- Don't expect to find all the subjects and topics within a course equally interesting. There may be parts that, for a whole variety of reasons, don't immediately fire your enthusiasm. There is nothing unusual about this; however do remember that something that may not appear particularly useful now may become crucial at some point in the future.
- However difficult things may get, don't be tempted to give up and don't be afraid to ask for assistance! Engineering is not, in itself, a difficult subject, rather it is a subject that *demands* logical thinking and an approach in which each new concept builds upon those that have gone before.
- Finally, don't be afraid to put your ideas into practice. Engineering is about *doing* – get out there and *do* it!

Good luck with your BTEC Engineering studies!

Chapter 1 Mechanical principles

Summary

The design, manufacture and servicing of engineered products are important to the nation's economy and well-being. One has only to think of the information technology hardware, aircraft, motor vehicles and domestic appliances we use in everyday life to realise how reliant we have become on engineered products. A product must be fit for its purpose. It must do the job for which it was designed for a reasonable length of time and with a minimum of maintenance. The term 'mechatronics' is often used to describe products which contain mechanical, electrical, electronic and IT systems. It is the aim of this unit to broaden your knowledge of the underpinning mechanical principles which are fundamental to engineering design, manufacturing and servicing.

Engineering structures

Loading systems

Forces whose lines of action lie in a single plane are called *coplanar forces*. If the lines of action pass through a single point, the forces are said to be *concurrent* and the point through which they pass is called the *point of concurrence* (Figure 1.1).

A concurrent system of coplanar forces can be reduced to a single force acting at the point of concurrence. This is called the *resultant force*. If a body is subjected to a system of concurrent coplanar forces and is not constrained, it will move in the direction of the resultant force. To prevent this from happening, a force must be applied which is equal and opposite to the resultant. This

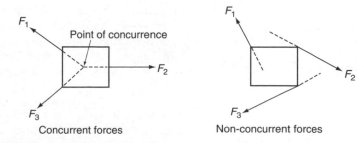

Figure 1.1 *Coplanar force systems*

balancing force, which will hold the body in a state of static equilibrium, is called the *equilibrant* of the system.

When a body is subjected to a system of non-concurrent coplanar forces, there is a tendency for the forces not only to make it move in a particular direction, but also to make it rotate. Such a non-concurrent system can be reduced to a single *resultant force* and a *resultant couple*. If the body is to be held at rest, an equilibrant must again be applied which is equal and opposite to the resultant force. This alone however will not be sufficient. A balancing couple or turning moment, must also be applied which is equal and opposite to the resultant couple.

If you have completed the core unit Science for Technicians, you will know how to find the resultant and equilibrant of a coplanar force system graphically by means of a force vector diagram. We will use this method again shortly but you now need to know how to do the same using mathematics.

Sign convention

When you are using mathematics to solve coplanar force system problems, you need to adopt a method of describing the action of the forces and couples. The following sign convention is that which is most often used (Figure 1.2):

(i) Upward forces are positive and downward forces are negative.
(ii) Horizontal forces acting to the right are positive and those acting to the left are negative.
(iii) Clockwise acting moments and couples are positive and anti-clockwise acting ones are negative.

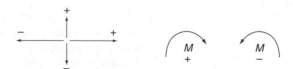

Figure 1.2 *Sign convention*

Resolution of forces

Forces which act at an angle exert a pull which is part horizontal and part vertical. They can be split into their horizontal and vertical parts or *components*, by the use of trigonometry. When you are doing this, it is a useful rule to always measure angles to the horizontal.

In Figure 1.3(a), the horizontal and vertical components are both acting in the positive directions and will be

$$F_H = +F\cos\theta \quad \text{and} \quad F_V = +F\sin\theta$$

In Figure 1.3(b), the horizontal and vertical components are both acting in the negative directions and will be

$$F_H = -F\cos\theta \quad \text{and} \quad F_V = -F\sin\theta$$

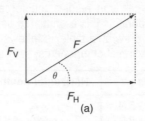 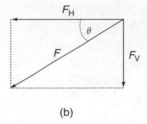

Figure 1.3 *Resolution of forces*

Forces which act upward to the left or downward to the right will have one component which is positive and one which is negative. Having resolved all of the forces in a coplanar system into their horizontal and vertical components, each set can then be added algebraically to determine the resultant horizontal pull, ΣF_H, and the resultant vertical pull, ΣF_V. The Greek letter Σ (sigma) means 'the sum or total' of the components. Pythagoras' theorem can then be used to find the single resultant force R, of the system.

That is,

$$R^2 = (\Sigma F_H)^2 + (\Sigma F_V)^2$$

$$R = \sqrt{(\Sigma F_H)^2 + (\Sigma F_V)^2} \tag{1.1}$$

The angle θ, which the resultant makes with the horizontal can also be found using

$$\tan \theta = \frac{\Sigma F_V}{\Sigma F_H} \tag{1.2}$$

With non-concurrent force systems, the algebraic sum of the moments of the vertical and horizontal components of the forces, taken about some convenient point, gives the resultant couple or turning moment. Its sign, positive or negative, indicates whether its direction is clockwise or anticlockwise. This in turn can be used to find the perpendicular distance of the line of action of the resultant from the chosen point.

Example 1.1

Find the magnitude and direction of the resultant and equilibrant of the concurrent coplanar force system shown in Figure 1.4.

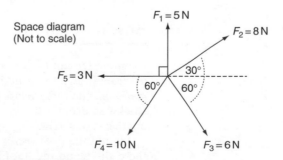

Figure 1.4

When you resolve the forces into their horizontal and vertical components it is essential to use the sign convention. A logical way is to draw up a table as follows with the forces, and their horizontal and vertical components, set out in rows and columns.

Force	Horizontal component	Vertical component
$F_1 = \ \ 5\,\text{N}$	0	$+5.0\,\text{N}$
$F_2 = \ \ 8\,\text{N}$	$+8\cos 30 = +6.93\,\text{N}$	$+8\sin 30 = +4.0\,\text{N}$
$F_3 = \ \ 6\,\text{N}$	$+6\cos 60 = +3.0\,\text{N}$	$-6\sin 60 = -5.2\,\text{N}$
$F_4 = 10\,\text{N}$	$-10\cos 60 = -5.0\,\text{N}$	$-10\sin 60 = -8.66\,\text{N}$
$F_5 = \ \ 3\,\text{N}$	$-\,3.0\,\text{N}$	0
Totals	$\Sigma F_H = +\mathbf{1.93\,N}$	$\Sigma F_V = -\mathbf{4.86\,N}$

The five forces have now been reduced to two forces, ΣF_H and ΣF_V. They can now be drawn as vectors (Figure 1.5).

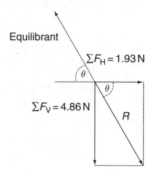

Figure 1.5

The resultant R, is found by Pythagoras as follows:

$$R = \sqrt{1.93^2 + 4.86^2}$$

$$R = \mathbf{5.23\,N}$$

The angle θ is found from

$$\tan\theta = \frac{\Sigma F_V}{\Sigma F_H} = \frac{4.86}{1.93} = 2.52$$

$$\theta = \tan^{-1} 2.52$$

$$\theta = \mathbf{68.3°}$$

The equilibrant, which is required to hold the system in a state of static equilibrium, is equal to the resultant but opposite in sense.

Bow's notation

Example 1.1 can be solved graphically by means of a force vector diagram drawn to a suitable scale. The process is known as *vector addition*. The force vectors are taken in order, preferably working clockwise around the system, and added nose to tail to produce *a polygon of forces*.

Should the final vector be found to end at the start of the first, there will be no resultant and the system will be in equilibrium. If however there is a gap between the two, this when measured from the start of the first vector to the end of the last, represents the magnitude, direction and sense of the resultant. The equilibrant will, of course, be equal and opposite. It must be remembered that when you solve problems graphically, the accuracy of the answers will depend on the accuracy of your measurement and drawing.

Bow's notation is a useful method of identifying the forces on a vector diagram and also the sense in which they act. In the space diagram, which shows the forces acting at the point of concurrence, the spaces between the forces are each given a capital letter. Wherever possible, the letters should follow a clockwise sequence around the diagram. In the solution shown in Figures 1.6 and 1.7, the force F_1 is between the spaces A and B, and when drawn on the vector diagram it is identified by the lower case letters as force ab.

The clockwise sequence of letters on the space diagram, i.e. A to B, gives the direction of the force on the vector diagram. The letter a is at the start of the vector and the letter b is at its end. Although arrows have been drawn to show the directions of the vectors, they are not really necessary and will be omitted in future graphical solutions.

> **Key point**
>
> Use capital letters and work in a clockwise direction when lettering space diagram using Bow's notation.

Example 1.1a Alternative graphical solution for Example 1.1

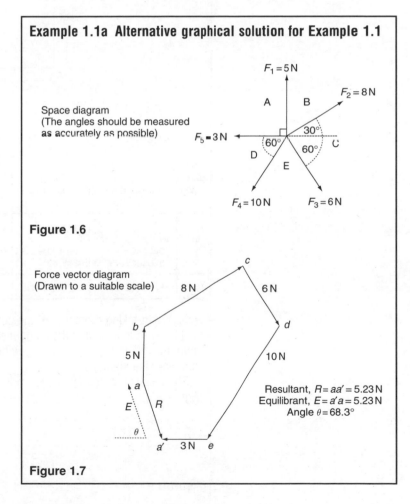

Figure 1.6

Figure 1.7

You can use the same graphical method to find the magnitude and direction of the resultant of a non-concurrent system of coplanar forces. Generally, however, it is best to use the analytical method as in Example 1.2. In problems, you are also usually asked to find the resultant couple or turning moment and this has to be done by calculation.

When drawing up a table for each force and its components, two extra columns are required. These are for the moments of the components, taken about some convenient point, and their total gives the resultant turning moment. With a little intuition, you can then determine the position of the line of action of the resultant.

Example 1.2

Determine the magnitude and direction of the resultant of the coplanar forces acting on the component shown in Figure 1.8. Determine also the perpendicular distance of its line of action from the corner A.

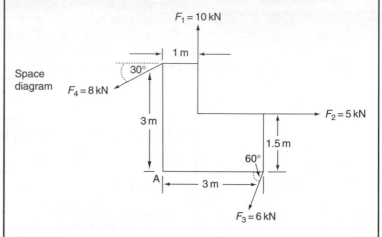

Figure 1.8

Force F (kN)	Horizontal component F_H (kN)	Vertical component F_V (kN)	Moment of F_H about A (kN m)	Moment of F_V about A (kN m)
10	0	+10	0	$-(1 \times 10) = -10$
5	+5	0	$+(5 \times 1.5) = +7.5$	0
6	$-6\cos 60 = -3$	$-6\sin 60 = -5.2$	0	$+(5.2 \times 3) = +15.6$
8	$-8\cos 30 = -6.93$	$-8\sin 30 = -4.0$	$-(6.93 \times 3) = -20.8$	0
Totals	$\Sigma F_H = -4.93$ kN	$\Sigma F_V = +0.8$ kN	$\Sigma M = -7.7$ kN m	

Note: When taking moments of the components about A, disregard the plus or minus sign in the components' columns. The sign of the moments is determined only by whether they are clockwise or anticlockwise about A.

The resultant of the force system is again found using Pythagoras (Figure 1.9).

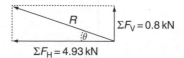

Figure 1.9

$$R = \sqrt{\Sigma F_H^2 + \Sigma F_V^2}$$
$$R = \sqrt{4.93^2 + 0.8^2}$$
$$R = \sqrt{4.99}\ \textbf{kN}$$

The angle θ is a again given by

$$\tan\theta = \frac{\Sigma F_V}{\Sigma F_H} = \frac{0.8}{4.93} = 0.162$$

$$\theta = \tan^{-1} 0.162$$

$$\theta = 9.22°$$

Let the perpendicular distance of the line of action of the resultant from the corner A be *a*. Because the resultant turning moment is negative, i.e. anticlockwise about A, the line of action of the resultant must be above A (Figure 1.10).

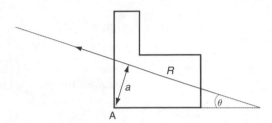

Figure 1.10

Now it must be that,

Resultant moment about A = moment of resultant about A

$$\Sigma M = R \times a$$

$$a = \frac{\Sigma M}{R} = \frac{7.7}{4.99}$$

$$a = 1.54 \, m$$

Test your knowledge 1.1

1. What are coplanar forces?
2. What are concurrent forces?
3. What are the conditions necessary for a body to be in static equilibrium under the action of a coplanar force system?
4. What are the resultant and equilibrant of a coplanar force system?
5. What can a non-concurrent coplanar force system be reduced to?

Activity 1.1

A connecting plate, which links the members in an engineering structure, is acted upon by four forces as shown in Figure 1.11. Calculate the magnitude, direction and sense of the resultant force, and the perpendicular distance of its line of action from the point A at which the 1 kN force acts.

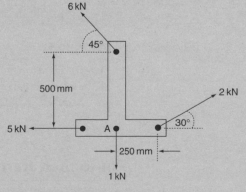

Figure 1.11

Problems 1.1

1. Determine the magnitude and direction of the resultant force for the coplanar force system shown in Figure 1.12.

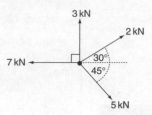

Figure 1.12

[1.79 kN, 15° to horizontal, upward to the left.]

2. Determine the magnitude and direction of the equilibrant required for the coplanar force system shown in Figure 1.13.

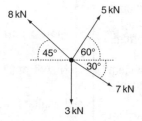

Figure 1.13

[4.54 kN, 50.2° to horizontal, downward to the left.]

3. Determine the magnitude and direction of the resultant force and turning moment for the force system shown in Figure 1.14.

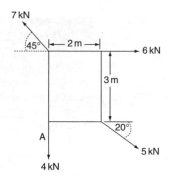

Figure 1.14

Determine also the perpendicular distance of the line of action of the resultant from the corner, A.

[5.8 kN, 7.53° to horizontal, downward to right, 6.57 kN m
clockwise about A, 1.13 m.]

Pin-jointed framed structures

Examples of framed structures which you see in everyday life are bicycles, roof trusses, electricity pylons and tower cranes. They are

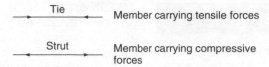

Figure 1.15 *Representation of structural members*

made up of members which are joined at their ends. Some of these are three-dimensional structures whose analysis is complex and it is only the two-dimensional, or coplanar structures, which we shall consider. There are three kinds of member in these structures (Figure 1.15).

1. *Ties*, which are in tension are shown diagrammatically with arrows pointing inwards. You have to imagine yourself in the place of a tie. You would be pulling inwards to stop yourself from being stretched. The arrows describe the force which the tie exerts on its neighbours to keep the structure in position.
2. *Struts*, which are in compression are shown diagrammatically with arrows pointing outwards. Once again, you have to imagine the way you would be pushing if you were in the place of a strut. You would be pushing outwards to keep the structure in position and to stop yourself from being squashed.
3. The third type is *redundant members*. A perfect framed structure is one which has just sufficient members to prevent it from becoming unstable. Any additional members, which may have been added to create a stiffer or stronger frame, are known as redundant members. Redundant members may be struts or ties or they may carry no load in normal circumstances. We shall avoid framed structures with redundant members as very often they cannot be solved by the ordinary methods of statics.

In reality the members are bolted, riveted or welded together at their ends but in our analysis we assume that they are pin-jointed or hinged at their ends, with frictionless pins. We further assume that because of this, the only forces present in the members are tensile and compressive forces. These are called *primary forces*. In practice there might also be bending and twisting forces present but we will leave these for study at a higher level.

When a structure is in a state of static equilibrium, the external active loads which it carries will be balanced by the reactions of its supports. The conditions for external equilibrium are:

1. The vector sum of the horizontal forces or horizontal components of the forces is zero.
2. The vector sum of the vertical forces or vertical components of the forces is zero.
3. The vector sum of the turning moments of the forces taken about any point in the plane of the structure is zero. That is,

$$\Sigma F_H = 0, \qquad \Sigma F_v = 0, \qquad \Sigma M = 0$$

We can also safely assume that if a structure is in a state of static equilibrium, each of its members will also be in equilibrium. It follows that the above three conditions can also be applied to individual members, groups of members and indeed to any internal part or section of a structure.

Key point

When a body or structure is in static equilibrium under the action of three concurrent external forces, the forces must be concurrent.

A full analysis of the external and internal forces acting on and within a structure can be carried out using mathematics or graphically by drawing a force vector diagram. We will use the mathematical method first and begin by applying the conditions for equilibrium to the external forces. This will enable us to find the magnitude and direction of the support reactions.

Next, we will apply the conditions for equilibrium to each joint in the structure, starting with one which has only two unknown forces acting on it. This will enable us to find the magnitude and direction of the force in each member.

Example 1.3

Determine the magnitude and direction of the support reactions at X and Y for the pin-jointed cantilever shown in Figure 1.16, together with the magnitude and nature of the force acting in each member.

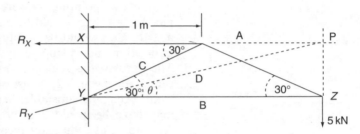

Figure 1.16

The wall support raction R_x must be horizontal because it is equal and opposite to the force carried by the top member CA. Also, the three external forces must be concurrent with lines of action meeting at the point P.

Begin by locating the point P and drawing in the support reactions at X and Y in the sense which you think they are acting. If you guess wrongly, you will obtain a negative answer from your calculations, telling you that your arrow should be pointing in the opposite direction. Letter the diagram using Bow's notation, with capital letters in the spaces between the external forces and inside the structure. Now you can begin the calculations.

Finding distances XY, YZ and the angle θ:

$$XY = 1 \times \tan 30 = 0.577\,\text{m}, \quad YZ = 2.0\,\text{m}$$

$$\tan \theta = \frac{XY}{YZ} = \frac{0.577}{2.0} = 0.289$$

$$\theta = \mathbf{16.1}°$$

Take moments about Y to find R_X. For equilibrium, $\Sigma M_Y = 0$,

$$(5 \times 2) - (R_X \times 0.577) = 0$$

$$10 - 0.577 R_X = 0$$

$$R_X = \frac{\mathbf{10}}{\mathbf{0.577}} = \mathbf{17.3\,kN}$$

The force in member AC will also be 17.3 kN because it is equal and opposite to R_x. It will be a tensile force, and this member will be a tie.

Take vertical components of external forces to find R_Y. For equilibrium, $\Sigma F_v = 0$,

$$R_Y \sin 16.1° - 5 = 0$$
$$0.277\, R_Y - 5 = 0$$
$$\mathbf{R_Y = \frac{5}{0.277} = 18.0\, kN}$$

You can now turn your attention to the forces in the individual members. You already know the force F_{AC} acting in member AC because it is equal and opposite to the support reaction R_X. Choose a joint where you know the magnitude and direction of one of the forces and the directions of the other two, i.e. where there are only two unknown forces. Joint ABD will be ideal. It is good practice to assume that the unknown forces are tensile, with arrows pulling away from the joint. A negative answer will tell you that the force is comperessive (Figure 1.17).

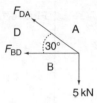

Figure 1.17

Take vertical components of the forces to find the force F_{DA} in member DA. For eqilibrium, $\Sigma F_v = 0$,

$$F_{DA} \sin 30 - 5 = 0$$
$$0.5\, F_{DA} - 5 = 0$$
$$\mathbf{F_{DA} = \frac{5}{0.5} = 10\, kN}$$

The positive answer denotes that the force F_{DA} in member DA is tensile and that the member is a tie. Now take horizontal components of the forces to find F_{BD} in member BD. For equilibrium, $\Sigma F_H = 0$,

$$-F_{DA} \cos 30 - F_{BD} = 0,$$
$$-10 \cos 30 - F_{BD} = 0$$
$$\mathbf{F_{BD} = -10 \cos 30 = -8.66\, kN}$$

The negative sign denotes that the force F_{BD} in member BD is compressive and that the member is a strut. Now go to joint ADC and take vertical components of the forces to find F_{DC}, the force in member DC (Figure 1.18). For equilibrium, $\Sigma F_v = 0$,

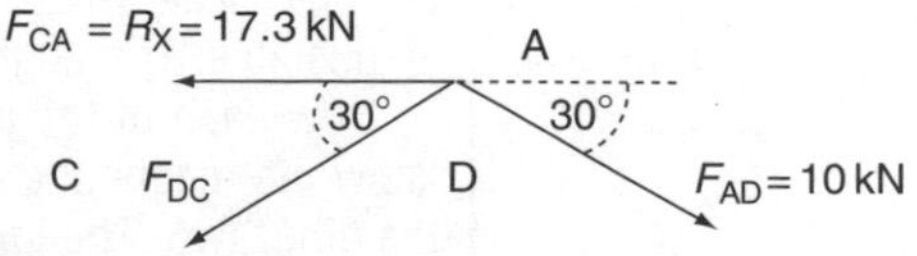

$F_{CA} = R_X = 17.3\ kN$ A

C F_{DC} 30°   D   30° $F_{AD} = 10\, kN$

Figure 1.18

$$-(10 \sin 30) - (F_{DC} \sin 30) = 0$$

$$-10 - F_{DC} = 0$$

$$\mathbf{F_{DC} = -10 \, kN}$$

The negative sign denotes that the force F_{DC} in member DC is compressive and that the member is a strut. You have now found the values of all the external and internal forces which can be tabulated and their directions indicated on the structure diagram as shown in Figure 1.19.

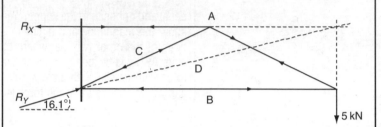

Figure 1.19

Reaction/Member	Force	Nature
R_x	17.3 kN	
R_y	18.0 kN	
AC	17.3 kN	Tie
DA	10.0 kN	Tie
BD	8.66 kN	Strut
DC	10.0 kN	Strut

You may also solve framed structure problems graphically. Sometimes this is quicker but the results may not be so accurate. As before, it depends on the accuracy of your drawing and measurement. An alternative solution to the above framed structure is given in Example 1.3a. Once again, use is made of Bow's notation to identify the members.

Example 1.3a Alternative graphical solution for Example 1.3

Begin by drawing the space diagram shown in Figure 1.20 to some suitable scale. Take care to measure the angles as accurately as possible. The angle of R_x can be measured, giving $\theta = 16.1°$.

Now each of the joints is in equilibrium. Choose one where you know the magnitude and direction of one force and the directions of the other two. The joint ABD is in fact the only one where you know these conditions.

Space diagram

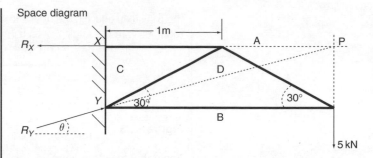

Figure 1.20

Draw the force vector diagram for this joint to a suitable scale, beginning with the 5 kN load. It will be a triangle of forces, which you can quite easily construct because you know the length of one side and the directions of the other two. The force vectors are added together nose to tail but don't put any arrows on the diagram. Bow's notation is sufficient to indicate the directions of the forces (Figure 1.21).

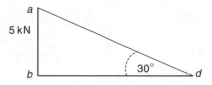

Figure 1.21

Accurate measurement gives
Force in DA = da = 10.0 kN
Force in BD = bd = 8.66 kN

Remember, the clockwise sequence of letters around the joint on the space diagram ABD gives the directions in which the forces act on the joint in the vector diagram, i.e. the force ab acts from a to b, the force bd acts from b to d and the force da acts from d to a.

Now go to another joint where you know the magnitude and direction of one of the forces, and the directions of the other two. Such a joint is ADC where the same 10 kN force da, which acts on joint ADB, also acts in the opposite direction on joint ADC. We now call it force ad, and the triangle of forces is as shown in Figure 1.22.

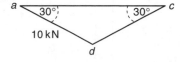

Figure 1.22

Accurate measurement gives
Force in DC = dc = 10.0 kN
Force in CA = ca = 17.3 kN

All of the internal forces have now been found and also the reaction R_X, which is equal and opposite to ca, and can thus be written as ac. A final triangle of forces ABC, can now be drawn, representing the three external forces (see Figure 1.23). Two of these, the 5 kN load and R_X, are now known in both magnitude and direction together with the angle θ which R_Y makes with the horizontal.

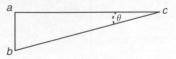

Figure 1.23

Accurate measurement gives
Reaction $R_Y = bc = 18.0$ kN
Angle $\theta = 16.10°$

You will note that the above three vector diagram triangles have sides in common and to save time it is usual to draw the second and third diagrams as additions to the first. The combined vector diagram appears as in Figure 1.24.

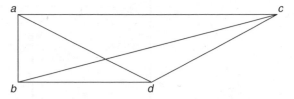

Figure 1.24

The directions of the forces, acting towards or away from the joints on which they act, can be drawn on the space diagram (Figure 1.25). This will immediately tell you which members are struts, and which are ties. The measured values of the support reactions and the forces in the members can then be tabulated.

Reaction/Member	Force	Nature
R_X	17.3 kN	
R_Y	18.0 kN	
AC	17.3 kN	Tie
DA	10.0 kN	Tie
BD	8.66 kN	Strut
DC	10.0 kN	Strut

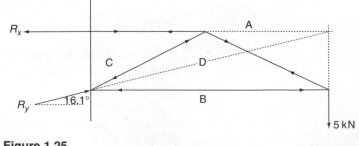

Figure 1.25

Activity 1.2

The jib-crane shown in Figure 1.26 carries a load of 10 kN. Making use of Bow's notation find the reactions of the supports at *X* and *Y*, and the magnitude and nature of the force in each member graphically by means of a force vector diagram. Apply the conditions for static equilibrium to the structure and check your results by calculation.

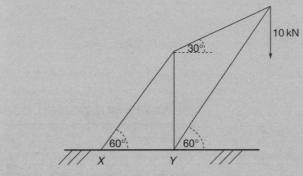

Figure 1.26

Problems 1.2

For each of the following framed structures, determine the support reactions, and the magnitude and nature of the force in each member (Figures 1.27, 1.28, and 1.29).

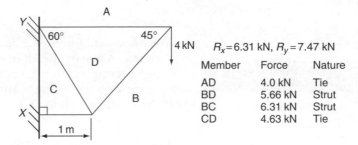

$R_x = 6.31$ kN, $R_y = 7.47$ kN

Member	Force	Nature
AD	4.0 kN	Tie
BD	5.66 kN	Strut
BC	6.31 kN	Strut
CD	4.63 kN	Tie

Figure 1.27

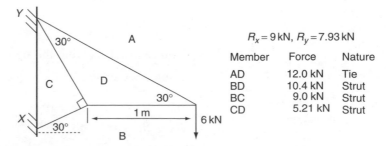

$R_x = 9$ kN, $R_y = 7.93$ kN

Member	Force	Nature
AD	12.0 kN	Tie
BD	10.4 kN	Strut
BC	9.0 kN	Strut
CD	5.21 kN	Strut

Figure 1.28

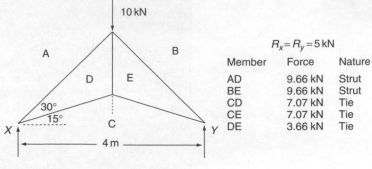

Member	Force	Nature
AD	9.66 kN	Strut
BE	9.66 kN	Strut
CD	7.07 kN	Tie
CE	7.07 kN	Tie
DE	3.66 kN	Tie

Figure 1.29

Simply supported beams

A simply supported beam is supported at two points in such a way that it is allowed to expand and bend freely. In practice the supports are often rollers. The loads on the beam may be concentrated at different points or uniformly distributed along the beam. Figure 1.30 shows concentrated loads only.

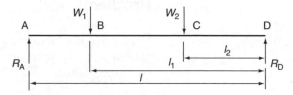

Figure 1.30 *Simply supported beam with concentrated loads*

The downward forces on a beam are said to be active loads, due to the force of gravity, whilst the loads carried by the supports are said to be reactive. When investigating the effects of loading, we often have to begin by calculating the supporting reactions. The beam is in static equilibrium under the action of these external forces, and so we proceed as follows:

1. Equate the sum of the turning moments, taken about the right hand support D, to zero. That is,

$$\Sigma M_D = 0$$
$$R_A l - W_1 l_1 - W_2 l_2 = 0$$

You can find R_A from this condition.

2. Equate vector sum of the vertical forces to zero. That is,

$$\Sigma F_V = 0$$
$$R_A + R_B - W_1 - W_2 = 0$$

You can find R_D from this condition.

Example 1.4

Calculate the support reactions of the simply supported beam shown in Figure 1.31.

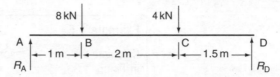

Figure 1.31

Take moments about the point D, remembering to use the sign convention that clockwise moments are positive and anticlockwise moments are negative. For equilibrium, $\Sigma M_D = 0$,

$$(R_A \times 4.5) - (8 \times 3.5) - (4 \times 1.5) = 0$$
$$4.5R_A - 28 - 6 = 0$$
$$R_A = \frac{28 + 6}{4.5} = 7.56 \text{ kN}$$

Equate the vector sum of the vertical forces to zero, remembering the sign convention that upward forces are positive and downward forces are negative. For equilibrium, $\Sigma F_V = 0$.

$$8 + 4 - 7.56 - R_D = 0$$
$$R_D = 8 + 4 - 7.56 = 4.44 \text{ kN}$$

Uniformly distributed loads (UDLs)

Uniformly distributed loads, or UDLs, are evenly spread out along a beam. They might be due to the beam's own weight, paving slabs or an asphalt surface. UDLs are generally expressed in kN per metre length, i.e. kN m^{-1}. This is also known as the 'loading rate'. Uniformly distributed loads are shown diagrammatically as in Figure 1.32.

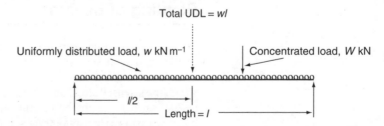

Figure 1.32 *Simply supported beam with concentrated and distributed loads*

The total UDL over a particular length l, of a beam is given by the product of the loading rate and the length. That is,

total UDL $= wl$

When you are equating moments to find the beam reactions, the total UDL is assumed to act at its centroid, i.e. at the centre of the length l. You can then treat it as just another concentrated load and calculate the support reactions in the same way as before.

Example 1.5

Calculate the support reactions of the simply supported beam shown in Figure 1.33.

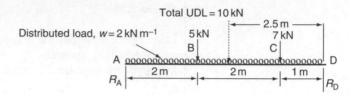

Figure 1.33

Begin by calculating the total UDL. Then, replace it by an equal concentrated load acting at its centroid. This is shown with a dotted line above.

Total UDL $= wl = 5 \times 2 = 10$ kN

You can now apply the conditions for static equilibrium. Begin by talking moments about the point D. For equilibrium, $\Sigma M_D = 0$,

$$(R_A \times 5) - (5 \times 3) - (10 \times 2.5) - (7 \times 1) = 0$$
$$5R_A - 15 - 25 - 7 = 0$$
$$R_A = \frac{15 + 25 + 7}{5} = 9.4 \text{ kN}$$

Now equate the vector sum of the vertical forces to zero. For equilibrium, $\Sigma F_V = 0$,

$$9.4 + R_D - 5 - 10 - 7 = 0$$
$$R_D = 5 + 10 + 7 - 9.4 = 12.6 \text{ kN}$$

Bending of beams

When structural engineers are designing beams to carry given loads, they have to make sure that the maximum allowable stresses will not be exceeded. On any transverse section of a loaded beam or cantilever, shear stress, tensile stress and compressive stress are all usually present.

As you can see in the cantilever in Figure 1.34(a), the load F, has a shearing effect and thus sets up shear stress at section Y–Y. The load also has a bending effect and at any section Y–Y, this produces tensile stress in the upper layers of the beam and compressive stress in the lower layers.

Generally, the stresses and deflection caused by bending greatly exceed those caused by shear, and the shearing effects of the loading are usually neglected for all but short and stubby cantilevers. Nevertheless, the distribution of shear loading is of importance

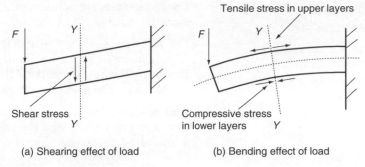

(a) Shearing effect of load   (b) Bending effect of load

Figure 1.34 *Stress in beams*

because it enables the likely positions of maximum bending stress to be pin-pointed, as will be shown.

Somewhere inside a beam there is a layer called the *neutral layer or neutral axis*. Although this becomes bent, like all the other layers, it is in neither tension nor compression and there is no tensile or compressive stress present. For elastic materials which have the same modulus of elasticity in tension and compression, it is found that the neutral axis is located at the centroid of the cross-section.

Shear force distribution

At any transverse section of a loaded horizontal beam the *shear force* is defined as being *the algebraic sum of the forces acting to one side of the section*. You might think of it as the upward or downward breaking force at that section. Usually, it is forces to the left of the section which are considered and the following sign convention is used.

1. Upward forces to the left of a section are positive and downward forces are negative.
2. Downward forces to the right of a section are positive and upward forces are negative.

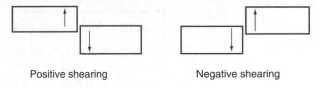

Positive shearing   Negative shearing

Figure 1.35 *Positive and negative shearing*

This gives rise to the idea of positive and negative shearing as shown in Figure 1.35, and the variation of the shear force along a loaded beam can be plotted on a shear force diagram.

Bending moment distribution

At any transverse section of a loaded beam, the *bending moment* is defined as being *the algebraic sum of the moments of the forces acting to one side of the section*. Usually it is moments to the left of a section which are considered and the following sign convention is used.

1. Clockwise moments to the left of a section are positive and anticlockwise moments are negative.
2. Anticlockwise moments to the right of a section are positive and clockwise moments are negative.

This gives rise to the idea of positive and negative bending as shown in Figure 1.36, and the variation of bending moment along a loaded beam can be plotted on a bending moment diagram.

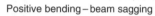

Positive bending – beam sagging   Negative bending – beam hogging

Figure 1.36 *Positive and negative bending*

When you are plotting shear force and bending moment diagrams you will see that

1. The maximum bending moment always occurs where the shear force diagram changes sign.
2. The area of the shear force diagram up to a particular section gives the bending moment at that section.
3. Under certain circumstances, the bending moment changes sign and this is said to occur at a point of contraflexure.

At a point of contraflexure, where the bending moment is zero, the deflected shape of a beam changes from sagging to hogging or vice versa. The location of these points is of importance to structural engineers since it is here that welded, bolted or riveted joints can be made which will be free of bending stress.

Example 1.6

Plot the shear force and bending moment distribution diagrams for the simple cantilever beam shown in Figure 1.37 and state the magnitude, nature and position of the maximum values of shear force, and bending moment.

shear force from A to B $= -5\,\text{kN}$

shear force from B to C $= -5 - 3 = -8\,\text{kN}$

   maximum shear force $= -8\,\text{kN}$ between B and C

bending moment at A $= 0$

bending moment at B $= -(5 \times 2) = -10\,\text{kN m}$

bending moment at C $= -(5 \times 3) - (3 \times 1) = -18\,\text{kN m}$

maximum bending moment $= -18\,\text{kN m}$ at C

As you can see, the shear force is negative over the whole length of the cantilever because there is a downward breaking force to the left of any section, i.e. negative shear.

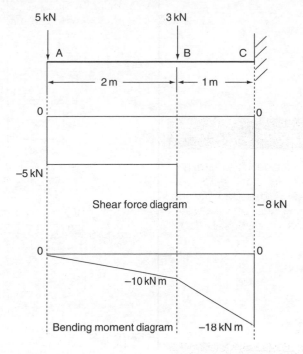

Figure 1.37

The bending moment is always zero at the free end of a simple cantilever and negative over the remainder of its length. You might think of a cantilever as being half of a hogging beam in which the bending moment at any section is anticlockwise, i.e. negative bending.

Example 1.7

Plot the shear force and bending moment distribution diagrams for the simply supported beam shown in Figure 1.38. State the magnitude, nature and position of the maximum values of shear force, and bending moment and the position of a point of contraflexure.

Begin by finding the support reactions:

For equilibrium, $\Sigma M_D = 0$,
$$(R_B \times 2) - (6 \times 3) - (10 \times 1) = 0$$
$$2R_B - 18 - 10 = 0$$
$$R_B = \frac{18 + 10}{2} = 14 \text{ kN}$$

Also, for equilibrium, $\Sigma F_V = 0$,
$$14 + R_D + 6 + 10 = 0$$
$$R_D = 6 + 10 - 14 = 2 \text{ kN}$$

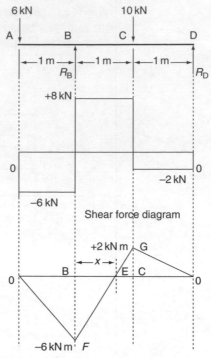

Figure 1.38

Next find the shear force values:

SF from A to B $= -6\,\text{kN}$
SF from B to C $= -6 + 14 = +8\,\text{kN}$
SF from C to D $= -6 + 14 - 10 = -2\,\text{kN}$

maximum shear force $= +8\,\text{kN}$ between B and C

Now find the bending moment values:

BM at A $= 0$
BM at B $= -(6 \times 1) = -6\,\text{kN}\,\text{m}$
BM at C $= -(6 \times 2) + (14 \times 1) = +2\,\text{kN}\,\text{m}$
BM at D $= 0$

maximum bending moment $= -6\,\text{kN}\,\text{m}$ at B

There is a point of contraflexure at E, where the bending moment is zero. To find its distance x, from B, consider the similar triangles BEF and ECG (Figure 1.39).

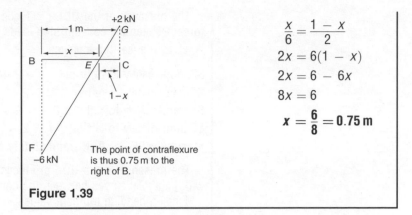

$$\frac{x}{6} = \frac{1-x}{2}$$
$$2x = 6(1-x)$$
$$2x = 6 - 6x$$
$$8x = 6$$
$$x = \frac{6}{8} = 0.75\,\text{m}$$

The point of contraflexure is thus 0.75 m to the right of B.

Figure 1.39

If you examine the shear force and bending moment diagrams in the above examples, you will find that there is a relationship between shear force and bending moment. Calculate the area under the shear force diagram from the left hand end, up to any point along the beam. You will find that this is equal to the bending moment at that point. Try it, but remember to use the sign convention that areas above the zero line are positive and those below are negative.

Example 1.8

Plot the shear force and bending moment distribution diagrams for the cantilever shown in Figure 1.40 and state the magnitude and position of the maximum values of shear force and bending moment.

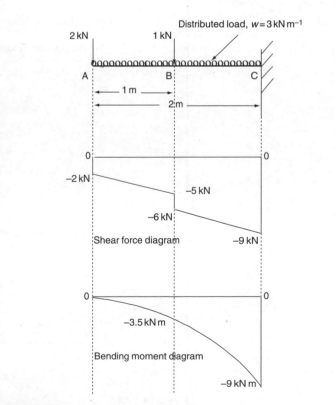

Figure 1.40

The presence of the UDL produces a gradually increasing shear force between the concentrated loads.

Finding shear force values:

SF immediately to right of A $= -2\,\text{kN}$

SF immediately to left of B $= -2 - (3 \times 1) = -5\,\text{kN}$

SF immediately to right of B $= -2 - 1 - (3 \times 1) = -6\,\text{kN}$

SF immediately to left of C $= -2 - 1 - (3 \times 2) = -9\,\text{kN}$

maximum SF $= -9\,\text{kN}$ immediately to left of C

The presence of the UDL produces a bending moment diagram with parabolic curves between the concentrated load positions.

Finding bending moment values:

BM at A $= 0$

BM at B $= -(2 \times 1) - (3 \times 1 \times 0.5) = -3.5\,\text{kN m}$

BM at C $= -(2 \times 1) - (1 \times 1) - (3 \times 2 \times 1) = -9\,\text{kN m}$

maximum bending moment $= -9\,\text{kN m}$ at C

Example 1.9

Plot the shear force and bending moment distribution diagrams for the simply supported beam shown in Figure 1.41. State the magnitude, nature and position of the maximum values of shear force and bending moment.

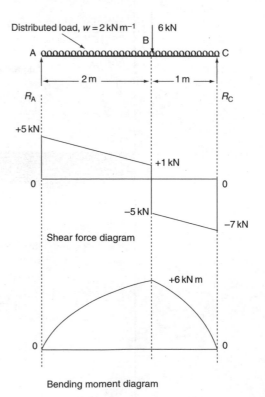

Figure 1.41

Finding support reactions:

For equilibrium, $\Sigma M_C = 0$,

$(R_A \times 3) - (6 \times 1) - (2 \times 3 \times 1.5) = 0$

$3R_A - 6 - 9 = 0$

$R_A = \dfrac{6+9}{3} = 5\,\text{kN}$

Also for equilibrium, $\Sigma F_V = 0$,

$5 + R_C - 6 - (2 \times 3) = 0$

$5 + R_C - 6 - 6 = 0$

$R_C = 6 + 6 - 5 = 7\,\text{kN}$

Finding the shear force values:

SF immediately to right of A $= +5\,\text{kN}$
SF immediately to left of B $= 5 - (2 \times 2) = +1\,\text{kN}$
SF immediately to right of B $= 5 - (2 \times 2) - 6 = -5\,\text{kN}$
SF immediately to left of C $= 5 - (2 \times 3) - 6 = -7\,\text{kN}$

maximum SF $= -7\,$kN immediately to left of C

Finding bending moment values:

BM at A $= 0$
BM at B $= (5 \times 2) - (2 \times 2 \times 1) = +6\,\text{kN m}$
BM at C $= 0$

maximum BM $= +6\,$kN m at B

As you can see from the diagrams in Figure 1.41, the effect of the UDL is to produce a shear force diagram that slopes between the supports and the concentrated load. Its slope $2\,\text{kN m}^{-1}$, which is the uniformly distributed loading rate. The effect on the bending moment diagram is to produce parabolic curves between the supports and the concentrated load.

Test your knowledge 1.3

1. What is the difference between an active and a reactive load?
2. How do you define the shear force at any point on a loaded beam?
3. How do you define the bending moment at any point on a loaded beam?
4. What is a point of contraflexure?

Activity 1.3

The simply supported beam shown in Figure 1.42 is made in two sections which will be joined together at some suitable point between the supports. Draw the shear force and bending moment diagrams for the beam. State the maximum values of shear force and bending moment, and the positions where they occur. Where would be the most suitable point to join the two sections of the beam together?

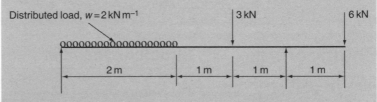

Distributed load, $w = 2\,\text{kN m}^{-1}$ 3 kN   6 kN

2 m   1 m   1 m   1 m

Figure 1.42

Problems 1.3

Sketch the shear force and bending moment diagrams for the simply supported beams and cantilevers shown in Figures 1.43–1.47. Indicate the magnitude and nature of the maximum shear force and bending moment and the positions where they occur. Indicate also the position of any point of contraflexure.

1.

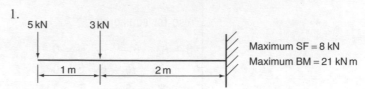

Maximum SF = 8 kN
Maximum BM = 21 kN m

Figure 1.43

2.

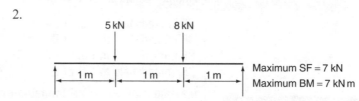

Maximum SF = 7 kN
Maximum BM = 7 kN m

Figure 1.44

3.

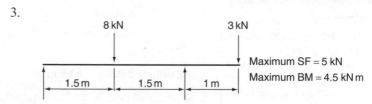

Maximum SF = 5 kN
Maximum BM = 4.5 kN m

Figure 1.45

4.

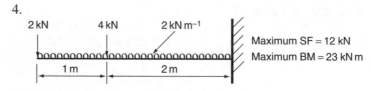

Maximum SF = 12 kN
Maximum BM = 23 kN m

Figure 1.46

5.

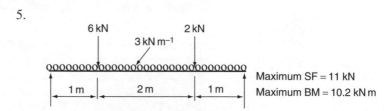

Maximum SF = 11 kN
Maximum BM = 10.2 kN m

Figure 1.47

Engineering components

Structural members

The primary forces acting in structural components such as ties and struts are direct tensile and compressive forces. Direct, or uni-axial loading occurs when equal and opposite tensile or compressive forces act on along the same line of action. The intensity of the loading in the component material is quantified as *direct stress*, and the deformation which it produces is quantified as *direct strain*.

Direct stress

Consider a component of original length *l*, and cross-sectional area *A*, which is subjected to a direct tensile load *F* as shown in Figure 1.48. Let the change of length be *x*.

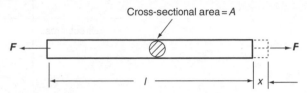

Figure 1.48 *Direct loading*

It is assumed that the load in the material is distributed evenly over the cross-sectional area *A*, of the component. The direct stress σ in the material is the load carried by each square millimetre or square metre of cross-sectional area.

$$\text{direct stress} = \frac{\text{direct load}}{\text{cross-sectional area}}$$

$$\sigma = \frac{F}{A} \; (\textbf{Pa or N m}^{-2}) \tag{1.3}$$

In Figure 1.47, the load and the stress are tensile and these are sometimes given a positive sign. Compressive loads produce compressive stress and these are sometimes given a negative sign. You will recall from the work you did in the core unit Science for Technicians, that the approved SI unit of stress is Pascal although you will often find its value quoted in $N m^{-2}$ and $N m m^{-2}$. These are, in fact, more convenient and because many engineers prefer them, you will still see them used in trade catalogues, British Standards and engineering publications.

Direct strain

Direct strain ε, is a measure of the deformation which the load produces. It is the change in length given as a fraction of the original length.

$$\text{direct strain} = \frac{\text{change in length}}{\text{original length}}$$

$$\varepsilon = \frac{x}{l} \; (\textbf{No units}) \tag{1.4}$$

Modulus of elasticity (Young's Modulus)

An elastic material is one in which the change in length is proportional to the load applied and in which the strain is proportional to the stress. Furthermore, a perfectly elastic material will immediately return to its original length when the load is removed.

A graph of stress σ, against strain ε, is a straight line whose gradient is always found to be the same for a given material. Figure 1.49 shows typical graphs for steel, copper and aluminium. The value of the gradient is a measure of the elasticity or 'stiffness' of the material, and is known as its Modulus of Elasticity, E.

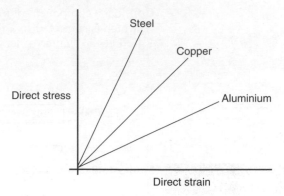

Figure 1.49 *Graph of stress v. strain*

$$\text{modulus of elasticity} = \frac{\text{direct stress}}{\text{direct strain}}$$

$$E = \frac{\sigma}{\varepsilon} \ (\textbf{Pa or N m}^{-2}) \tag{1.5}$$

Substituting the expressions for stress and strain from equations (1.3) and (1.4) gives an alternative formula.

$$E = \frac{\frac{F}{A}}{\frac{x}{l}}$$

$$E = \frac{F}{A} \times \frac{l}{x} \tag{1.6}$$

> **Key point**
>
> The modulus of elasticity of a material is a measure of its stifness, i.e. its resistance to being stretched or compressed.

Factor of safety

Engineering components should be designed so that the working stress which they are likely to encounter is well below the ultimate stress at which failure occurs.

$$\textbf{factor of safety} = \frac{\textbf{ultimate stress}}{\textbf{working stress}} \tag{1.7}$$

A factor of safety of at least 2 is generally applied for static structures. This ensures that the working stress will be no more than half of that at which failure occurs. Much lower factors of safety are applied in aircraft design where weight is at a premium, and with some of the major components it is likely that failure will eventually occur due to metal fatigue. These are rigorously tested at the prototype stage to predict their working life and replaced periodically in service well before this period has elapsed.

Example 1.9

A strut of diameter 25 mm and length 2 m carries an axial load of 20 kN. The ultimate compressive stress of the material is 350 MPa and its modulus of elasticity is 150 GPa. Find (a) the compressive stress in the material, (b) the factor of safety in operation, (c) the change in length of the strut.

(a) Finding cross-sectional area of strut:

$$A = \frac{\pi d^2}{4} = \frac{\pi \times 0.025^2}{4}$$

$$\boldsymbol{A = 491 \times 10^{-6}\,m^2}$$

Finding compressive stress:

$$\sigma = \frac{F}{A} = \frac{20 \times 10^3}{491 \times 10^6}$$

$$\boldsymbol{\sigma = 40.7 \times 10^6\,Pa} \quad \text{or} \quad \boldsymbol{40.7\,MPa}$$

(b) Finding factor of safety in operation:

$$\text{factor of safety} = \frac{\text{ultimate stress}}{\text{working stress}} = \frac{350 \times 10^6}{40.7 \times 10^6}$$

$$\boldsymbol{\text{factor of safety} = 8.6}$$

(c) Finding compressive strain:

$$E = \frac{\sigma}{\varepsilon}$$

$$\varepsilon = \frac{\sigma}{E} = \frac{40.7 \times 10^6}{150 \times 10^9}$$

$$\varepsilon = 271 \times 10^{-6}$$

Finding change in length:

$$\varepsilon = \frac{x}{l}$$

$$x = \varepsilon\,l = (271 \times 10^{-6}) \times 2$$

$$\boldsymbol{x = 0.543 \times 10^{-3}\,m} \quad \text{or} \quad \boldsymbol{0.543\,mm}$$

Test your knowledge 1.4

1. What is meant by uni-axial loading?
2. What is the definition of an elastic material?
3. A tie bar of cross-sectional area 50 mm^2 carries a load of 10 kN. What is the tensile stress in the material measured in Pascals?
4. If the ultimate tensile stress at which failure occurs in a material is 550 MPa and a factor of safety of 4 is to apply, what will be the allowable working stress?

Activity 1.4

A tube of length 1.5 m has an inner diameter 50 mm and a wall thickness of 6 mm. When subjected to a direct tensile load of 75 kN the length is seen to increase by 0.55 mm. Determine (a) the tensile stress in the material, (b) the factor of safety in operation if the ultimate tensile strength of the material is 350 MPa, (c) the modulus of elasticity of the material.

Thermal loading

Most materials expand when their temperature rises. This is certainly true of the more commonly used metals in engineering and the effect is measured as the *linear expansivity* of a material. It is defined as *the change in length per unit of length per degree of temperature change* and its symbol is α. It is also known as the *coefficient of linear expansion* and its units are $°C^{-1}$.

To find the change in length x, of a component of original length l and linear expansivity α, when its temperature changes by $t°C$, we use the formula,

$$x = l\alpha t \tag{1.8}$$

Some typical values of linear expansivity α, are shown in Table 1.1.

Table 1.1 *Linear expansivity values*

Material	Linear expansivity ($°C^{-1}$)
Aluminium	24×10^{-6}
Brass and bronze	19×10^{-6}
Copper	17×10^{-6}
Nickel	13×10^{-6}
Carbon steel	12×10^{-6}
Cast iron	10×10^{-6}
Platinum	9×10^{-6}
Invar	1.6×10^{-6}

Invar is an alloy steel containing around 36% nickel. The combination results in a material with a very low expansivity. It is used in applications such as instrumentation systems, where expansion of the components could result in output errors.

Example 1.10

A steel bar of length 2.5 m and linear expansivity $12 \times 10^{-6}\ °C^{-1}$ undergoes a rise in temperature from 15 °C to 250 °C. What will be its change in length?

$x = l\alpha t$

$x = 2.5 \times 12 \times 10^{-6} \times (250 - 15)$

$x = 7.05 \times 10^{-3}$ m   or   7.05 mm

Equation (1.8) can be rearranged to give the thermal strain ε, which has resulted from the temperature change. It is measured in just the same way as the strain due to direct loading.

$$\varepsilon = \frac{x}{l} = \alpha t \tag{1.9}$$

If a material is allowed to expand freely, there will be no stress produced by the temperature change. If, however, the material is securely held and the change in length is prevented, thermal stress σ, will be induced. The above equation gives the virtual mechanical strain to which it is proportional and if the modulus of elasticity, E, for the material is known, the value of the stress can be calculated as follows.

$$E = \frac{\sigma}{\varepsilon} = \frac{\sigma}{\alpha t}$$

$$\sigma = E\alpha t \tag{1.10}$$

You will see from the above equation that thermal stress depends only on the material properties and the temperature change.

Key point

Thermal stress is independent of the dimensions of a restrained component. It is dependant only on its modulus of elasticity, its linear expansivity and the temperature change which takes place.

Example 1.11

A steam pipe made from steel is fitted at a temperature of $20\,°C$. If expansion is prevented, what will be the compressive stress in the material when steam at a temperature of $150\,°C$ flows through it? Take $\alpha = 12 \times 10^{-6}\,°C^{-1}$ and $E = 200\,GN\,m^{-2}$.

$$\sigma = E\alpha t$$

$$\sigma = 200 \times 10^9 \times 12 \times 10^{-6} \times (150 - 20)$$

$$\boldsymbol{\sigma = 312 \times 10^6\,Pa} \quad \text{or} \quad \textbf{312 MPa}$$

You should note that this is quite a high level of stress which could very easily cause the pipe to buckle. This is why expansion loops and expansion joints are included in steam pipe systems, to reduce the stress to an acceptably low level.

Combined direct and thermal loading

It is of course quite possible to have a loaded component which is rigidly held and which also undergoes temperature change. This often happens with aircraft components and with components in process plant. Depending on whether the temperature rises or falls, the stress in the component may increase or decrease.

When investigating the effects of combined direct and thermal stress, it is useful to adopt the sign convention that tensile stress and strain are positive, and compressive stress and strain are negative. The resultant strain and the resultant stress are the algebraic sum of the direct and thermal values.

resultant strain = direct strain + thermal strain

$$\sigma = \sigma_D + \sigma_T$$

$$\sigma = \frac{F}{A} + E\alpha t \tag{1.11}$$

resultant strain = direct strain + thermal strain

$$\varepsilon = \varepsilon_D + \varepsilon_T$$

$$\varepsilon = \frac{x}{l} + \alpha t \tag{1.12}$$

Having calculated either the resultant stress or the resultant strain, the other can be found from the modulus of elasticity of the material.

$$\text{modulus of elasticity} = \frac{\text{resultant stress}}{\text{resultant strain}}$$

$$E = \frac{\sigma}{\varepsilon} \tag{1.13}$$

Key point

Residual thermal stresses are sometimes present in cast and forged components which have cooled unevenly. They can cause the component to become distorted when material is removed during machining. Residual thermal stresses can be removed by heat treatment and this will be described in Chapter 3.

Example 1.12

A rigidly held tie bar in a heating chamber has a diameter of 15 mm and is tensioned to a load of 150 kN at a temperature of $15\,°C$. What is the initial stress in the bar and what will be the resultant stress when the temperature in the chamber has risen to $50\,°C$? Take $E = 200\,GN\,m^{-2}$ and $\alpha = 12 \times 10^{-6}\,°C^{-1}$.

Note: The initial tensile stress will be positive but the thermal stress will be compressive and negative. It will thus have a cancelling effect.

Finding cross-sectional area:

$$A = \frac{\pi d^2}{4} = \frac{\pi \times 0.015^2}{4}$$

$$A = 2.83 \times 10^{-3}\,\text{m}^2$$

Finding initial direct tensile stress at 15 °C:

$$\sigma_D = \frac{F}{A} = \frac{+150 \times 10^3}{2.83 \times 10^{-3}}$$

$$\sigma_D = +53.0 \times 10^6\,\text{Pa} \quad \text{or} \quad +53.0\,\text{MPa}$$

Finding thermal compressive stress at 50 °C:

$$\sigma_T = E\alpha t = -200 \times 10^9 \times 12 \times 10^{-6} \times (50 - 15)$$

$$\sigma_T = -84.0 \times 10^6\,\text{Pa} \quad \text{or} \quad -84.0\,\text{MPa}$$

Finding resultant stress at 180 °C:

$$\sigma = \sigma_D + \sigma_T = +53.0 + (-84.0)$$

$$\sigma = -31.0\,\text{MPa}$$

That is, the initial tensile stress has been cancelled out during the temperature rise and the final resultant stress is compressive.

Test your knowledge 1.5

1. How do you define the linear expansivity of a material?
2. If identical lengths of steel and aluminium bar undergo the same temperature rise, which one will expand the most?
3. What effect do the dimensions of a rigidly clamped component have on the thermal stress caused by temperature change?

Activity 1.5

A rigidly fixed strut in a refrigeration system carries a compressive load of 50 kN when assembled at a temperature of 20 °C. The initial length of the strut is 0.5 m and its diameter is 30 mm. Determine the initial stress in the strut and the amount of compression under load. At what operating temperature will the stress in the material have fallen to zero? Take $E = 150\,\text{GN m}^{-2}$ and $\alpha = 16 \times 10^{-6}$.

Problems 1.4

1. A steel spacing bar of length 0.5 m is assembled in a structure whilst the temperature is 20 °C. What will be its change in length if the temperature rises to 50 °C? What will be the stress in the bar if free expansion is prevented? Take $E = 200\,\text{GPa}$ and $\alpha = 12 \times 10^{-1}\,°\text{C}^{-1}$.

[0.18 mm, 72 MPa]

2. A copper bar is rigidly fixed at its ends at a temperature of 15 °C. Determine the magnitude and nature of the stress induced in the material (a) if the temperature falls to −25 °C, (b) if the temperature rises to 95 °C. Take $E = 96\,\text{GPa}$ and $\alpha = 17 \times 10^{-1}\,°\text{C}^{-1}$.

[66 MPa, 131 MPa]

3. A steel tie bar of length 1 m and diameter of 30 mm is tensioned to carry a load of 180 kN at a temperature of 20 °C. What is the stress in the bar and its change in length when tensioned? What

will be the stress in the bar if the temperature rises to 80 °C and any change in length is prevented? Take $E = 200\,\text{GN}\,\text{m}^{-2}$ and $\alpha = 12 \times 10^{-6}\,{}^{\circ}\text{C}^{-1}$.

[255 MPa, 1.27 mm, 111 MPa]

4. A metal component of initial length 600 mm undergoes an increase in length of 0.05 mm when loaded at a temperature of 18 °C. Determine, (a) the tensile stress in the component when first loaded, (b) the temperature at which the stress will be zero if the component is rigidly held at its extended length. Take $E = 120\,\text{GPa}$ and $\alpha = 15 \times 10^{-6}\,{}^{\circ}\text{C}^{-1}$ for the component material.

[10 MPa, 73.6 °C]

5. A brittle cast iron bar is heated to 140 °C and then rigidly clamped so that it cannot contract when cooled. The bar is seen to fracture when the temperature has fallen to 80 °C. What is the ultimate tensile strength of the cast iron? Take $E = 120\,\text{GPa}$ and $\alpha = 10 \times 10^{-6}\,{}^{\circ}\text{C}^{-1}$.

[72 MPa]

Compound members

Engineering components are sometimes fabricated from two different materials. Materials joined end to end form series compound members. Materials which are sandwiched together or contained one within another form parallel compound members.

With a series compound member as shown on Figure 1.50, the load F is transmitted through both materials.

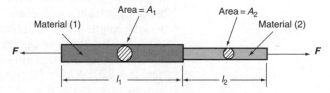

Figure 1.50 *Series compound member*

Let the moduli of elasticity or the materials be E_1 and E_2.
Let the cross-sectional areas be A_1 and A_2.
Let the stresses in the materials be σ_1 and σ_2.
Let the strains in the materials be ε_1 and ε_2.
Let the changes in length be x_1 and x_2.

The stress in each material is given by

$$\sigma_1 = \frac{F}{A_1} \tag{1.14}$$

and

$$\sigma_2 = \frac{F}{A_2} \tag{1.15}$$

If the modulus of elasticity of the material (1) is E_1, the strain in the material will be

$$\varepsilon_1 = \frac{\sigma_1}{E_1}$$

and the change in length will be

$$x_1 = \varepsilon_1 l_1$$

$$x_1 = \frac{\sigma_1 l_1}{E_1} \qquad (1.16)$$

If the modulus of elasticity of the material (2) is E_2, the strain in the material will be,

$$\varepsilon_2 = \frac{\sigma_2}{E_2}$$

and the change in length will be

$$x_2 = \varepsilon_2 l_2$$

$$x_2 = \frac{\sigma_2 l_2}{E_2} \qquad (1.17)$$

The total change in length will thus be

$$x = x_1 + x_2 \qquad (1.18)$$

Substituting for x_1 and x_2 from equations (1.16) and (1.17) gives

$$x = \frac{\sigma_1 l_1}{E_1} + \frac{\sigma_2 l_2}{E_2} \qquad (1.19)$$

Key point

The load is transmitted through both materials in series compound members, i.e. both materials carry the same load.

Example 1.13

The compound member, shown in Figure 1.51, consists of a steel bar of length 2 m and diameter 30 mm, to which is brazed a copper tube of length 1.5 m, outer diameter 30 mm and inner diameter 15 mm. If the member carries a tensile load of 15 kN, determine the stress in each material and the overall change in length. For steel, $E = 200\,\text{GN m}^{-2}$, for copper $E = 120\,\text{GN m}^{-2}$.

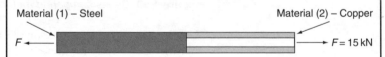

Material (1) – Steel   Material (2) – Copper

$F \leftarrow$ $\rightarrow F = 15\,\text{kN}$

Figure 1.51

Finding cross-sectional area of steel:

$$A_1 = \frac{\pi D_1^2}{4} = \frac{\pi \times 0.03^2}{4}$$

$$\boldsymbol{A_1 = 707 \times 10^{-6}\ \text{m}^2}$$

Finding cross-sectional area of copper:

$$A_2 = \frac{\pi (D_1^2 - d_2^2)}{4} = \frac{\pi \times (0.03^2 - 0.015^2)}{4}$$

$$\boldsymbol{A_2 = 530 \times 10^{-6}\ \text{m}^2}$$

Finding stress in steel:

$$\sigma_1 = \frac{F}{A_1} = \frac{15 \times 10^3}{707 \times 10^{-6}}$$

$\sigma_1 = 21.2 \times 10^6\,\text{Pa}$ or $21.2\,\text{MPa}$

Finding stress in copper:

$$\sigma_2 = \frac{F}{A_2} = \frac{15 \times 10^3}{530 \times 10^{-6}}$$

$\sigma_2 = 28.3 \times 10^6\,\text{Pa}$ or $28.3\,\text{MPa}$

Finding overall change in length using equation (1.19):

$$x = \frac{\sigma_1 l_1}{E_1} + \frac{\sigma_2 l_2}{E_2} = \frac{(21.2 \times 10^6 \times 2)}{200 \times 10^9} + \frac{(28.3 \times 10^6 \times 1.5)}{120 \times 10^9}$$

$$x = (212 \times 10^{-6}) + (354 \times 10^{-6})$$

$x = 0.566 \times 10^{-3}\,\text{m}$ or $0.566\,\text{mm}$

With parallel compound members, the loads in each material will most likely be different but when added together, they will equal the external load. The change in length of each material will be the same. When you are required to find the stress in each material, this information enables you to write down two equations, each of which contains the unknown stresses. You can then solve these by substitution and use one of the stress values to find the common change in length.

Consider a member made up of two plates of the same material between which is bonded a plate of another material as shown in Figure 1.52. It is a good idea to let the material with the larger modulus of elasticity be material (1) and that with the lower value be material (2).

Let the moduli of elasticity of the materials be E_1 and E_2.
Let the loads carried be F_1 and F_2.
Let the cross-sectional areas be A_1 and A_2.
Let the stresses in the materials be σ_1 and σ_2.

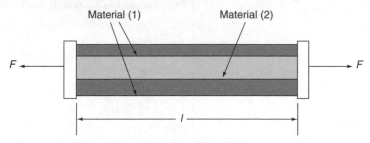

Figure 1.52 *Parallel compound member*

The external load is equal to the sum of the loads carried by the two materials.

$$F = F_1 + F_2$$

Now, $F_1 = \sigma_1 A_1$ and $F_2 = \sigma_2 A_2$

$$F = \sigma_1 A_1 + \sigma_2 A_2 \tag{1.20}$$

Key point

Both materials undergo the same strain in parallel compound members and the sum of the loads carried by each material is equal to the external load.

The strain in each material is the same, and so

$\varepsilon_1 = \varepsilon_2$

Now, $\varepsilon_1 = \sigma_1/E_1$ and $\varepsilon_2 = \sigma_2/E_2$

$$\frac{\sigma_1}{E_1} = \frac{\sigma_2}{E_2} \tag{1.21}$$

Equations (1.20) and (1.21) can be solved simultaneously to find σ_1 and σ_2. The common strain can then be found from either of the above two expressions for strain, and this can then be used to find the common change in length.

Example 1.14

A concrete column 200 mm square is reinforced by nine steel rods of diameter 20 mm. The length of the column is 3 m and it is required to support an axial compressive load of 500 kN. Find the stress in each material and the change in length of the column under load. For steel $E = 200\,\text{GN m}^{-2}$ and for concrete, $E = 20\,\text{GN m}^{-2}$.

Because the steel has higher modulus of elasticity, let it be material (1) and the concrete be material (2).

Begin by finding the cross-sectional areas of the two materials:

$$A_1 = 9 \times \frac{\pi d^2}{4} = 9 \times \frac{\pi \times 0.02^2}{4}$$

$$A_1 = 2.83 \times 10^{-3}\,\text{m}^2$$

$$A_2 = (0.2 \times 0.2) - A_1 = 0.04 - (2.83 \times 10^{-3})$$

$$A_2 = 37.2 \times 10^{-3}\,\text{m}^2$$

Now equate the external force to the sum of the forces in the two materials using equation (1.20):

$$F = \sigma_1 A_1 + \sigma_2 A_2$$

$$500 \times 10^3 = (2.83 \times 10^{-3})\sigma_1 + (37.2 \times 10^{-3})\sigma_2$$

This can be simplified by dividing both sides by 10^{-3}:

$$500 \times 10^6 = 2.83\sigma_1 + 37.2\sigma_2 \tag{1}$$

Now equate the strains in each material using equation (1.21) :

$$\frac{\sigma_1}{E_1} = \frac{\sigma_2}{E_2}$$

$$\sigma_1 = \frac{E_1 \sigma_2}{E_2}$$

$$\sigma_1 = \frac{200 \times 10^9}{20 \times 10^9} \times \sigma_2$$

$$\sigma_1 = 10\,\sigma_2 \tag{2}$$

Substitute in equation (1) for σ_1:

$$500 \times 10^6 = 2.83(10\sigma_2) + 37.2\sigma_2$$

$$500 \times 10^6 = 28.3\sigma_2 + 37.2\sigma_2 = 65.5\sigma_2$$

$$\sigma_2 = \frac{500 \times 10^6}{65.5}$$

$$\sigma_2 = 7.63 \times 10^6\,\text{Pa} \quad \text{or} \quad 7.63\,\text{MPa}$$

Test you knowledge 1.6

1. What is the parameter that the two materials in a series compound member have in common when it is under load?
2. What is the parameter that the two materials in a parallel compound member have in common when it is under load?
3. On which material property does the ratio of the stresses in the materials of a parallel compound member depend?

Finding σ_1 from equation (2):

$$\sigma_1 = 10 \times 7.63 = 7.63\,\text{MPa}$$

Now find the common strain using the stress and modulus of elasticity for material (2):

$$\varepsilon = \frac{\sigma_2}{E_2} = \frac{7.63 \times 10^6}{20 \times 10^9}$$

$$\varepsilon = 382 \times 10^{-6}$$

Finally find the common change in length:

$$\varepsilon = \frac{x}{l}$$

$$x = \varepsilon\,l = 382 \times 10^{-6} \times 3$$

$$x = 1.15 \times 10^3\,\text{m} \quad \text{or} \quad 1.15\,\text{mm}$$

Activity 1.6

The tie bar shown in Figure 1.53 consists of a steel core 25 mm in diameter on to which is cast an aluminium outer case of outer diameter 40 mm. The compound section 1.5 m long, carries a tensile load of 30 kN. The load is applied to the protruding steel core as shown below.

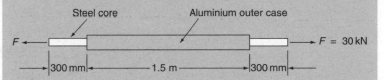

Figure 1.53

Find (a) the stress in the protruding steel which carries the full load, (b) the stress in the two materials in the compound section, (c) the overall change in length. For steel, $E = 200\,\text{GN}\,\text{m}^{-2}$ and for aluminium, $E = 90\,\text{GN}\,\text{m}^{-2}$.

Problems 1.5

1. A duralumin tie bar of diameter 40 mm and length 600 mm carries a tensile load of 180 kN. If the bar contains a 30 mm diameter hole along 100 mm of its length, determine the stress in the two sections and the overall change in length. Take $E = 180\,\text{GPa}$.

    [143 MPa, 327 MPa, 1.24 mm]

2. A compound strut consists of a steel tube of length 1 m, outer diameter 35 mm and inner diameter 25 mm which is brazed to a copper tube of length 1.5 m, outer diameter 35 mm and inner diameter 20 mm. If the member carries a compressive load of 20 kN, determine the stress in each material and the overall change in length.

    For steel, $E = 200\,\text{GN}\,\text{m}^{-2}$, for copper $E = 120\,\text{GN}\,\text{m}^{-2}$.

    [42.4 MPa, 30.9 MPa, 0.6 mm]

3. A rectangular timber strut of cross-section 125 mm × 105 mm is reinforced by two aluminium bars of diameter 60 mm. Determine the stress in each material when the member carries a load of 300 kN. For aluminium, $E = 90\,\text{GN}\,\text{m}^{-2}$, for timber $E = 15\,\text{GN}\,\text{m}^{-2}$.

[103 MPa, 17.2 MPa]

4. A compression member is made up of a mild steel bar, 38 mm in diameter, which is encased in a brass tube of inner diameter 38 mm and outer diameter 65 mm. Determine the stress in the two materials and the reduction in length when the member carries a load of 200 kN. For steel, $E = 200\,\text{GPa}$, for brass $E = 96\,\text{GPa}$.

[92.9 MPa, 43.3 MPa, 0.068 mm]

5. A steel reinforced concrete column of height 3 m has a square cross-section with sides of 375 mm. If the column contains four steel reinforcing rods of diameter 25 mm, determine the stress in each material and the amount of compression when carrying a compressive load of 600 kN. For steel, $E = 200\,\text{GPa}$, for concrete $E = 13.8\,\text{GPa}$.

[53.6 MPa, 3.56 MPa, 0.77 mm]

Fastenings

Engineering fastenings such as rivets, bolts, the different kinds of machine screws and setscrews, self-tapping screws and hinge pins are frequently subjected to shear loading. Shear loading occurs when equal and opposite parallel forces act on a component. Direct loading tends to cause failure perpendicular to the direction of loading whereas shear loading tends to cause failure parallel to the direction of loading. Direct loading and its effects have already been dealt with and we will now examine the effects of shear loading (Figure 1.54).

Key point

Direct loading tends to cause failure in a plane which is at right angles to the direction of loading whereas shear loading tends to cause failure in a plane which is parallel to the direction of loading.

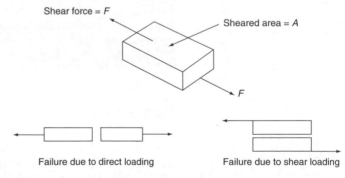

Figure 1.54 *Direct and shear loading*

Shear stress

Shear stress τ, is a measure of the intensity of loading over the sheared area A.

$$\text{shear stress} = \frac{\text{shear force}}{\text{sheared area}}$$

$$\tau = \frac{F}{A} \ (\textbf{Pa or N}\,\textbf{m}^{-2}) \tag{1.22}$$

Shear strain

Shearing forces tend to distort the shape of a component as shown in Figure 1.55.

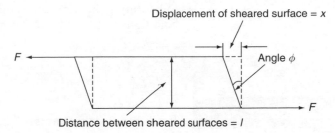

Figure 1.55 *Deformation due to shearing*

Shear strain γ, is a measure of the deformation which the shearing force produces. It is the ratio of the displacement of the sheared surfaces to the distance between them.

$$\text{shear strain} = \frac{\text{displacement of sheared surfaces}}{\text{distance between sheared surfaces}}$$

$$\gamma = \frac{x}{l} \text{ (No units)} \qquad (1.23)$$

The angle ϕ, is called the angle of shear. Its tangent is equal to the shear strain.

$$\tan \phi = \frac{x}{l}$$

$$\mathbf{\tan \phi = \gamma} \qquad (1.24)$$

Shear modulus or (Modulus of rigidity)

When an elastic material is subjected to shear loading, the displacement x of the sheared surfaces is proportional to the load F, which is applied. Also the shear stress τ is proportional to the shear strain γ.

A graph of shear stress against shear strain is a straight line, as shown in Figure 1.56, whose gradient for a given material is always found to be the same. It gives a measure of the elasticity or

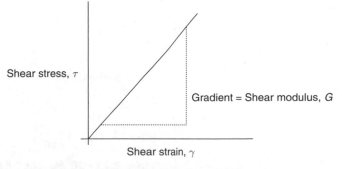

Figure 1.56 *Graph of shear stress v. shear strain*

'stiffness' of the material in shear and is known as its *Shear Modulus, G*. In older text books you might find that it is called the *Modulus of Rigidity*.

$$G = \frac{\tau}{\gamma} \ (\mathbf{Pa \ or \ Nm^{-2}}) \tag{1.25}$$

Substituting the expressions for shear stress and shear strain from equations (1.22) and (1.23) gives an alternative formula

$$G = \frac{\frac{F}{A}}{\frac{x}{l}}$$

$$G = \frac{F}{A} \times \frac{l}{x} \tag{1.26}$$

It will be noted that several of the above formulae are similar to those derived for direct stress and strain and the Modulus of Elasticity but they should not be confused. The symbols F, A, l and x have different meanings when used to calculate shear stress, shear strain and shear modulus. Furthermore, the values of Modulus of Elasticity E, and Shear Modulus G, are not the same for any given material. With mild steel, for example, $E = 210 \, \text{GN m}^{-2}$ whilst $G = 85 \, \text{GN m}^{-2}$.

Key point

Shear modulus is a measure of the shear stiffness of a material, i.e. its resistance to being deformed by shearing forces.

Example 1.15

A block of an elastic material is subjected to a shearing force of 50 kN which deforms its shape as shown in Figure 1.57. Find (a) the shear stress, (b) the shear strain, (c) the shear modulus for the material.

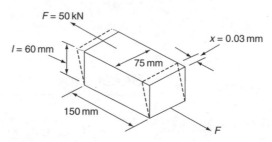

Figure 1.57

(a) Finding shear stress:

$$\tau = \frac{F}{A} = \frac{50 \times 10^3}{0.15 \times 0.075}$$

$$\tau = \mathbf{4.44 \times 10^6 \ Pa} \quad \text{or} \quad \mathbf{4.44 \ MPa}$$

(b) Finding shear strain:

$$\gamma = \frac{x}{l} = \frac{0.03}{60}$$

$$\gamma = \mathbf{500 \times 10^{-6}}$$

(c) Finding shear modulus:

$$G = \frac{\tau}{\gamma} = \frac{4.44 \times 10^6}{500 \times 10^{-6}}$$

$$G = \mathbf{8.88 \times 10^9 \ Pa} \quad \text{or} \quad \mathbf{8.88 \ GPa}$$

Fastenings in single shear

Riveted lap joints and joints employing screwed fastenings are often subjected to shearing forces. Tensile forces may also be present and these are very necessary to hold the joint surfaces tightly together. It is very likely, however, that the external loads will have a shearing effect and it is assumed that this will be carried entirely by the fastenings. The tendency is to shear the fastenings at the joint interface and this is known as *single shear* as shown in Figure 1.58.

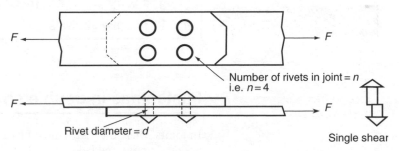

Figure 1.58 *Lap joint with rivets in single shear*

The total sheared cross-sectional area of the fastenings is

$$A = \frac{n\pi d^2}{4}$$

The shear stress in the fastenings will be

$$\tau = \frac{F}{A} = \frac{F}{\frac{n\pi d^2}{4}}$$

$$\tau = \frac{4F}{n\pi d^2} \qquad (1.27)$$

In design problems you will probably know the safe working stress and the recommended rivet or bolt diameter for the thickness of the materials being joined. The task will then be to calculate the number of fastenings required and to decide on their spacing. Transposing the above formula gives

$$n = \frac{4F}{\tau\pi d^2} \qquad (1.28)$$

The fastenings should not be too close together or too near to the edge of the material being joined. You can find the rules for particular applications in British and international standard specifications, and also in design code handbooks which are based on them.

Key point

The allowable working stress is the stress at which a component is considered to have failed divided by the factor of safety which is required. It is calculated for shear loading in exactly the same way as for direct loading.

Example 1.16

A lap joint is required to join plates using rivets of diameter 6 mm. The shearing force to be carried by the joint is 12 kN and the shear strength of the rivet material is 300 MPa . If a factor of safety of 8 is to apply, find the number of rivets required for the joint.

Finding allowable shear stress in rivets:

$$\tau = \frac{\text{shear strength}}{\text{factor of safety}} = \frac{300}{8}$$

$\tau = \mathbf{37.5\ MPa}$

Finding number of rivets required:

$$n = \frac{4F}{\tau \pi d^2}$$

$$n = \frac{4 \times 12 \times 10^3}{37.5 \times 10^6 \times \pi \times (6 \times 10^{-3})^2}$$

$n = \mathbf{11.3}$ i.e. use **12 rivets**

Fastenings in double shear

In joints where the plates must be butted together, connecting plates are used above and below the joint. There is then a tendency to shear the rivets in two places and they are said to be in *double shear*.

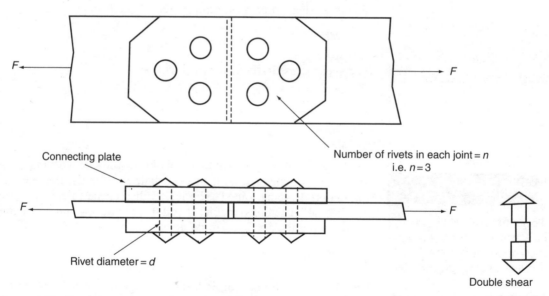

Figure 1.59 *Butt joint with rivets in double shear*

There are in fact two joints in Figure 1.59, where each of the butted plates is riveted to the connecting plates. As a result, the number of rivets per joint is half the total number shown. That is, $n = 3$ rivets in this particular example.

The total sheared cross-sectional area of the fastenings is double that for single shear. That is,

$$A = \frac{2n\pi d^2}{4}$$

The shear stress in the fastenings will be

$$\tau = \frac{F}{A} = \frac{F}{\frac{2n\pi d^2}{4}}$$

$$\tau = \frac{2F}{n\pi d^2} \tag{1.29}$$

If you compare this with equation (1.27) you will note that for the same load, diameter and number of rivets, the shear stress for double shear is half the value for single shear. A joint in double shear can therefore carry twice as much load as the equivalent joint in single shear. If it is required to find the number of rivets required, the above equation can be transposed to give

$$n = \frac{2F}{\tau \pi d^2} \tag{1.30}$$

Example 1.17

Two aluminium plates are to be joined by means of a double strap riveted butt joint. The joint is required to support a load of 15 kN and the total number of rivets is 12. The rivet material has a shear strength of 200 MPa and a factor of safety of 6 is to apply. Determine the required diameter of the rivets.

Finding allowable shear stress in rivets:

$$\tau = \frac{\text{shear strength}}{\text{factor of safety}} = \frac{200}{6}$$

$$\tau = \mathbf{33.3\ MPa}$$

Finding required rivet diameter:

$$\tau = \frac{2F}{n\pi d^2}$$

$$d^2 = \frac{2F}{n\pi\tau}$$

$$d = \sqrt{\frac{2F}{n\pi\tau}} \quad \text{where } n = \frac{12}{2} = 6 \text{ rivets}$$

$$d = \sqrt{\frac{2 \times 10 \times 10^3}{6 \times \pi \times 33.3 \times 10^6}}$$

$$d = \mathbf{5.64 \times 10^{-3}\ m} \quad \text{or} \quad \mathbf{5.64\ mm} \quad \text{i.e. use } \mathbf{6.0\ mm\ revets}$$

Activity 1.7

Two steel plates are riveted together by means of a lap joint. Ten rivets of diameter 9 mm are used, and the load carried is 20 kN. What is the shear stress in the rivet material and the factor of safety in operation if the shear strength is 300 MPa? If the lap joint were replaced by a double strap butt joint with ten rivets per joint and each carrying the same shear stress as above, what would be the required rivet diameter?

Problems 1.6

1.

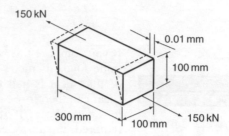

Figure 1.60

A block of elastic material is loaded in shear as shown in Figure 1.60. Determine (a) the shear stress, (b) the shear strain, (c) the shear modulus of the material.

$$[5\,\text{MPa},\ 100 \times 10^{-6},\ 50\,\text{GPa}]$$

2. A mild steel plate of thickness 15 mm has a 50 mm diameter hole punched in it. If the ultimate shear stress of the steel is 275 MPa, determine (a) the punching force required, (b) the compressive stress in the punch.

$$[648\,\text{kN},\ 330\,\text{MPa}]$$

3.

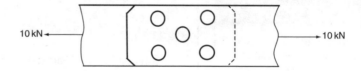

Figure 1.61

Figure 1.61 shows a riveted lap joint. Determine a suitable diameter for the rivets if the ultimate shear strength of the rivet material is 325 MPa and a factor of safety of 6 is to apply.

$$[6.86\,\text{mm, i.e. use 7 mm dia rivets}]$$

4.

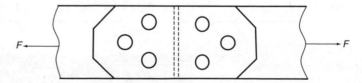

Figure 1.62

Determine the load F which the double strap butt joint shown in Figure 1.62 can carry. The rivets are 6 mm in diameter with an ultimate shear strength of 325 MPa. A factor of safety of 8 is to apply.

$$[6.9\,\text{kN}]$$

Centripetal acceleration and centripetal force

Newton's first law of motion states that *a moving body will continue in a straight line unless acted upon by some external force*. It follows that when a body is travelling in a circular path, there must be some sideways force pushing, or pulling it, towards the centre of rotation. This is known as *centripetal* force, which means 'centre seeking'. With a car travelling round a curve, the centripetal force is provided by friction between the tyres and the road surface, and with an orbiting satellite, it is provided by the earth's gravitational pull.

Vector change of velocity

You will recall that velocity is a vector quantity. This means that it has both magnitude and direction. For a body travelling in a straight path we very often only state the magnitude of its velocity, i.e. its speed. We take its direction for granted but strictly speaking, we should state both magnitude and direction when we write down the velocity of a body.

Because velocity is a vector quantity, it can be represented in magnitude direction and sense by a line v_1 drawn to a suitable scale on a vector diagram. If either the speed or the direction of a body should change, then its velocity will change. The new vector v_2 will be drawn from the same starting point but to a different length or in a different direction. There is then said to have been a vector change of velocity which can be measured as the distance between the end of the initial vector to the end of the final vector.

> **Key point**
>
> Velocity is a vector quantity and if either the speed or the direction of a body changes, its velocity will change.

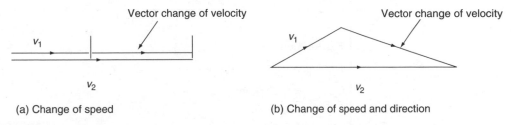

(a) Change of speed   (b) Change of speed and direction

Figure 1.63 *Vector change of velocity*

You will also recall that acceleration is the rate of change of velocity, i.e., the change of velocity per second. This applies not only for changes of speed, as in Figure 1.63(a), but also for vector changes of velocity as in Figure 1.63(b). If the time taken for the vector change is known, the acceleration which has taken place can be calculated. This is also a vector quantity whose direction is the same as the vector change of velocity.

> **Key point**
>
> Acceleration is the rate of change of velocity which can result from a change of speed or a change of direction.

$$\text{acceleration} = \frac{\text{vector change of velocity}}{\text{time taken for change}}$$

$$a = \frac{\Delta v}{t} \tag{1.31}$$

When a body picks up speed or changes direction, there must be some external force acting to cause the change. This force is given by Newton's second law of motion which states that *the rate of change of momentum of a body is equal to the force which is causing the change.* It is from this law that we obtain the formula

force = mass × acceleration

$$F = ma \qquad (1.32)$$

Having found the vector change of velocity of a body from a vector diagram, you can then calculate the acceleration which has occurred using expression (1.31). If you know the mass m of the body, you can then calculate the force which has produced the change using expression (1.32). The direction in which the force acts on the body is the same as that of the vector change of velocity and the acceleration it produces.

Example 1.18

A vehicle of mass 750 kg travelling east at a speed of 50 km h^{-1} accelerates around a bend in the road and emerges 10 s later at a speed of 90 km h^{-1} travelling in a north easterly direction. What is its vector change of velocity, the acceleration and the force which produces the change?

The vector change of velocity Δv, is 65.1 km h^{-1} in a direction which is 77.8° north of west as shown in Figure 1.64.

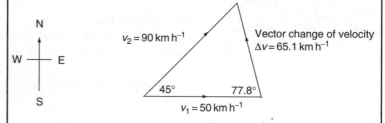

Figure 1.64

Finding velocity change in ms^{-1}:

$$\Delta v = \frac{65.1 \times 1000}{60 \times 60}$$

$$\Delta v = 18.1 \, \text{ms}^{-1}$$

Finding acceleration produced:

$$a = \frac{\Delta v}{t} = \frac{18.1}{10}$$

$$a = 1.81 \, \text{m s}^{-2}$$

Finding force which has produced the change of speed and direction:

$$F = ma = 750 \times 1.81$$

$$F = 13.6 \times 10^3 \, \text{N} \quad \text{or} \quad 13.6 \, \text{kN}$$

This force is the resultant of the force between the driving wheels and the road, which is producing the increase in speed, and the side thrust on the vehicle, which results from turning the steering wheel into the bend. Its direction is the same as the vector change which it produces.

If the vehicle in Example 1.18 had been travelling around the bend without increasing its speed, there would still have been a vector change in velocity and a resulting acceleration. It is thus possible to have acceleration at constant speed. This may sound a bit odd, but you must remember that velocity is a vector quantity and if either the speed or the direction of a body changes, there will be a change of velocity and an acceleration.

Centripetal acceleration and force

Consider now what happens when a body is travelling at a constant speed $v\,\mathrm{m\,s^{-1}}$ and angular velocity $\omega\,\mathrm{rad\,s^{-1}}$ around a circular path of radius r. Its direction, and also its velocity, is continually changing.

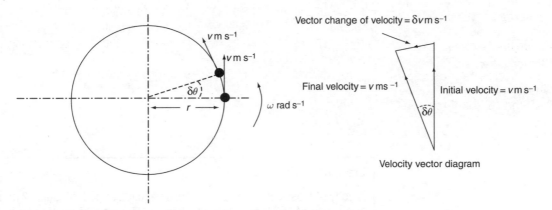

Figure 1.65 *Uniform circular motion*

Let the body move through a small angle $\delta\theta$ radians in time δt s, as shown in Figure 1.65. There will be a vector change in velocity during this period, as shown on the velocity vector diagram. If the angle $\delta\theta$ is very small, the velocity vector diagram will be a long, thin isosceles triangle and the vector change in velocity will be given by

$$\delta v = v\delta\theta$$

The acceleration which occurs will be

$$a = \frac{\delta v}{\delta t} = v\frac{\delta\theta}{\delta t}$$

But $\frac{\delta\theta}{\delta t} = \omega$, the angular velocity of the body,

$$a = v\omega \tag{1.33}$$

This is known as the *centripetal acceleration* of the body and whatever its position, its centripetal acceleration is always directed towards the centre of the circular path.

Now $v = \omega r$ and $\omega = \dfrac{v}{r}$. Substituting for v and ω in equation (1.33) gives

$$a = \omega^2 r \tag{1.34}$$

and

$$a = \frac{v^2}{r} \tag{1.35}$$

It is more usual to use these expressions when calculating centripetal acceleration than that given by equation (1.33).

Wherever acceleration occurs, there must be a force present. For a body travelling in a circular path, this is the centripetal force, pulling or pushing it towards the centre of the path. If the body shown in Figure 1.65 were attached to the centre of its path by a cord, this would all the time be pulling it inwards and the tension in the cord would be the centripetal force. For a body of mass m kg, the centripetal force is obtained using expression (1.32) which comes from Newton's second law of motion.

$$F = ma$$
$$F = m\omega^2 r \tag{1.36}$$

or

$$F = \frac{mv^2}{r} \tag{1.37}$$

Example 1.19

A body of mass 5 kg travels in a horizontal circular path of radius 2 m at a speed of 100 rpm and is attached to the centre of rotation by a steel rod of diameter 3 mm. Determine (a) the centripetal acceleration of the mass, (b) the centripetal force, (c) the tensile stress in the rod.

(a) Finding angular velocity of rotation:

$$\omega = N(\text{rpm}) \times \frac{2\pi}{60} = 100 \times \frac{2\pi}{60}$$
$$\omega = \mathbf{10.5\,rad\,s^{-1}}$$

Finding centripetal acceleration of mass:

$$a = \omega^2 r = 10.5^2 \times 2$$
$$a = \mathbf{219\,m\,s^{-2}}$$

(b) Finding centripetal force:

$$F = ma = 5 \times 219$$
$$F = \mathbf{1.10 \times 10^3\,N} \quad \text{or} \quad \mathbf{1.10\,kN}$$

(c) Finding cross-sectional area of rod:

$$A = \frac{\pi d^2}{4} = \frac{\pi \times 0.003^2}{4}$$
$$A = \mathbf{7.07 \times 10^{-6}\,m^2}$$

Test your knowledge 1.8

1. What is a vector quantity?
2. How is it possible for a body to have acceleration whilst travelling at constant speed?
3. In what direction does centripetal force act?
4. What is centrifugal reaction?

Finding tensile stress in rod:

$$\sigma = \frac{F}{A} = \frac{1.10 \times 10^3}{7.07 \times 10^{-6}}$$

$$\sigma = 155 \times 10^6 \text{ Pa} \quad \text{or} \quad 155 \text{ kPa}$$

Activity 1.8

A body of mass 0.6 kg rotates on the end of a helical spring of stiffness 1.2 kN m^{-1} in a horizontal plane. The initial distance from the centre of rotation to the centre of the mass is 200 mm. What will be the radius of rotation of the mass when the rotational speed is 300 rpm?

Centrifugal clutches

Centrifugal clutches are to be found on motor-driven equipment such as lawn mowers, mixers and pumps. They engage automatically when the motor speed reaches a pre-determined level, and do not depend on the skill of the operator.

In its simplest form a centrifugal clutch consists of a drive shaft to which are attached two or more spring loaded masses. These are lined with a friction material on their outer surfaces and rotate inside a drum which is fixed to the output shaft (see Figure 1.66).

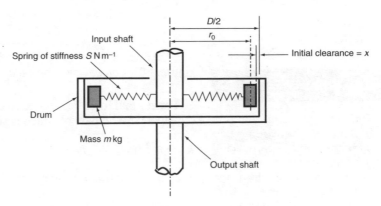

Figure 1.66 *Centrifugal clutch*

The masses, or 'bobs', are free to slide outwards in guides which are rigidly fixed to the input shaft. These are not shown in the above diagram. As the angular velocity of the input shaft increases, the masses slide outwards and eventually make contact with the drum. A further increase in speed causes the drive to be transmitted to the output shaft through friction between the masses and the drum.

Let initial clearance between the masses and the drum be x.
Let the initial radius of rotation from the centre of the shafts to the centre of gravity of the masses be r_0.
Let the diameter of the drum be D.

Let the coefficient of friction between the drum and the masses be μ. Let the angular velocity at which the masses just make contact with the drum be ω_0.

When the masses just make contact with the drum, the tension F_0 in the springs is given by

spring tension = spring stiffness × extension

$$F_0 = Sx \tag{1.38}$$

This is also the centripetal force acting on the masses whose radius of rotation is now $r_0 + x$. Equating spring tension and centripetal force enables the angular velocity ω_0, at which the masses engage with the drum, to be found

$$F_0 = m\omega_0^2 (r_0 + x)$$

$$\omega_0 = \sqrt{\frac{F_0}{m(r_0 + x)}} \tag{1.39}$$

Consider now the normal force and the tangential force between the masses and the drum at some higher speed ω, when the drive is being transmitted to the output shaft. These are shown in Figure 1.67.

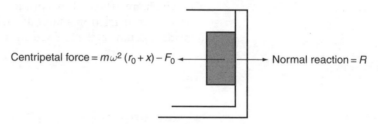

Centripetal force = $m\omega^2 (r_0 + x) - F_0$ Normal reaction = R

Figure 1.67 *Forces acting on clutch drum*

The normal force between each mass and the drum is equal to the centripetal force, part of which is supplied by the spring tension F_0.

$$R = m\omega^2 (r_0 + x) - F_0 \tag{1.40}$$

The tangential friction force between each mass and the drum will be

$$F = \mu R$$

$$F = \mu[m\omega^2 (r_0 + x) - F_0] \tag{1.41}$$

If the number of masses is n, the torque transmitted will be

$$T = nF\frac{D}{2}$$

$$T = n\mu[m\omega^2 (r_0 + x) - F_0]\frac{D}{2} \tag{1.42}$$

Having calculated the torque transmitted, the power transmitted can be calculated.

$$\text{power} = T\omega \tag{1.43}$$

Example 1.20

A centrifugal clutch has two rotating contact masses of 0.25 kg each. These are able to move in guides attached to the input shaft and are held by springs of stiffness $8\,\text{kN}\,\text{m}^{-1}$. The initial clearance between the masses and the internal surface of a drum on the output shaft is 5 mm. The inner diameter of the drum is 300 mm and the distance from the centre of rotation to the centre of gravity of the masses is 135 mm. The coefficient of friction between the masses and the drum is 0.35. Determine (a) the rotational speed in revs per minute at which the clutch begins to transmit power, (b) the torque and power transmitted at a speed of 1000 rpm.

(a) Finding tension in springs as masses make contact with drum:

$$F_0 = Sx = 8 \times 10^3 \times 5 \times 10^{-3}$$
$$\boldsymbol{F_0 = 40\,\text{N}}$$

Finding angular velocity at which masses just make contact with the drum:

$$\omega_0 = \frac{F_0}{\sqrt{m(r_0 + x)}} = \frac{40}{\sqrt{0.25(0.135 + 0.005)}}$$
$$\omega_0 = 33.8\,\text{rad s}^{-1}$$

Changing to revs per minute,

$$N_0 = \omega_0 \times \frac{60}{2\pi} = 33.8 \times \frac{60}{2\pi}$$
$$\boldsymbol{N_0 = 323\ \text{rpm}}$$

(b) Finding angular velocity at rotational speed of 1000 rpm:

$$\omega = N \times \frac{2\pi}{60} = 1000 \times \frac{2\pi}{60}$$
$$\boldsymbol{\omega = 105\,\text{rad s}^{-1}}$$

Finding torque transmitted at this speed:

$$T = n\mu[m\omega^2(r_0 + x) - F_0]\frac{D}{2}$$
$$T = 2 \times 0.35 \times [\{0.25 \times 105^2 \times (0.135 + 0.005)\} - 40] \times 0.15$$
$$\boldsymbol{T = 36.3\,\text{N m}}$$

Finding power transmitted at this speed:

$$\text{Power} = T\omega = 36.3 \times 105$$
$$\boldsymbol{\text{Power} = 3.81 \times 10^3\ \text{W}}\quad\text{or}\quad\boldsymbol{3.81\ \text{kW}}$$

Test your knowledge 1.9

1. What are the units in which spring stiffness is measured?
2. If the normal force between two surfaces is 100 N and their coefficient of friction is 0.3, what force will be required to make them slide over each other?
3. What determines the rotational speed at which a centrifugal clutch starts to engage?
4. How do you convert rotational speed in revolutions per minute to angular velocity measured in radians per second?

Activity 1.9

A centrifugal clutch has four bobs of mass 0.25 kg each which can slide outwards in guides and are attached to the input shaft by restraining springs. The drum attached to the output shaft has an inner diameter of 350 mm. The static clearance between the bobs and the drum is 6 mm, and the distance from the centre of rotation to the centre of gravity of the bobs is initially 155 mm. The coefficient of friction between the bobs and the drum is 0.3. If the clutch starts to engage at a speed of 500 rpm, what is the stiffness of the restraining springs? What will be the power transmitted by the clutch at a speed of 1200 rpm?

Problems 1.7

1. A body of mass 3 kg is whirled round in a horizontal plane at the end of a steel wire. The distance from the centre of rotation to the centre of the mass is 1.5 m and the diameter of the wire is 2 mm. If the ultimate tensile strength of the steel is 500 MPa, determine the rotational speed at which the wire will break.

[178 rpm]

2. A body of mass 1 kg is whirled in a horizontal plane at the end of a spring. The spring stiffness is $5 \, \text{kN m}^{-1}$ and the initial distance from the centre of rotation to the centre of the body is 200 mm. What will be the extension of the spring at a rotational speed of 300 rpm?

[49.1 mm]

3. A centrifugal clutch has four shoes, each of mass 2 kg which are held on retaining springs of stiffness $10 \, \text{kN m}^{-1}$. The internal diameter of the clutch drum is 600 mm and the radius of the centre of gravity of the shoes when in contact with the drum is 250 mm. The stationary clearance between the shoes and the drum is 20 mm, and the coefficient of friction is 0.25. Determine (a) the speed at which the clutch starts to engage, (b) the power transmitted at a speed of 1500 rpm.

[191 rpm, 2.29 MW]

4. A centrifugal clutch has four shoes, each of mass 0.3 kg. When at rest, the radial clearance between the shoes and the clutch drum is 5 mm and the radius to the centre of gravity of the shoes is 195 mm. The shoes are held on retaining springs of stiffness $8 \, \text{kN m}^{-1}$ and the internal diameter of the clutch drum is 500 mm. The coefficient of friction between the shoes and the drum is 0.2. Determine for a rotational speed of 1000 rpm (a) the radial force which each shoe exerts on the drum, (b) the torque transmitted, (c) the power transmitted.

[622 N, 124 N m, 13.1 kW]

Stability of vehicles

When you are travelling around a bend on a motor cycle or in a car the centripetal force is supplied by friction between the wheels and the road surface as you turn into the curve. If you are travelling too fast for the road conditions, or the condition of your vehicle, one of two things may happen.

1. The friction force between the wheels and the road surface will be insufficient and you will skid. The direction of your skid will be at a tangent to the curve.
2. If the centre of gravity of your vehicle is high it may overturn before starting to skid.

With a motor cycle only the first option is possible. You may of course skid into the kerb and then overturn but it is skidding which will initiate the problem. Cars, buses and trucks are designed so that even when fully loaded, they should skid before reaching the speed at which overturning would occur. Large heavy items stacked on a roof rack may raise the centre of gravity of a car to such a height that overturning is a possibility and of course this would also increase the possibility of overturning should the vehicle skid into the kerb.

Consider a four wheeled vehicle of mass m, travelling at a speed v, round a level unbanked curve of radius r as shown in Figure 1.68.

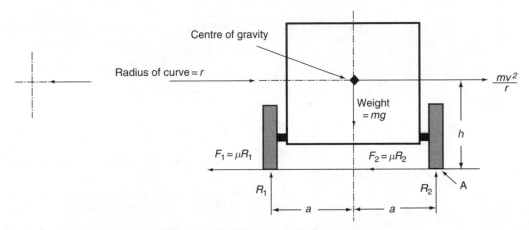

Figure 1.68 *Vehicle on an unbanked horizontal curve*

Although the vehicle is moving, it can be considered to be in a state of dynamic equilibrium. In the vertical direction, the active force is the weight W acting downwards, which is balanced by the upward reactions R_1 and R_2 of the wheels. The active forces in the horizontal direction are the friction forces F_1 and F_2 which provide the centripetal force. The equal and opposite reaction to them is the centrifugal force given by mv^2/r. This method of equating the forces acting on a moving object is known as D'Alembert's principle which will be explained more fully in Chapter 2.

Let the height of the centre of gravity be h.
Let the track width be $2a$.
Let the limiting coefficient of friction between the wheels and the road be μ.

To find the speed at which skidding is likely to occur, equate horizontal forces as the vehicle is about to skid.

centripetal force = centrifugal reaction

$$F_1 + F_2 = \frac{mv^2}{r}$$

Key point

When a motor vehicle is travelling round a curve, centripetal force is provided by friction between its tyres and the road surface. When skidding occurs, the vehicle continues in a straight line at a tangent to the curve.

In the limit as the vehicle is about to skid, $F_1 = \mu R_1$ and $F_2 = \mu R_2$. Substituting for these gives

$$\mu R_1 + \mu R_2 = \frac{mv^2}{r}$$

$$\mu(R_1 + R_2) = \frac{mv^2}{r}$$

But $R_1 + R_2 = m\,g$, the weight of the vehicle

$$\mu \cancel{m} g = \frac{\cancel{m} v^2}{r}$$

$$\mu r g = v^2$$

$$v = \sqrt{\mu r g} \tag{1.44}$$

To find the speed at which overturning is likely to occur, equate moments about the point A in the limit as the vehicle is about to overturn. In this condition, R_1 is zero as the nearside wheels are about to lift off the road and all the weight will be carried on the offside wheels.

clockwise overturning moment = anticlockwise righting moment

$$\frac{\cancel{m} v^2 h}{r} = \cancel{m} g a$$

$$v^2 = \frac{g a r}{h}$$

$$v = \sqrt{\frac{a r g}{h}} \tag{1.45}$$

Whichever of the equations (1.44) and (1.45) gives the lower value of velocity, that will be the limiting value. You should note that both values of limiting velocity are independent of the mass of the vehicle. Comparison of the two equations shows that

If $\mu < \dfrac{a}{h}$, the vehicle will skid before the overturning speed is reached.

If $\mu > \dfrac{a}{h}$, the vehicle will overturn before the skidding speed is reached.

The reaction of the outer wheels will always be greater than that of the inner wheels. To find these reactions at any speed below that at which overturning is likely, take moments about the point A again but this time include the moment of the reaction R_1.

clockwise overturning = anticlockwise righting
    moments   moment

$$\frac{mv^2}{r} h + R_1 2a = mga$$

$$R_1 2a = mga - \frac{mv^2}{r} h = m\left(ga - \frac{v^2 h}{r} \right)$$

$$R_1 = \frac{m}{2a}\left(ga - \frac{v^2}{r} h \right) \tag{1.46}$$

Now equate vertical forces to find the reaction of the outer wheels:

$$R_1 + R_2 = mg$$

$$R_2 = mg - R_1 \tag{1.47}$$

Key point

Properly loaded motor vehicles will skid rather than overturn if their speed around a curve is excessive. Overloading a roof rack can raise the centre of gravity to such a degree that overturning becomes a danger.

Example 1.21

A car of mass 900 kg travels round an unbanked horizontal curve of radius 30 m. The track width of the car is 1.5 m, and its centre of gravity is central and at a height of 0.9 m above the road surface. The limiting coefficient of friction between the tyres and the road is 0.7. Show that the car is likely to skid rather than overturn if the speed is excessive and calculate the maximum speed in km h^{-1} at which it can travel round the curve. Determine also the reactions of the inner and outer wheels at this speed.

Finding the ratio of half the track width to height of centre of gravity:

$$\frac{a}{h} = \frac{0.75}{0.9} = 0.833$$

Comparing this with the coefficient of friction, whose value is $\mu = 0.7$, shows that,

$$\mu < \frac{a}{h}$$

This indicates that the car will skid before the overturning speed is reached.

Finding speed at which car can travel round the curve, i.e. when skidding is likely to occur:

$$v = \sqrt{\mu r g} = \sqrt{0.7 \times 30 \times 9.81}$$

$$v = \textbf{14.35 m s}^{-1}$$

Changing to km h^{-1}

$$v = \frac{14.4 \times 60 \times 60}{1000}$$

$$v = \textbf{51.7 km h}^{-1}$$

Finding reaction R_1 of inner wheels by taking moments about point of contact of outer wheels and road:

$$R_1 = \frac{m}{2a}\left(ga - \frac{v^2}{r}h\right)$$

$$R_1 = \frac{900}{1.5}\left[(9.81 \times 0.75) - \frac{(14.35^2 \times 0.9)}{30}\right]$$

$$R_1 = \textbf{708 N}$$

Finding R_2 by equating vertical forces:

$$R_2 = mg - R_1 = (900 \times 9.81) - 708$$

$$R_2 = \textbf{8121 N} \quad \text{or} \quad \textbf{8.12 kN}$$

In practice the bends on major roads are banked to reduce the tendency for side-slip. When driving at the speed where there is no side-slip, the forces acting on a vehicle are as shown in Figure 1.69.

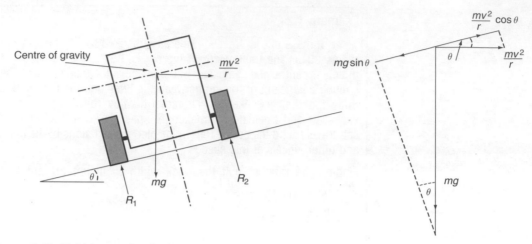

Figure 1.69 *Vehicle on a banked curve*

The speed at which there is no side-slip can be found by resolving forces parallel to the road surface. The two wheel reactions are perpendicular to the road surface, and are not involved. This just leaves the component of the weight down the slope, which is the active centripetal force, and the component of the centrifugal reaction up the slope. For no side-slip, these must be equal and opposite.

$$\frac{mv^2}{r}\cos\theta = mg\sin\theta$$

$$v^2 = rg\frac{\sin\theta}{\cos\theta} = rg\tan\theta$$

$$v = \sqrt{rg\tan\theta} \qquad (1.48)$$

Key point

In the condition where there is no tendency for side-slip, the reactions of the inner and outer wheels are equal.

Example 1.22

A vehicle has a track width of 1.4 m and the height of its centre of gravity is 1.2 m above the road surface. The limiting coefficient of friction between the wheels and the road surface is 0.7.

(a) What is the maximum speed at which the vehicle can travel round a curve of radius 75 m?

(b) To what angle would the curve have to be banked for the vehicle to travel round it at a speed of 50 km h^{-1} without any tendency to side-slip?

(a) Finding the ratio of half the track width to height of centre of gravity:

$$\frac{a}{h} = \frac{0.7}{1.2} = 0.583$$

Comparing this with the coefficient of friction whose value is $\mu = 0.7$ shows that

$$\mu > \frac{a}{h}$$

This indicates that the car will overturn before the skidding speed is reached.

(b) Finding speed at which car can travel round the curve, i.e. when overturning is likely to occur:

$$v = \sqrt{\frac{arg}{h}} = \sqrt{\frac{0.7}{1.2} \times 75 \times 9.81}$$
$$v = 20.7 \, \text{m s}^{-1}$$

Changing to km h^{-1}

$$v = \frac{20.7 \times 60 \times 60}{1000}$$
$$v = 74.6 \, \text{km h}^{-1}$$

Changing $50 \, \text{km h}^{-1}$ to m s^{-2},

$$v = \frac{50 \times 1000}{60 \times 60}$$
$$v = 13.9 \, \text{m s}^{-1}$$

Finding angle to which curve would need to be banked for no side-slip at this speed:

$$v = \sqrt{rg \tan \theta}$$
$$v^2 = rg \tan \theta$$
$$\tan \theta = \frac{v^2}{rg} = \frac{13.9^2}{75 \times 9.81}$$
$$\tan \theta = 0.262$$
$$\theta = 14.7°$$

Activity 1.10

A vehicle of mass 1 tonne travels round a horizontal unbanked curve of radius 35 m. The track width of the vehicle is 1.8 m and the height of its centre of gravity above the road surface is 0.95 m. The coefficient of friction between the tyres and the road is 0.65.

(a) Show that if the speed of the vehicle is excessive, it is likely to skid rather than overturn.
(b) Calculate the speed at which skidding is likely to occur.
(c) Calculate the reactions of the inner and outer wheels at this speed.
(d) If the curve were to be banked at an angle of 3.5° what would be the speed at which the vehicle could travel around it without any tendency for side-slip?

Problems 1.8

1. A four-wheeled vehicle has a track width of 1.15 m and a mass of 750 kg. Its centre of gravity is 1.2 m above the road surface and equidistant from each of the wheels. What will be the reactions of the inner and outer wheels when travelling at a speed of $60 \, \text{km h}^{-1}$ around a level curve of radius 80 m.

[6.4 kN, 961 N]

2. What is the maximum speed at which a car can travel over a hump-backed bridge whose radius of curvature is 18 m?

[47.8 km h^{-1}]

3. A railway wagon has a mass of 12 tonne and the height of its centre of gravity is 1.5 m above the rails. The track width is 1.37 m and the wagon travelling at 70 km h^{-1}. What is the minimum radius of level curve that the wagon can negotiate at this speed without overturning?

[84.4 m]

4. A car of mass 850 kg travels round an unbanked horizontal curve of radius 35 m. The limiting coefficient of friction between the tyres and the road is 0.65. The track width of the car is 1.65 m and its centre of gravity is central at a height of 0.8 m above the road surface. Show that the car is likely to skid rather than overturn if the speed is excessive and calculate the maximum speed in km h^{-1} at which it can travel round the curve. What will be the reactions of the inner and outer wheels at this speed?

[53.8 km h^{-1}, 1541 N, 4149 N]

5. The height of the centre of gravity of a four-wheeled vehicle is 1.1 m above the road surface and equidistant from each of the wheels. The track width is 1.5 m and the limiting coefficient of friction between the wheels and the road surface is 0.7. Determine (a) the maximum speed at which the vehicle can travel round a curve of radius 90 m, (b) the angle to which the curve would have to be banked for the vehicle to travel round it at a speed of 60 km h^{-1} without any tendency to side-slip?

[88.3 km h^{-1}, 17.5°]

Simple machines

A simple machine is an arrangement of moving parts whose purpose is to transmit motion and force. The ones which we will consider are those in which a relatively small input force is used to raise a heavy load. They include lever systems, inclined planes, screw jacks, wheel and axle arrangements, and gear trains.

For all simple machines the *mechanical advantage* or *force ratio* is the ratio of the load W, raised to the input effort E (Figure 1.70).

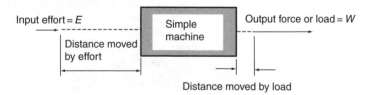

Figure 1.70 *Block diagram of a simple machine*

$$\text{mechanical advantage} = \frac{\text{load}}{\text{effort}}$$

$$MA = \frac{W}{E} \tag{1.49}$$

A characteristic of a simple machine is that the distance moved by the load is much smaller than the distance moved by the input effort. The *velocity ratio* or *movement ratio* is used to measure this effect.

$$\text{velocity ratio} = \frac{\textbf{distance moved by effort}}{\textbf{distance moved by load}} \qquad (1.50)$$

or

$$\text{velocity ratio} = \frac{\textbf{velocity at which effort moves}}{\textbf{velocity at which load moves}} \qquad (1.51)$$

The efficiency η of a simple machine is the ratio of the work output to the work input. It is usually given as a percentage.

$$\text{efficiency} = \frac{\text{work output}}{\text{work input}}$$

$$\eta = \frac{\text{load} \times \text{distance moved by load}}{\text{effort} \times \text{distance moved by effort}} \times 100\%$$

$$\eta = MA \times \frac{1}{VR} \times 100\%$$

$$\eta = \frac{MA}{VR} \times 100\% \qquad (1.52)$$

There is always some friction between the moving parts of a machine. Some of the work input must be used to overcome friction and so the work output is always less than the input. As a result, the efficiency can never be 100% and it is very often a great deal less. If there were no friction present, the mechanical advantage would be equal to the velocity ratio. In practice it is always a lower figure.

Example 1.23

A jack requires an input effort of 25 N to raise a load of 150 kg. If the distance moved by the effort is 500 mm and the load is raised through a height of 6 mm, find (a) the mechanical advantage, (b) the velocity ratio, (c) the efficiency, (d) the work done in overcoming friction.

(a) Finding mechanical advantage:

$$MA = \frac{\text{load}}{\text{effort}} = \frac{150 \times 9.81}{25}$$

$$\textbf{\textit{MA}} = \textbf{58}.\textbf{9}$$

(b) Finding velocity ratio:

$$VR = \frac{\text{Distance moved by effort}}{\text{Distance moved by load}} = \frac{500}{6}$$

$$\textbf{\textit{VR}} = \textbf{83}.\textbf{3}$$

(c) Finding efficiency:

$$\eta = \frac{MA}{VR} = \frac{58.9}{83.3} \times 100\%$$

$$\eta = \textbf{70}.\textbf{7\%}$$

(d) Finding work done in overcoming friction:

The work output is 70.7% of the work input. The remaining 29.3% of the work input is the work done in overcoming friction.

Friction work $=$ work input $\times \frac{29.3}{100}$
Friction work $=$ effort $\times$ distance effort moves $\times \frac{29.3}{100}$
Friction work $= 25 \times 0.5 \times \frac{29.3}{100}$
Friction work $= 3.66\,J$

Velocity ratio formulae

The expressions (1.50) and (1.51) are general formulae for calculating velocity ratio, and these can now be applied to a range of devices. Each will then have its own particular formula for velocity ratio which is dependent upon its dimensions and its mode of operation. For each device let the distance moved by the effort be a, and the corresponding distance moved by the load be b. That is,

$$\text{velocity ratio} = \frac{\text{distance moved by effort}}{\text{distance moved by load}}$$

$$VR = \frac{a}{b} \tag{1.53}$$

Simple lever systems

For both types of lever shown in Figure 1.71, the distances moved by the effort and the load are proportional to the distances x and y from the fulcrum of the lever.

$$VR = \frac{a}{b} = \frac{x}{y} \tag{1.54}$$

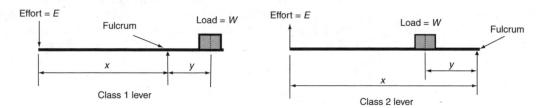

Figure 1.71 *Lever systems*

Inclined plane

In pulling the load up the incline, the effort moves through distance a whilst lifting the load through a vertical height b (Figure 1.72).

$$VR = \frac{a}{b} = \frac{\cancel{a}}{\cancel{a} \sin \theta}$$

$$VR = \frac{1}{\sin \theta} \tag{1.55}$$

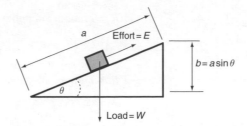

Figure 1.72 *Inclined plane*

Screw jack

The screw jack is a practical form of the inclined plane. The plane is now in the form of a spiral and the effort is applied horizontally to the end of the operating handle. One complete turn of the handle causes the load to rise through a distance equal to the pitch of the screw thread (Figure 1.73).

$$VR = \frac{\text{distance moved by effort in one revolution of screw thread}}{\text{pitch of screw thread}}$$

$$VR = \frac{2\pi r}{p} \tag{1.56}$$

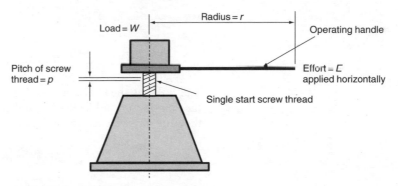

Figure 1.73 *Screw jack*

Pulley blocks

Here the velocity ratio is equal to the number of pulleys in operation or alternatively, the number of rope lengths connecting the pulleys, excluding the effort rope (Figure 1.74).

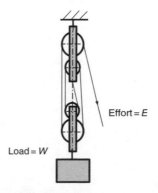

Figure 1.74 *Pulley blocks*

$$VR = \text{number of pulleys} \tag{1.57}$$

or

$$VR = \text{number of connecting rope lengths} \tag{1.58}$$

Weston differential pulley block

This comprises a compound pulley with diameters D_1 and D_2, and a snatch block which carries the load. They are connected by an endless chain to which the effort is applied as shown in Figure 1.75.

For one rotation of the compound pulley, the chain is wound onto the larger diameter and off the smaller diameter. That part of the chain passing round the snatch block shortens by a length equal to the difference between the circumferences of the compound pulley. The load is raised by half of this distance.

$$VR = \frac{\text{circumference of larger compound pulley}}{\text{half the difference of the compound pulley circumferences}}$$

$$VR = \frac{\pi D_1}{\frac{\pi D_1 - \pi D_2}{2}} = \frac{\cancel{\pi} D_1}{\frac{\cancel{\pi}(D_1 - D_2)}{2}}$$

$$VR = \frac{2D_1}{(D_1 - D_2)} \tag{1.59}$$

The links of the chain engage on teeth or 'flats' which are cut on the compound pulley. If the numbers of flats n_1 and n_2, are known, they too can be used to calculate the velocity ratio.

$$VR = \frac{2n_1}{(n_1 - n_2)} \tag{1.60}$$

You will note that it is the difference between the diameters on the compound pulley which determines the velocity ratio. The closer the two diameters, the higher will be the velocity ratio.

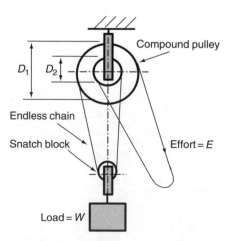

Figure 1.75 *Differential pulley block*

Simple wheel and axle

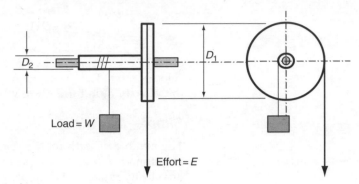

Figure 1.76 *Simple wheel and axle*

For one revolution of the wheel and axle, the effort moves a distance equal to the wheel circumference and the load is raised through a distance equal to the axle circumference (Figure 1.76).

$$VR = \frac{\text{circumference of wheel}}{\text{circumference of axle}} = \frac{\cancel{\pi}D_1}{\cancel{\pi}D_2}$$

$$VR = \frac{D_1}{D_2} \tag{1.61}$$

Differential wheel and axle

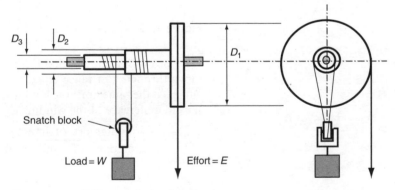

Figure 1.77 *Differential wheel and axle*

Here the wheel, whose diameter is D_1, has a compound axle with two diameters, D_2 and D_3. The two axle diameters are wound in opposite directions with the same cord which also passes round the snatch block. For one revolution of the wheel, the effort moves a distance equal to the wheel's circumference. At the same time the length of cord around the snatch block shortens by a length equal to the difference between the two axle circumferences, and the load is raised by a distance which is half of this difference (Figure 1.77).

$$VR = \frac{\text{circumference of wheel}}{\text{half the difference of the axle circumferences}}$$

$$VR = \frac{\pi D_1}{\frac{\pi D_2 - \pi D_3}{2}} = \frac{\not{\pi} D_1}{\frac{\not{\pi}(D_2 - D_3)}{2}}$$

$$VR = \frac{2D_1}{(D_2 - D_3)} \tag{1.62}$$

You will note that it is not only the wheel diameter, but also the difference between the two axle diameters which determines the velocity ratio. The closer the two axle diameters, the higher will be the velocity ratio for a given wheel size.

Simple gear winch

The number of teeth on the gears A and B in the speed reduction gear train are t_A and t_B respectively (Figure 1.78). The gear ratio, or velocity ratio, of the simple gear train alone will be

$$\text{gear ratio} = \frac{t_A}{t_B}$$

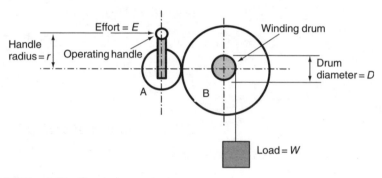

Figure 1.78 *Simple gear winch*

For one complete revolution of the operating handle, the effort will move through a distance equal to the circumference of its turning circle. At the same time, the load will rise through a distance equal to the circumference of the winding drum multiplied by the gear ratio.

$$VR = \frac{\text{circumference of handle turning circle}}{\text{circumference of winding drum} \times \text{gear ratio}}$$

$$VR = \frac{2\not{\pi} r}{\not{\pi} D \times \left(\frac{t_A}{t_B}\right)}$$

$$VR = \frac{2rt_B}{Dt_A} \tag{1.63}$$

Compound gear winch or crab winch

The numbers of teeth on gears A, B, C and D are t_A, t_B, t_C and t_D respectively. Gears B and C are keyed on the same shaft and rotate together to form a compound gear (Figure 1.79). The gear ratio or velocity ratio of the compound reduction gear train alone will be

$$\text{gear ratio} = \frac{t_A}{t_B} \times \frac{t_C}{t_D}$$

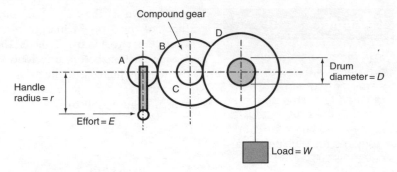

Figure 1.79 *Crab winch*

As with the simple gear winch, one complete revolution of the operating handle will cause the effort to move through a distance equal to the circumference of its turning circle. At the same time the load will rise through a distance equal to the circumference of the winding drum multiplied by the gear ratio.

$$VR = \frac{\text{circumference of handle turning circle}}{\text{circumference of winding drum} \times \text{gear ratio}}$$

$$VR = \frac{2\pi r}{\pi D \times \left(\dfrac{t_A}{t_B} \times \dfrac{t_C}{t_D}\right)}$$

$$VR = \frac{2r}{D}\left(\frac{t_B}{t_A} \times \frac{t_D}{t_C}\right) \tag{1.64}$$

Law of a machine

When a range of load and the corresponding effort values are tabulated for any of the above machines, a graph of effort against load is found to have the straight line form shown in Figure 1.80(a).

The equation of graph in Figure 1.80(a) is

$$E = aW + b \tag{1.65}$$

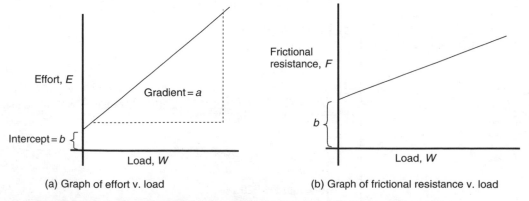

(a) Graph of effort v. load   (b) Graph of frictional resistance v. load

Figure 1.80 *Graphs of effort and frictional resistance v. load*

66 *Mechanical principles*

Key point

Frictional resistance increases with load and some of the effort is required to overcome it. As a result, the work output is always less than the work input and the efficiency can never be 100%.

This is known as the *law of the machine*. The constant a, is the gradient of the straight line graph. The constant b, is the intercept on the effort axis. This is the effort initially required to overcome friction before any load can be lifted. It is found that as the load is increased, the frictional resistance in the mechanism increases from the initial value b, in a linear fashion as shown in Figure 1.80(b).

Example 1.24

A differential wheel and axle has a wheel diameter of 275 mm and axle diameters of 50 mm and 100 mm. The law of the machine is $E = 0.11\,W + 4.5$. Determine (a) the velocity ratio of the device, (b) mechanical advantage and efficiency when raising a load of 25 kg, (c) the work input which is required to raise the load through a height of 1.5 m.

(a) Finding velocity ratio:

$$VR = \frac{2D_1}{(D_2 - D_3)} = \frac{2 \times 275}{(100 - 50)}$$

$$\boldsymbol{VR = 11}$$

(b) Finding effort required to raise a mass of 25 kg:

$$E = 0.15\,W + 4.5 = (0.11 \times 25 \times 9.81) + 4.5$$

$$\boldsymbol{E = 31.5\,N}$$

Finding mechanical advantage when raising this load:

$$MA = \frac{W}{E} = \frac{25 \times 9.81}{31.5}$$

$$\boldsymbol{MA = 7.79}$$

Finding efficiency when raising this load:

$$\eta = \frac{MA}{VR} = \frac{7.79}{11}$$

$$\boldsymbol{\eta = 0.708} \quad \text{or} \quad \boldsymbol{70.8\%}$$

(c) Finding work input:

$$\eta = \frac{\text{work output}}{\text{work input}}$$

$$\text{work input} = \frac{\text{work output}}{\eta} = \frac{\text{load} \times \text{distance moved by load}}{\eta}$$

$$\text{work input} = \frac{25 \times 9.81 \times 1.5}{0.708}$$

$$\boldsymbol{\text{work input} = 520\,J}$$

Limiting efficiency and mechanical advantage

It is found that the efficiency of a simple machine increases with load but not in a linear fashion.

As can be seen in Figure 1.81, the efficiency eventually levels off at a limiting value. This is found to depend on the constant a, in the law of the machine, and its velocity ratio.

$$\text{efficiency} = \frac{\text{mechanical advantage}}{\text{velocity ratio}}$$

$$\eta = \frac{MA}{VR} = \frac{W}{E \times VR}$$

Now, from the law of the machine, $E = aW + b$

$$\eta = \frac{W}{(aW + b) \times VR}$$

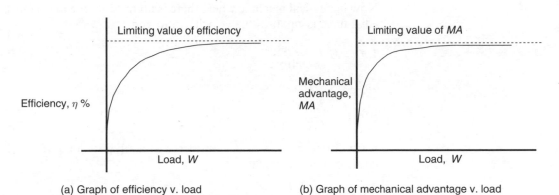

(a) Graph of efficiency v. load   (b) Graph of mechanical advantage v. load

Figure 1.81 *Variation of efficiency and mechanical advantage with load*

Dividing numerator and denominator by W gives

$$\eta = \frac{1}{\left(a + \dfrac{b}{W}\right) \times VR}$$

$$\eta = \frac{1}{a\,VR + \dfrac{b}{W}\,VR} \qquad (1.66)$$

Examination of this expression shows that as the load W, increases, the second term in the denominator becomes smaller and smaller. When the load becomes very large, this term tends to zero and the limiting value of efficiency is

$$\eta = \frac{1}{a\,VR} \qquad (1.67)$$

In this limiting condition the mechanical advantage levels off to a value given by

$$\eta = \frac{MA}{VR}$$

$$MA = \eta \times VR$$

$$MA = \frac{1}{a} \qquad (1.68)$$

The limiting value of mechanical advantage is thus the reciprocal of the gradient of the effort v. load graph.

Key point

The mechanical advantage and the efficiency of a simple machine increase with load but eventually level off at a limiting value.

Overhauling

A simple machine is said to *overhaul* if when the effort is removed, the load falls under the effects of gravity. If a machine does not overhaul, then friction alone must be sufficient to support the load. If you have changed a wheel on a car you probably used a screw jack to raise the wheel off the road surface. Friction in the screw thread is then sufficient to support the car. This means, of course that there must be quite a lot of friction present which results in a low value of efficiency. This however is the price that often has to be paid for safety in lifting devices such as car jacks and engine hoists.

Frictional resistance can be considered as an additional load which the effort must overcome.

$$\frac{\text{total}}{\text{effort}} = \frac{\text{effort to overcome}}{\text{friction load, } F} + \frac{\text{effort to overcome}}{\text{actual load, } W}$$

Now in an ideal machine where there is no friction, the mechanical advantage is equal to the velocity ratio and

$$VR = MA = \frac{\text{load}}{\text{effort}}$$

and

$$\text{effort} = \frac{\text{load}}{VR} = \frac{W}{VR}$$

If friction force, F is being considered as a separate load, the total effort can be written as

$$\text{total effort} = \frac{F}{VR} + \frac{W}{VR}$$

$$E = \frac{F + W}{VR} \tag{1.69}$$

Now, the efficiency is given by

$$\eta = \frac{MA}{VR} = \frac{W}{E \times VR}$$

Substituting for E gives

$$\eta = \frac{W}{\left(\frac{F + W}{VR}\right) VR}$$

$$\eta = \frac{W}{F + W} \tag{1.70}$$

When the effort is removed there is only the frictional resistance F, to oppose the load W. To stop it overhauling the friction force must be equal or greater than the load. In the limit when $F = W$, the efficiency will be

$$\eta = \frac{W}{2W} = 0.5 \quad \text{or} \quad \mathbf{50\%} \tag{1.71}$$

It follows that a machine will overhaul if its efficiency is greater than 50%.

Key point

A simple machine will overhaul if the load falls under the effects of gravity when the effort is removed. Overhauling is likely to occur if the efficiency of a machine is greater than 50%.

Example 1.25

The law of the gear winch shown in Figure 1.82 is $E = 0.0105\,W + 5.5$ and it is required to raise a load of 150 kg. Determine, (a) its velocity ratio, (b) the effort required at the operating handle, (c) the mechanical advantage and efficiency when raising this load, (d) the limiting mechanical advantage and efficiency and (e) state whether the winch is likely to overhaul when raising the 150 kg load.

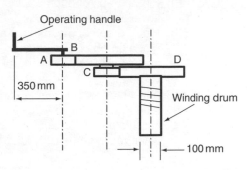

Teeth on gear A = 16
Teeth on gear B = 72
Teeth on gear C = 18
Teeth on gear D = 60

Figure 1.82

(a) Finding velocity ratio:

$$VR = \frac{2r}{D}\left(\frac{t_B}{t_A} \times \frac{t_D}{t_C}\right) = \frac{2 \times 350}{100}\left(\frac{72 \times 60}{16 \times 18}\right)$$

$$\boldsymbol{VR = 105}$$

(b) Finding effort required:

$$E = 0.0105\,W + 5.5 = (0.0105 \times 150 \times 9.81) + 5.5$$

$$\boldsymbol{E = 21.0\,N}$$

(c) Finding mechanical advantage:

$$MA = \frac{W}{E} = \frac{150 \times 9.81}{21.0}$$

$$\boldsymbol{MA = 70.1}$$

Finding efficiency:

$$\eta = \frac{MA}{VR} = \frac{70.1}{105}$$

$$\boldsymbol{\eta = 0.668} \quad \text{or} \quad \boldsymbol{66.8\%}$$

(d) Finding limiting mechanical advantage:

limiting $MA = \dfrac{1}{a}$ (where $a = 0.0105$, from the law of the machine)

$$\textbf{limiting } \boldsymbol{MA} = \frac{1}{\textbf{0.0105}} = \textbf{95.2}$$

Finding limiting efficiency:

limiting $\eta = \dfrac{1}{aVR}$

$$\textbf{limiting } \boldsymbol{\eta} = \frac{1}{\textbf{0.0105} \times \textbf{105}} = \textbf{0.907} \quad \text{or} \quad \textbf{90.7\%}$$

(e) When raising the 150 kg load the efficiency is greater than 50% and so the winch will overhaul under these conditions.

Test your knowledge 1.11

1. How does the mechanical advantage of a simple machine vary with the load raised?
2. How does the frictional resistance in a simple machine vary with the load raised?
3. How is the efficiency of a simple machine defined?
4. What information does the law of a machine contain?
5. What is meant by overhauling and how can it be predicted?

Activity 1.11

A screw jack has a single start thread of pitch 6 mm and an operating handle of radius 450 mm. The following readings of load and effort were taken during a test on the jack.

Load (kN)	0	1	2	3	4	5	6	7	8	9	10
Effort (N)	2.2	6.6	11.8	17.0	22.2	27.2	32.2	36.8	41.4	46.9	52.4

(a) Plot a graph of effort against load and from it, determine the law of the machine.
(b) Plot a graph of mechanical advantage against load.
(c) Plot a graph of efficiency against load and state whether the machine is likely to overhaul.
(d) Calculate the theoretical limiting values of mechanical advantage and efficiency, and state whether your graphs tend towards these values.
(e) Calculate the work input and the work done against friction when raising the 10 kN load through a height of 50 mm.

Problems 1.9

1. A screw jack has a single start thread of pitch 10 mm and is used to raise a load of 900 kg. The required effort is 75 N, applied at the end of an operating handle 600 mm long. For these operating conditions, determine (a) the mechanical advantage, (b) the velocity ratio, (c) the efficiency of the machine.

   [118, 377, 31.3%]

2. The top pulleys of a Weston differential pulley block have diameters of 210 mm and 190 mm. Determine the effort required to raise a load of 150 kg if the efficiency of the system is 35%. What is the work done in overcoming friction when the load is raised through a height of 2.5 m?

   [200 N, 6.83 kJ]

3. A differential wheel and axle has a wheel diameter of 300 mm and axle diameters of 100 mm and 75 mm. The effort required to raise a load of 50 kg is 55 N and the effort required to raise a load of 200 kg is 180 N. Determine (a) the velocity ratio, (b) the law of the machine, (c) the efficiency of the machine when raising a load of 100 kg.

   [24, $E = 0.085\,W + 13.3$, 42.3%]

4.

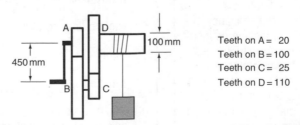

Teeth on A =	20
Teeth on B =	100
Teeth on C =	25
Teeth on D =	110

Figure 1.83

The crab winch shown in Figure 1.83 has a law of the form $E = 0.01W + 7.5$. Determine (a) its velocity ratio, (b) its mechanical advantage and efficiency when raising a load of 50 kg, (c) the limiting values of its mechanical advantage and efficiency.

[198, 20%, 100, 50.5%]

Belt drives

Power is often transmitted between parallel shafts by means of a belt running on pulleys attached to the shafts. Belt drives are friction drives. They depend on the frictional resistance around the arcs of contact between the belt and the pulleys in order to transmit torque. You will find many examples of belt drives in everyday life. The alternator and water pump on most car engines are driven by a belt from a pulley on the crankshaft. Washing machine drums and tumble dryers are also driven by a belt from an electric motor.

Torque is transmitted from the driving to the driven pulley as a result of the difference in tension which is present on opposite sides of the belt. As more and more torque is transmitted, a point is reached where the frictional resistance has a limiting value. Here the belt begins to slip. Slipping always occurs on the smaller of the two pulleys where the arc of contact, or 'angle of lap' is least. Compared with geared systems, chain drives and shaft drives, belt drives have certain advantages.

- They are automatically protected against overload because slipping occurs if the load exceeds the maximum which can be sustained by friction.
- The initial tension to which the belt is set can be adjusted to determine the load at which slipping occurs.
- The length of the belt can be selected to suit a wide range of shaft centre distances.
- Different sizes of pulley may be used to step up or step down the rotational speed.
- The cost is less than for the equivalent gear train or chain drive.

Consider a typical belt drive in which torque is transmitted from a driving pulley A, to a driven pulley B as in Figure 1.84.

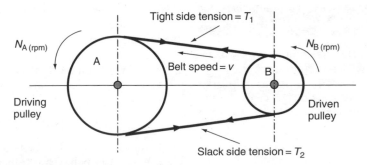

Figure 1.84 *Belt drive*

The velocity ratio of the belt drive is given by

$$VR = \frac{\text{input speed}}{\text{output speed}}$$

$$VR = \frac{N_A}{N_B} = \frac{\omega_A}{\omega_B} \qquad (1.72)$$

Alternatively, the velocity ratio may be found from the pulley dimensions. Now, pulley speed is inversely proportional to size so if you use the pulley diameters or radii, remember to invert the subscript letters as follows:

$$VR = \frac{D_B}{D_A} = \frac{r_B}{r_A} \qquad (1.73)$$

The speed of the belt can be found from either of the pulley speeds. First, change the speed in revolutions per minute to angular velocity in radians per second. Then, multiply by the appropriate pulley radius to change angular velocity to belt speed in metres per second.

For pulley A,

$$\omega_A = N_A \times \frac{2\pi}{60} \quad \text{and} \quad v = \omega_A \, r_A \qquad (1.74)$$

For pulley B,

$$\omega_B = N_B \times \frac{2\pi}{60} \quad \text{and} \quad v = \omega_B \, r_B \qquad (1.75)$$

The power transmitted from A to B is given by

power = resultant belt tension × belt speed

$$\textbf{power} = (T_1 - T_2)\, v \qquad (1.76)$$

Power loss in the shaft bearings is usually negligible compared to the power transmitted and it can be assumed that

input power = output power

input torque × input angular velocity = output torque × output angular velocity

$$T_A \times \omega_A = T_B \times \omega_B$$
$$\frac{T_A}{T_B} = \frac{\omega_B}{\omega_A} = \frac{1}{VR} \qquad (1.77)$$

The material from which a belt is made, is assumed to be perfectly elastic so that when the drive is taken up, the increase in tension on the tight side is equal to the decrease in tension on the slack side. If the initial tension setting is T_0, then

increase in tension on tight side = decrease in tension on slack side

$$T_1 - T_0 = T_0 - T_2$$
$$T_1 + T_2 = 2T_0 \qquad (1.78)$$

Key point

The maximum power which a belt drive can transmit before slipping occurs depends on its initial tension setting.

Both flat and V-section belts are used in belt drives. It is found that V-section belts can transmit more power before slipping occurs because of the wedge action between the belt and the sides of the V-shaped pulley. It is possible to calculate the ratio of the tight and slack side tensions at which slipping takes place. If you continue your studies to a higher level you will derive the formulae used.

For flat belts, slipping typically occurs when $\dfrac{T_1}{T_2} = 2 - 2.5$

For V-section belts, slipping typically occurs when $\dfrac{T_1}{T_2} = 6 - 15$

Example 1.26

In a belt drive, the smaller driving pulley A has a diameter of 110 mm and rotates at a speed of 500 rpm. The larger driven pulley B, has a diameter of 275 mm. The tension in the belt is not to exceed 2 kN and the initial tension setting is 1.2 kN. Find (a) the angular velocity of the driven pulley, (b) the belt speed, (c) the limiting slack side tension, (d) the maximum power that the belt can transmit, (e) the maximum output torque.

(a) Finding velocity ratio:

$$VR = \frac{D_B}{D_A} = \frac{275}{110}$$
$$\textbf{\textit{VR} = 2.5}$$

Finding output speed:

$$VR = \frac{N_A}{N_B}$$
$$N_B = \frac{N_A}{VR} = \frac{500}{2.5}$$
$$\textbf{\textit{N}}_B = \textbf{200 rpm}$$

Changing to angular velocity in rads s^{-1}

$$\omega_B = N_B \times \frac{2\pi}{60} = 200 \times \frac{2\pi}{60}$$
$$\boldsymbol{\omega_B = 20.9\,\text{rad s}^{-1}}$$

(b) Finding belt speed from angular velocity of pulley B:

$$v = \omega_B r_B = 20.9 \times \frac{275 \times 10^{-3}}{2}$$
$$\textbf{\textit{v} = 2.87 m s}^{-1}$$

(c) Finding the slack side tension when the tight side tension is 2 kN:

$$2\,T_0 = T_1 + T_2$$
$$2 \times 1200 = 2000 + T_2$$
$$2400 = 2000 = T_2$$
$$T_2 = 2400 - 2000$$
$$\textbf{\textit{T}}_2 = \textbf{400 N}$$

1. Where is slipping likely to occur on a belt drive?
2. How can a belt drive be adjusted to alter the load at which slipping occurs?
3. Why is a V-section belt drive able to transmit more power before slipping than a flat belt of the same material and tension setting?
4. What is the relationship between the velocity ratio of a belt drive and the ratio of the input and output torques?

(d) Finding maximum power that the belt can transmit:

$$\text{maximum power} = (T_1 - T_2)v$$
$$\text{maximum power} = (2000 - 400) \times 2.87$$
$$\textbf{maximum power} = \textbf{4.59} \times \textbf{10}^3 \textbf{ W} \quad \text{or} \quad \textbf{4.59 kW}$$

(e) Finding maximum output torque T_B:

$$\text{maximum power} = \text{output torque} \times \text{output angular velocity}$$
$$\text{maximum power} = T_B \omega_B$$
$$T_B = \frac{\text{maximum power}}{\omega_B} = \frac{4.59 \times 10^3}{20.9}$$
$$T_B = \textbf{220 N m}$$

Activity 1.12

A flat belt connects a driving pulley of diameter 350 mm and a driven pulley of diameter 150 mm. The driving pulley rotates at 250 rpm and the belt is given an initial tension of 1 kN. When the belt is on the point of slipping, the ratio of the tight and slack side tensions is 2.5. Find (a) the angular velocity of the driven pulley, (b) the belt speed, (c) the tight and slack side tensions when the belt is on the point of slipping, (d) the maximum power which can be transmitted, (e) the maximum output torque.

Problems 1.10

1. A belt drive for speed reduction has pulley diameters of 180 mm and 320 mm. The tight side tension is to be limited to 500 N and it is required to transmit 1.5 kW with an output speed of 400 rpm. Determine (a) the input torque, (b) belt speed, (c) the limiting slack side tension, (d) the initial tension to which the belt should be set. Friction losses may be neglected.

   [20.1 N m, 6.71 m s^{-1}, 276 N, 388 N]

2. A belt drive is set to an initial tension of 350 N. The driving pulley is 300 mm in diameter and rotates at 500 rpm. The driven pulley is 200 mm in diameter and friction losses may be neglected. If the tight side tension is to be limited to 600 N, determine (a) the limiting slack side tension, (b) the belt speed, (c) the maximum power that the belt can transmit, (d) the maximum values of input and output torque.

   [100 N, 7.85 m s^{-1}, 3.93 kW, 75 Nm, 50 N m]

3. The input pulley of a belt drive is 150 mm in diameter and rotates at a speed of 600 rpm. The output pulley is 350 mm in diameter and it can be assumed that there are no friction losses in the bearings. The belt is initially set to a tension of 200 N

and slipping occurs when the tight and slack side tensions are in the ratio 10:1. Determine (a) the limiting tight and slack side tensions, (b) the maximum power which can be transmitted, (c) the output torque.

[36 N, 400 N, 1.71 kW, 81.8 N m]

Plane linkage mechanisms

Machines are devices in which the input work, or energy, is converted into a more useful form in order to do a particular job of work. The screw jacks, pulley systems and gear winches, which we have already examined, are examples of simple machines. Here the input power, which may be manual or from a motor, is converted into a lifting force.

Machines may contain a number of parts. Levers, transmission shafts, pulleys, lead screws, gears, belts and chains are all typical machine components. Machines may also contain rigid links. The crank, connecting rods and pistons in an internal combustion engine are links. Car suspension units are made up of links and they are also to be found in machine tools, photocopiers and printers. A *linkage mechanism* may be defined as a device which transmits or transfers motion from one point to another.

Two of the most common plane linkage mechanisms are the slider-crank and the four-bar chain shown in Figure 1.85.

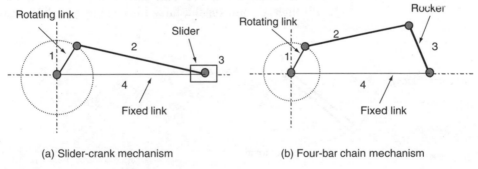

(a) Slider-crank mechanism   (b) Four-bar chain mechanism

Figure 1.85 *Linkage mechanisms*

With the slider-crank mechanism, the crank (1) rotates at a uniform speed and imparts a reciprocating translational motion to the piston (3). It is said to have an input rotation and output translation. The connecting link (2) has both a rotational and translational motion and link (4) is formed by the stationary machine frame or cylinder block.

With the four-bar linkage the crank (1) rotates at a uniform speed and imparts a rocking rotational motion to the output link (3). It has input rotation and output rotation. Here again, the connecting link (2) has both a rotational and translational motion and link (4) is formed by the stationary machine frame.

The nature of the output rotation can be altered by changing the lengths of the links. Different output characteristics can also be obtained by fixing a different link in the chain. Such mechanisms are called *inversions* of the slider crank and four-bar chain. Whatever the arrangement, there are three types of link in the above mechanisms.

Key point

A link may have linear motion, rotational motion, or a combination of the two. Translational motion is another name for linear motion.

1. *Links which have translational motion*
   These are links such as the piston (4) in the slider-crank. Here, all points on the link have the same linear velocity. The velocity vector diagram for such a link is shown in Figure 1.86.

Figure 1.86 *Slider velocity vector*

The velocity vector *ab* gives the magnitude and direction of the piston velocity. No arrow is necessary. The sequence of the letters, *a* to *b*, gives the direction. The notation v_{BA}, indicates that this is the velocity of the piston B, relative to a point A on the fixed link or machine frame. The notation is similar to that which we have already used on force vector diagrams, where upper case letters are used on the space diagram and lower case letters on the vector diagram.

2. *Links with rotational motion*
   These are links where one end rotates about a fixed axis through a point on the link. Link (1) in the slider crank and

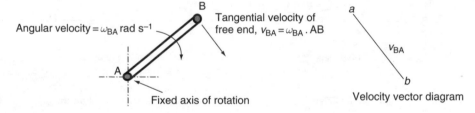

Figure 1.87 *Rotating link velocity vector*

links (1) and (4) in the four-bar chain are of this type. The velocity vector diagram for such a link is shown in Figure 1.87.

   Here the velocity vector *ab*, gives the magnitude and direction of the velocity of the free end B, as it rotates about the fixed axis through the end A, i.e. in the direction *a* to *b*. The notation v_{BA} again indicates that this is the velocity of B relative to A.

3. *Links which have a combined translational and rotational motion*
   Links such as the connecting link (2) in both the slider crank and the four-bar chain are of this type. The velocity vector diagram for such a link is shown in Figure 1.88.

   Here the two ends of the link BC are moving in different directions at the instant shown. This gives both a rotational and translational motion to the link. The two velocity vectors *ab* and *ac* are drawn from the same point and the vector *bc* is the velocity of C relative to B. In other words, if you sat on the end B, looking at C, that is what its velocity v_{CB}, would appear to be. You should

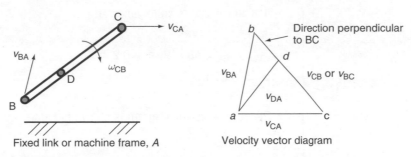

Figure 1.88 *Velocity vector diagram for link with translational and rotational motion*

note that the direction of the vector *bc* is always perpendicular to the link BC.

If you sat on the end B looking at C, it would also appear to be rotating about you with angular velocity ω_{CB}. This angular velocity can be calculated by dividing v_{CB} by the radius of rotation which is the length BC. That is,

$$\omega_{CB} = \frac{v_{CB}}{CB} \tag{1.79}$$

Alternatively, if you sat on the end C, looking at B, it would appear to be moving with velocity v_{BC} which is given by the vector *cb*. The velocity vector diagram also enables the velocity of any point D, on the link to be measured. Suppose that the point D is one third of the way along the link from the end B. The point *d*, on the vector diagram is similarly located at one third of the distance *bc* measured from *b*. The vector *ad* then gives the velocity v_{DA} of the point D relative to the fixed point A.

Key point

When the two ends of a link are travelling in different directions, the velocity of one end relative to the other is always at right angles to the link.

Example 1.27

In the slider-crank mechanism shown in Figure 1.89, the crank AB is of length 200 mm and the connecting rod BC is of length 800 mm. The crank rotates clockwise at a steady speed of 240 rpm. At the instant shown, determine (a) the velocity of the crank pin at B, (b) the velocity of the piston, (c) the velocity of the point D which is at the centre of the connecting rod, (d) the angular velocity of the connecting rod.

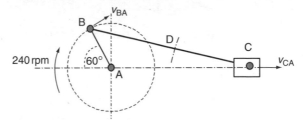

Figure 1.89

(a) Finding angular velocity of crank:

$$\omega_{BA} = N \times \frac{2\pi}{60} = 240 \times \frac{2\pi}{60}$$

$$\omega_{BA} = 25.1 \text{ rad s}^{-1}$$

Finding velocity of crank pin at B:

$$v_{BA} = \omega_{BA} \times AB = 25.1 \times 0.2$$
$$\boldsymbol{v_{BA} = 5.03\ m\ s^{-1}}$$

The velocity vector diagram is shown in Figure 1.90.

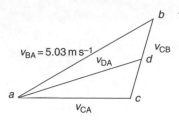

Figure 1.90

1. Draw vector *ab* for v_{BA} to a suitable scale.
2. From the point *a*, draw in a horizontal construction line for the direction of vector *ac*.
3. From the point *b*, draw a construction line perpendicular to the connecting rod BC for the vector *bc*. The intersection of the two construction lines fixes the point *c*.
4. Locate the point *d* which is the mid-point of vector *bc* and draw in the vector *ad*.

(b) Finding velocity of piston v_{CA}:
    From diagram,

$$\boldsymbol{v_{CA} = ac = 3.8\ m\ s^{-1}}$$

(c) Finding v_{DA}, which is the velocity of D, the mid-point of the connecting rod:
    From diagram,

$$\boldsymbol{v_{DA} = ad = 4.3\ m\ s^{-1}}$$

(d) Finding v_{CB}, which is the velocity of C relative to B:
    From diagram,

$$\boldsymbol{v_{CB} = bc = 2.6\ m\ s^{-1}}$$

Finding v_{CB}, which is the velocity of the conecting rod:

$$\omega_{CB} = \frac{v_{CB}}{BC} = \frac{2.6}{0.8}$$
$$\boldsymbol{\omega_{CB} = 3.25\ m\ s^{-1}}$$

Example 1.28

In the four-bar linkage shown in Figure 1.91, AB = 200 mm, BC = 250 mm, CD = 200 mm and the distance between the axes of rotation at A and D is 500 mm. At the instant shown, the crank AB is rotating clockwise at a speed of 300 rpm. Determine (a) the angular velocity of the crank AB and the velocity of the point B, (b) the velocity of the point C and the angular velocity of the rocker CD, (c) the velocity of the point E on the connecting rod BC which is 100 mm from B, (d) the angular velocity of the connecting rod.

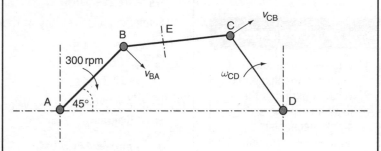

Figure 1.91

(a) Finding angular velocity of crank:

$$\omega_{BA} = N \times \frac{2\pi}{60} = 300 \times \frac{2\pi}{60}$$

$$\omega_{BA} = 31.4 \text{ rad s}^{-1}$$

Finding velocity of B:

$$v_{BA} = \omega_{BA} \times AB = 31.4 \times 0.2$$

$$\boldsymbol{v_{BA} = 6.28 \text{ m s}^{-1}}$$

The vector diagram is shown in Figure 1.92.

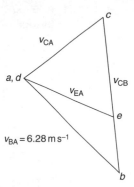

Figure 1.92

1. Draw vector *ab* for v_{BA} to a suitable scale.
2. From the point *a*, draw in a construction line in a direction at right angles to DC for the vector *dc*.

3. From the point b, draw a construction line perpendicular to the connecting rod BC for the vector bc. The intersection of the two construction lines fixes the point c.

4. Locate the point e, on vector bc such that

$$\frac{BE}{EC} = \frac{be}{ec} = \frac{100}{150} = \frac{2}{3}$$

5. Draw in the vector ae which gives the velocity of the point E.

(b) Finding velocity v_{CD}, of the point C:
From diagram,

$$v_{AD} = ac = \mathbf{4.5\,m\,s^{-1}}$$

Finding angular velocity of rocker CD:

$$\omega_{CD} = \frac{v_{CD}}{CD} = \frac{4.5}{0.2}$$

$$\omega_{CD} = \mathbf{22.5\,rad\,s^{-1}}$$

(c) Finding v_{EA}, which is the velocity of E, on the connecting rod:
From diagram,

$$v_{EA} = ae = \mathbf{4.4\,m\,s^{-1}}$$

(d) Finding v_{CB}, which is the velocity of C relative of B:
From diagram,

$$v_{CB} = bc = \mathbf{7.1\,m\,s^{-1}}$$

Finding ω_{CB}, the angular velocity of the connecting rod:

$$\omega_{CB} = \frac{v_{CB}}{BC} = \frac{7.1}{0.25}$$

$$\omega_{CB} = \mathbf{28.4\,rad\,s^{-1}}$$

Activity 1.13

In the plane mechanism shown in Figure 1.93, the crank AB is rotating clockwise at 250 rpm at the instant shown. The links have the following dimensions:

AB = 165 mm
BC = 240 mm
CD = 200 mm
CE = 350 mm

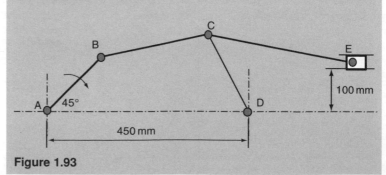

Figure 1.93

Determine (a) the angular velocity of the crank AB and the velocity of the point B, (b) the velocity of the point C and the angular velocity of the rocker CD, (c) the angular velocity of the connecting rod BC, (d) the velocity of the piston E.

Problems 1.11

1. In the slider-crank mechanism shown in Figure 1.94, the crank OA is 150 mm long and the connecting rod AB is 300 mm long. The crank rotates clockwise at a speed of 3600 rpm, and C is the mid-point of the connecting rod. At the instant shown, determine (a) the velocity of the piston, (b) the velocity of the point C, (c) the angular velocity of the connecting rod.

$$[40.5\,\mathrm{m\,s^{-1}},\ 48.6\,\mathrm{m\,s^{-1}},\ 252\,\mathrm{rad\,s^{-1}}]$$

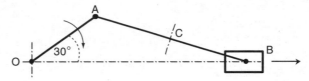

Figure 1.94

2. The crank OA of the engine mechanism shown in Figure 1.95 rotates clockwise at a speed of 3000 rpm. The length of the crank is 200 mm and the length of the connecting rod AB is 500 mm. The point C on the connecting rod is 200 mm from the crankpin at A. At the position shown, determine (a) the velocity of the piston, (b) the velocity of the point C, (c) the angular velocity of the connecting rod.

$$[66\,\mathrm{m\,s^{-1}},\ 61.8\,\mathrm{m\,s^{-1}},\ 66\,\mathrm{rad\ s^{-1}}]$$

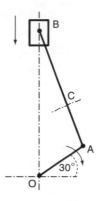

Figure 1.95

3. A four-bar linkage is shown in Figure 1.96. The crank OA, the connecting rod AB and the rocker BC are 20 mm, 120 mm and 60 mm long respectively. The crank rotates at a speed of 60 rpm and the centre distance OC is 90 mm. At the instant shown,

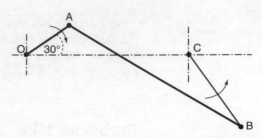

Figure 1.96

determine (a) the velocity of C, (b) the angular velocity of the
rocker BC, (c) the angular velocity of the connecting rod AB.

[0.22 m s^{-1}, 2.12 rad s^{-1}, 3.72 rad s^{-1}]

Chapter 2 Further mechanical principles

Summary

The aim of this unit is to build on the work that you have done in the core units *Science for Technicians* and *Mechanical Principles*. The section of this chapter which deals with the selection of standard section structural members, builds on the work which you have done on shear force and bending moment distribution in simply supported beams. The section on frictional resistance in mechanisms builds on the work which you have done on power transmission systems and the section on combined linear and angular motion aims to increase your knowledge of rotational dynamics. The final section investigates the occurrence of natural vibrations in mechanical systems. It will introduce you to the concept of simple harmonic motion and the characteristics of some common mechanical systems in which a disturbing force has produced mechanical oscillations.

Selection of standard section structural members

Selection of beams

In Chapter 1, we investigated the distribution of shear force and bending moment in simply supported beams. The diagrams that you plotted showed the value and position of the maximum bending moment. When selecting beams for a particular purpose, structural engineers need to know the *maximum bending moment* and the *maximum allowable stress* in the beam material. It is usual to apply a *factor of safety* of at least two against elastic failure for static structures.

$$\text{F.O.S} = \frac{\text{stress at which failure occurs}}{\text{maximum allowable working stress}}$$

$$\textbf{maximum allowable} \atop \textbf{working stress} = \frac{\textbf{stress at which failure occurs}}{\textbf{F.O.S}} \qquad (2.1)$$

Key point

The stress at which failure is considered to have occurred may be the UTS or the yield stress. The design specifications for a beam will tell you which to use when calculating factor of safety.

Elastic section modulus (z)

The *elastic section modulus* of a beam is the ratio of the maximum bending moment, M to the maximum allowable stress, σ. That is,

$$\text{elastic section modulus} = \frac{\text{maximum bending moment}}{\text{maximum allowable stress}}$$

$$z = \frac{M}{\sigma}\ (\text{m}^3) \tag{2.2}$$

Structural grade mild steel is the most common metal used for beams in static engineering structures. It is available in a variety of standard hot-rolled cross-sections. The range covers I-sections, T-sections and L-sections. Box, tubular and channel sections are also available. Their dimensions are specified by the British Standards Institution. The I-section is perhaps the most widely used for beams. Its range of sizes is covered by BS4: Part 1: 1980 which is entitled, Universal beams: dimensions and properties. An extract from it, which we will use for our selections, is shown in Table 2.1.

The different cross-sections are grouped in serial sizes and for each one the overall depth, h and the overall width, b is given together with the web and flange thickness and the mass per metre length. The web of an I-section beam is the central vertical part and the flanges are the horizontal parts at the top and bottom. When loaded, the flanges carry the greater part of the bending forces. The elastic section modulus for each section is also given and you will note that its units are cm^3. It is found that these are the most convenient units and you will have to convert your values from m^3 to these units. The selection procedure is as follows.

1. Calculate the elastic section modulus, z (m^3).
2. Convert the units of z from m^3 to cm^3. The conversion factor which you need to multiply by is 10^6.
3. Go to the column headed Elastic modulus X–X axis and select the next higher value which has the least mass per metre or the least possible depth or width. You are generally told which criteria to adopt in the design specifications for a beam.

Example 2.1

Select a standard rolled steel I-section for the simply supported beam shown in Figure 2.1. A factor of safety of 6 is to apply and the ultimate tensile strength of the material is 500 MPa. The selected section must have the least possible weight. The weight of the beam itself may be neglected when calculating the maximum bending moment.

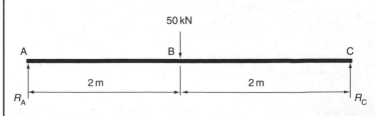

Figure 2.1 *Simply supported beam with central concentrated load*

Finding support reactions:
The beam is symmetrically loaded and so the support reactions R_A and R_C will be equal. That is,

$$R_A = R_C = \frac{50}{2}$$

$$R_A = R_C = 25 \text{ kN}$$

Finding maximum bending moment, M which will be at the centre of the beam:

$$M = R_A \times 2 = 25 \times 10^3 \times 2$$

$$M = 50 \times 10^3 \text{ N m}$$

Finding maximum allowable stress, σ:

$$\sigma = \frac{\text{ultimate tensile strength}}{\text{factor of safety}} = \frac{500 \times 10^6}{6}$$

$$\sigma = 83.3 \times 10^6 \text{ Pa}$$

Finding elastic section modulus, z:

$$z = \frac{M}{\sigma} = \frac{50 \times 10^3}{83.3 \times 10^6}$$

$$z = 600 \times 10^{-6} \text{ m}^3$$

Multiply by 10^6 to convert section modulus to cm^3:

$$z = 600 \times 10^{-6} \times 10^6$$

$$z = 600 \text{ cm}^3$$

Finding section with next higher value and least mass per metre from standard section table:

elastic section modulus, $z = 626.9 \text{ cm}^3$
serial size $= 406 \times 140$ mm

    depth, $h = 398$ mm
    width, $b = 141.8$ mm

Example 2.2

A standard rolled steel I-section is required for the simply supported beam shown in Figure 2.2. A factor of safety of 12 is to apply and the ultimate tensile strength of the material is 510 MPa. The selected section must have the least possible depth. The weight of the beam itself may be neglected when calculating the maximum bending moment.

Finding support reactions:
The beam is symmetrically loaded and so the support reactions R_A and R_C will be equal. That is,

$$R_A = R_C = \frac{3 \times 20}{2}$$

$$R_A = R_C = 30 \text{ kN}$$

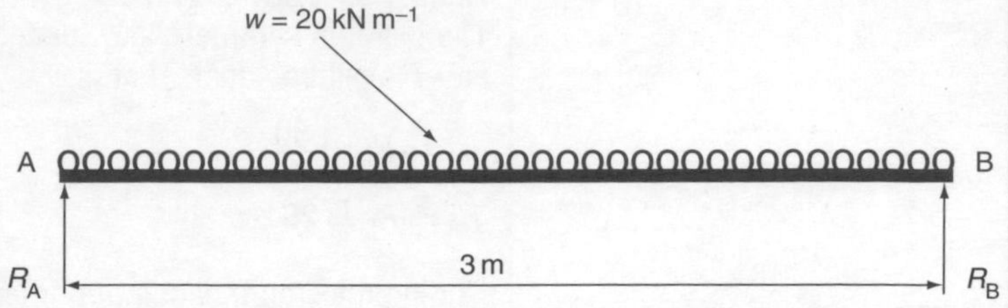

Figure 2.2 *Simply supported beam with uniformly distributed load*

Finding maximum bending moment, *M* which will be at the centre of the beam:

$M = (R_A \times 1.5) - (w \times 1.5 \times 0.75)$

$M = (30 \times 10^3 \times 1.5) - (20 \times 10^3 \times 1.5 \times 0.75)$

$\boldsymbol{M = 22.5 \times 10^3}$ **N m**

Finding maximum allowable stress, σ:

$$\sigma = \frac{\text{ultimate tensile strength}}{\text{factor of safety}} = \frac{510 \times 10^6}{12}$$

$\boldsymbol{\sigma = 42.4 \times 10^6}$ **Pa**

Finding elastic section modulus, *z*:

$$z = \frac{M}{\sigma} = \frac{22.5 \times 10^3}{42.4 \times 10^6}$$

$\boldsymbol{z = 530.7 \times 10^{-6}}$ $\mathbf{m^3}$

Multiply by 10^6 to convert section modulus to cm^3:

$z = 530.7 \times 10^{-6} \times 10^6$

$\boldsymbol{z = 530.7}$ $\mathbf{cm^3}$

Finding section with next higher value and least depth, *D* from standard section table:

elastic section modulus, $\boldsymbol{z = 561.2}$ $\mathbf{cm^3}$

serial size = 305 × 165 mm

depth, $\boldsymbol{h = 303.4}$ **mm**

width, $\boldsymbol{b = 165.0}$ **mm**

Test your knowledge 2.1

1. What is *factor of safety*?
2. What is the *elastic section modulus* of a beam?
3. How do you convert elastic section modulus values from m^3 to cm^3?
4. Which part of an I-section beam is known as the web?
5. Which part of an I-section beam carries the bulk of the bending forces?

Activity 2.1

A simply supported beam of span 2 m carries a central concentrated load of 200 kN and a uniformly distributed load of 50 kN m^{-1}. A standard rolled steel I-section is to be used with a factor of safety of 8. The ultimate tensile strength of the steel is 500 MN m^{-2}. Select a suitable section from standard tables which (a) has the least weight and depth, (b) has the least possible weight and width.

Table 2.1 Universal beams: dimensions and properties (To BS 4: Part 1: 1993)

Designation	Mass per metre (kg/m)	Depth of section h (mm)	Width of section b (mm)	Thickness web s (mm)	Thickness flange t (mm)	Root radius r (mm)	Depth between fillets d (mm)	Ratios for local buckling Flange b/2t	Ratios for local buckling Web d/s	Second moment of area Axis x-x (cm⁴)	Second moment of area Axis y-y (cm⁴)	Radius of gyration Axis x-x (cm)	Radius of gyration Axis y-y (cm)	Elastic modulus Axis x-x (cm³)	Elastic modulus Axis y-y (cm³)	Plastic modulus Axis x-x (cm³)	Plastic modulus Axis y-y (cm³)
914 × 419 × 388	388.0	921.0	420.5	21.4	36.6	24.1	799.6	5.74	37.4	719600	45440	38.2	9.59	15630	2161	17607	3341
914 × 419 × 343	343.3	911.8	418.5	19.4	32.0	24.1	799.6	6.54	41.2	625800	39160	37.8	9.46	13730	1871	15480	2890
914 × 305 × 289	289.1	926.6	307.7	19.5	32.0	19.1	824.4	4.81	42.3	504200	15600	37.0	6.51	10880	1014	12570	1601
914 × 305 × 253	253.4	918.4	305.5	17.3	27.9	19.1	824.4	5.47	47.7	436300	13300	36.8	6.42	9501	871	10940	1371
914 × 305 × 224	224.2	910.4	304.1	15.9	23.9	19.1	824.4	6.36	51.8	376400	11240	36.3	6.27	8269	739	9535	1163
914 × 305 × 201	200.9	903.0	303.3	15.1	20.2	19.1	824.4	7.51	54.6	325300	9423	35.7	6.07	7204	621	8351	982
838 × 292 × 226	226.5	850.9	293.8	16.1	26.8	17.8	761.7	5.48	47.3	339700	11360	34.3	6.27	7985	773	9155	1212
838 × 292 × 194	193.8	840.7	292.4	14.7	21.7	17.8	761.7	6.74	51.8	279200	9066	33.6	6.06	6641	620	7640	974
838 × 292 × 176	175.9	834.9	291.7	14.0	18.8	17.8	761.7	7.76	54.4	246000	7799	33.1	5.90	5893	535	6808	842
762 × 267 × 197	196.8	769.8	268.0	15.6	25.4	16.5	686.0	5.28	44.0	240000	8175	30.9	5.71	6234	610	7176	959
762 × 267 × 173	173.0	762.2	266.7	14.3	21.6	16.5	686.0	6.17	48.0	205300	6850	30.5	5.58	5387	514	6198	807
762 × 267 × 147	146.9	754.0	265.2	12.8	17.5	16.5	686.0	7.58	53.6	168500	5455	30.0	5.40	4470	411	5156	647
762 × 267 × 134	133.9	750.0	264.4	12.0	15.5	16.5	686.0	8.53	57.2	150700	4788	29.7	5.30	4018	362	4644	570
686 × 254 × 170	170.2	692.9	255.8	14.5	23.7	15.2	615.1	5.40	42.4	170300	6630	28.0	5.53	4916	518	5631	811
686 × 254 × 152	152.4	687.5	254.5	13.2	21.0	15.2	615.1	6.06	46.6	150400	5784	27.8	5.46	4374	455	5000	710
686 × 254 × 140	140.1	683.5	253.7	12.4	19.0	15.2	615.1	6.68	49.6	136300	5183	27.6	5.39	3987	409	4558	638
686 × 254 × 125	125.2	677.9	253.0	11.7	16.2	15.2	615.1	7.81	52.6	118000	4383	27.2	5.24	3481	346	3994	542
610 × 305 × 238	238.1	635.8	311.4	18.4	31.4	16.5	540.0	4.96	29.3	209500	15840	26.3	7.23	6589	1017	7486	1574
610 × 305 × 179	179.0	620.2	307.1	14.1	23.6	16.5	540.0	6.51	38.3	153000	11410	25.9	7.07	4935	743	5547	1144
610 × 305 × 149	149.1	612.4	304.8	11.8	19.7	16.5	540.0	7.74	45.8	125900	9308	25.7	7.00	4111	611	4594	937

Table 2.1 *Continued*

Designation	Mass per metre (kg/m)	Depth of section h (mm)	Width of section b (mm)	Thickness web s (mm)	Thickness flange t (mm)	Root radius r (mm)	Depth between fillets d (mm)	Ratios for local buckling		Second moment of area		Radius of gyration		Elastic modulus		Plastic modulus	
								Flange b/2t	Web d/s	Axis x–x (cm⁴)	Axis y–y (cm⁴)	Axis x–x (cm)	Axis y–y (cm)	Axis x–x (cm³)	Axis y–y (cm³)	Axis x–x (cm³)	Axis y–y (cm³)
610 × 229 × 140	139.9	617.2	230.2	13.1	22.1	12.7	547.6	5.21	41.8	111800	4505	25.0	5.03	3622	391	4142	611
610 × 229 × 125	125.1	612.2	229.0	11.9	19.6	12.7	547.6	5.84	46.0	98610	3932	24.9	4.97	3221	343	3676	535
610 × 229 × 113	113.0	607.6	228.2	11.1	17.3	12.7	547.6	6.60	49.3	87320	3434	24.6	4.88	2874	301	3281	469
610 × 229 × 101	101.2	602.6	227.6	10.5	14.8	12.7	547.6	7.69	52.2	75780	2915	24.2	4.75	2515	256	2881	400
533 × 210 × 122	122.0	544.5	211.9	12.7	21.3	12.7	476.5	4.97	37.5	76040	3388	22.1	4.67	2793	320	3196	500
533 × 210 × 109	109.0	539.5	210.8	11.6	18.8	12.7	476.5	5.61	41.1	66820	2943	21.9	4.60	2477	279	2828	436
533 × 210 × 101	101.0	536.7	210.0	10.8	17.4	12.7	476.5	6.03	44.1	61520	2692	21.9	4.57	2292	256	2612	399
533 × 210 × 92	92.1	533.1	209.3	10.1	15.6	12.7	476.5	6.71	47.2	55230	2389	21.7	4.51	2072	228	2360	356
533 × 210 × 82	82.2	528.3	208.8	9.6	13.2	12.7	476.5	7.91	49.6	47540	2007	21.3	4.38	1800	192	2059	300
457 × 191 × 98	98.3	467.2	192.8	11.4	19.6	10.2	407.6	4.92	35.8	45730	2347	19.1	4.33	1957	243	2232	379
457 × 191 × 89	89.3	463.4	191.9	10.5	17.7	10.2	407.6	5.42	38.8	41020	2089	19.0	4.29	1770	218	2014	338
457 × 191 × 82	82.0	460.0	191.3	9.9	16.0	10.2	407.6	5.98	41.2	37050	1871	18.8	4.23	1611	196	1831	304
457 × 191 × 74	74.3	457.0	190.4	9.0	14.5	10.2	407.6	6.57	45.3	33320	1671	18.8	4.20	1458	176	1653	272
457 × 191 × 67	67.1	453.4	189.9	8.5	12.7	10.2	407.6	7.48	48.0	29380	1452	18.5	4.12	1296	153	1471	237
457 × 152 × 82	82.1	465.8	155.3	10.5	18.9	10.2	407.6	4.11	38.8	36590	1185	18.7	3.37	1571	153	1811	240
457 × 152 × 74	74.2	462.0	154.4	9.6	17.0	10.2	407.6	4.54	42.5	32670	1047	18.6	3.33	1414	136	1627	213
457 × 152 × 87	67.2	458.0	153.8	9.0	15.0	10.2	407.6	5.13	45.3	28930	913	18.4	3.27	1263	119	1453	187
457 × 152 × 60	59.8	454.6	152.9	8.1	13.3	10.2	407.6	5.75	50.3	25500	795	18.3	3.23	1122	104	1287	163
457 × 152 × 52	52.3	449.8	152.4	7.6	10.9	10.2	407.6	6.99	53.6	21370	645	17.9	3.11	950	84.6	1096	133
406 × 178 × 74	74.2	412.8	179.5	9.5	16.0	10.2	360.4	5.61	37.9	27310	1545	17.0	4.04	1323	172	1501	267
406 × 178 × 67	67.1	409.4	178.8	8.8	14.3	10.2	360.4	6.25	41.0	24330	1365	16.9	3.99	1189	153	1346	237
406 × 178 × 60	60.1	406.4	177.9	7.9	12.8	10.2	360.4	6.95	45.6	21600	1203	16.8	3.97	1063	135	1199	209
406 × 178 × 54	54.1	402.6	177.7	7.7	10.9	10.2	360.4	8.15	46.8	18720	1021	16.5	3.85	930	115	1055	178
406 × 140 × 46	46.0	403.2	142.2	6.8	11.2	10.2	360.4	6.35	53.0	15690	538	16.4	3.03	778	75.7	888	118
406 × 140 × 39	39.0	398.0	141.8	6.4	8.6	10.2	360.4	8.24	56.3	12510	410	15.9	2.87	629	57.8	724	90.8

Table 2.1 *Continued*

Designation	Mass per metre (kg/m)	Depth of section h (mm)	Width of section b (mm)	Thickness web s (mm)	Thickness flange t (mm)	Root radius r (mm)	Depth between fillets d (mm)	Ratios for local buckling Flange b/2t	Ratios for local buckling Web d/s	Second moment of area Axis x-x (cm⁴)	Second moment of area Axis y-y (cm⁴)	Radius of gyration Axis x-x (cm)	Radius of gyration Axis y-y (cm)	Elastic modulus Axis x-x (cm³)	Elastic modulus Axis y-y (cm³)	Plastic modulus Axis x-x (cm³)	Plastic modulus Axis y-y (cm³)
356 × 171 × 67	67.1	363.4	173.2	9.1	15.7	10.2	311.6	5.52	34.2	19460	1362	15.1	3.99	1071	157	1211	243
356 × 171 × 57	57.0	358.0	172.2	8.1	13.0	10.2	311.6	6.62	38.5	16040	1108	14.9	3.91	896	129	1010	199
356 × 171 × 51	51.0	355.0	171.5	7.4	11.5	10.2	311.6	7.46	42.1	14140	968	14.8	3.86	796	113	896	174
356 × 171 × 45	45.0	351.4	171.1	7.0	9.7	10.2	311.6	8.82	44.5	12070	811	14.5	3.76	687	94.8	775	147
356 × 127 × 39	39.1	353.4	126.0	6.6	10.7	10.2	311.6	5.89	47.2	10170	358	14.3	2.68	576	56.8	659	89.1
356 × 127 × 33	33.1	349.0	125.4	6.0	8.5	10.2	311.6	7.38	51.9	8249	280	14.0	2.58	473	44.7	543	70.3
305 × 165 × 54	54.0	310.4	166.9	7.9	13.7	8.9	265.2	6.09	33.6	11700	1063	13.0	3.93	754	127	846	196
305 × 165 × 46	46.1	306.6	165.7	6.7	11.8	8.9	265.2	7.02	39.6	9899	896	13.0	3.90	646	108	720	166
306 × 165 × 40	40.3	303.4	165.0	6.0	10.2	8.9	265.2	8.09	44.2	8503	764	12.9	3.86	560	92.6	623	142
305 × 127 × 48	48.1	311.0	125.3	9.0	14.0	8.9	265.2	4.47	29.5	9575	461	12.5	2.74	616	73.6	711	116
305 × 127 × 42	41.9	307.2	124.3	8.0	12.1	8.9	265.2	5.14	33.2	8196	389	12.4	2.70	534	62.6	614	98.4
305 × 127 × 37	37.0	304.4	123.3	7.1	10.7	8.9	265.2	5.77	37.4	7171	336	12.3	2.67	471	54.5	539	85.4
305 × 102 × 33	32.8	312.7	102.4	6.6	10.8	7.6	275.9	4.74	41.8	6501	194	12.5	2.15	416	37.9	481	60.0
305 × 102 × 28	28.2	308.7	101.8	6.0	8.8	7.6	275.9	5.78	46.0	5366	155	12.2	2.08	348	30.5	403	48.5
305 × 102 × 25	24.8	305.1	101.6	5.8	7.0	7.6	275.9	7.26	47.6	4455	123	11.9	1.97	292	24.2	342	38.8
254 × 146 × 43	43.0	259.6	147.3	7.2	12.7	7.6	219.0	5.80	30.4	6544	677	10.9	3.52	504	92.0	566	141
254 × 146 × 37	37.0	256.0	146.4	6.3	10.9	7.6	219.0	6.72	34.8	5537	571	10.8	3.48	433	78.0	483	119
254 × 146 × 31	31.1	251.4	146.1	6.0	8.6	7.6	219.0	8.49	36.5	4413	448	10.5	3.36	351	61.3	393	94.1
254 × 102 × 28	28.3	260.4	102.2	6.3	10.0	7.6	225.2	5.11	35.7	4005	179	10.5	2.22	308	34.9	353	54.8
254 × 102 × 25	25.2	257.2	101.9	6.0	8.4	7.6	225.2	6.07	37.5	3415	149	10.3	2.15	266	29.2	306	46.0
254 × 102 × 22	22.0	254.0	101.6	5.7	6.8	7.6	225.2	7.47	39.5	2841	119	10.1	2.06	224	23.5	259	37.3
203 × 133 × 30	30.0	206.8	133.9	6.4	9.6	7.6	172.4	6.97	26.9	2896	385	8.71	3.17	280	57.5	314	88.2
203 × 133 × 25	25.1	203.2	133.2	5.7	7.8	7.6	172.4	8.54	30.2	2340	308	8.56	3.10	230	46.2	258	70.9
203 × 102 × 23	23.1	203.2	101.8	5.4	9.3	7.6	169.4	5.47	31.4	2105	164	8.46	2.36	207	32.2	234	49.8
178 × 102 × 19	19.0	177.8	101.2	4.8	7.9	7.6	146.8	6.41	30.6	1356	137	7.48	2.37	153	27.01	171	41.6
152 × 89 × 16	16.0	152.4	88.7	4.5	7.7	7.6	121.8	5.76	27.1	834	89.8	6.41	2.10	109	20.2	123	31.2
127 × 76 × 13	13.0	127.0	76.0	4.0	7.6	7.6	96.6	5.00	24.1	473	55.7	5.35	1.84	74.6	14.7	84.2	22.6

Second moment of area

A beam can be considered to be made up of an infinite number of layers, rather like the pages of a book. It was stated in Chapter 1 that when bending occurs, some of the layers are in tension and some are in compression. In between these is a layer which, although bent like the others, is in neither tension nor compression. It is called the *neutral layer* or *neutral axis* of the beam and generally passes through the centroid of its cross-section.

There are two parameters which determine the stiffness of a beam and its resistance to bending. One of these will be quite familiar to you. It is the *modulus of elasticity* of the beam material. The greater the modulus of elasticity, the greater the stiffness and resistance to bending. Suppose that you have two beams with the same dimensions, one of mild steel whose modulus of elasticity is 200 GPa and one of an aluminium alloy whose modulus of elasticity is 100 GPa. The values indicate that the alloy beam has only half the stiffness of the steel beam. If the two beams are subjected to the same load, it will be found that the deflection of the aluminium beam is double that of the steel beam.

The other parameter concerns the shape and orientation of the beam's cross-section. It is quite easy to bend a ruler when the bending forces are applied to its flat faces. When it is turned through 90° however, and the bending forces are applied to its edges, it becomes very difficult to bend. The dimensional parameter which governs the stiffness of a beam is the *second moment of area* of its cross-section, taken about the neutral axis of bending. It is given the symbol, I and its units are m^4. Consider a rectangular section beam of width, B and depth, D as shown in Figure 2.3.

The second moment of area of the elemental strip, δI about the neutral axis is its area, δA multiplied by the square of its distance, x from the neutral axis. That is,

$$\delta I = x^2 \, \delta A$$

The total second moment of area, I of the whole cross-section is the sum of all such elements.

$$I = \Sigma(x^2 \delta A) \tag{2.3}$$

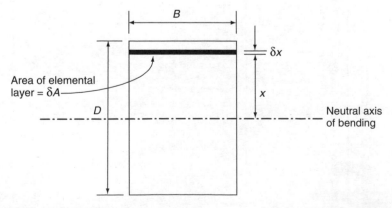

Figure 2.3 *Rectangular cross-section*

Now $\delta A = B\delta x$ and so

$$I = \Sigma(x^2 B\delta x)$$

This sum can be found by integration between the limits of $x = +D/2$ and $x = -D/2$. As $\delta x \longrightarrow 0$, the expression becomes

$$I = \int_{-D/2}^{+D/2} x^2 B\,\mathrm{d}x$$

$$I = B\left(\frac{x^3}{3}\right)_{-D/2}^{+D/2} = \frac{B}{3}\left(\left(\frac{D}{2}\right)^3 - \left(\frac{-D}{2}\right)^3\right]$$

$$I = \frac{B}{3}\left(\frac{D^3}{8} + \frac{D^3}{8}\right) = \frac{B}{3}\left(\frac{D^3}{4}\right)$$

$$I = \frac{BD^3}{12}\ (\mathrm{m}^4) \tag{2.4}$$

The same procedure can be applied to circular cross-sections, although the integration is a little more difficult. The formula which results for a circular section of diameter, D is,

$$I = \frac{\pi D^4}{64}\ (\mathrm{m}^4) \tag{2.5}$$

Equation (2.4) shows that if you double the width, B of a rectangular section beam you will double its stiffness. If however you double its depth, D you will increase its stiffness by a factor of eight because of the D^3 term, i.e. $2^3 = 8$. This is why your ruler is so much more difficult to bend when the bending forces are applied to its edges. Equation (2.5) for a circular section beam shows that if you double the diameter, the stiffness will increase by a factor of 16 because of the D^4 term, i.e. $2^4 = 16$.

The second moment of area of a section is also sometimes called its *moment of inertia*. This is traditional but rather misleading because the term really applies to rotating bodies, as we shall see later in this chapter. It is, however, the term used for second moment of area in British Standard tables and in Table 6.1 you will see listed the values for *moment of inertia about the x–x neutral axis and moment of inertia about the y–y neutral axis*. When the section is loaded as an I-section, with the web vertical, the x–x axis is the neutral axis of bending. The section can also be turned through 90° with the web horizontal and loaded as an H-section. The y–y axis then becomes the neutral axis of bending. As you can see from the moment of inertia values, the beams are much stiffer when used as an I-section than as an H-section.

Radius of gyration

We shall encounter this term again when we investigate the dynamics of rotating bodies later in the chapter. In that context, the *radius of gyration* is the radius of rotation at which the mass of a rotating body can be considered to be concentrated. The term is also used in structural engineering and you can see its values, about the x–x and y–y neutral axes of bending, listed in Table 2.1. The name is a little misleading in this context since no rotation is actually taking place. You will recall that the second moment of area of a section, as given by equation (2.3), is

$$I = \Sigma(x^2 \delta A)$$

If there is an average value of all the x^2 terms whose value we may call k^2, the second moment of area can be written as

I = average value of $x^2 \times$ total cross-sectional area

$$\mathbf{I = k^2 A} \tag{2.6}$$

This term k, is the radius of gyration of the cross-section measured from the neutral axis of bending. It may be correctly described as a *root-mean square radius*, i.e. it is the square root of the average of all the x^2 values. You may have come across a similar thing in the study of alternating current. Here a current of say 3 amps, is the root-mean square value of an alternating current which has peak values of ± 4.24 amps.

Example 2.3

Determine the second moment of area and radius of gyration of a rectangular beam section of width 50 mm and depth 100 mm. By how much would its depth need to be increased in order to double its stiffness?

Finding initial second moment of area, I_1, about the neutral axis of bending for depth $D_1 = 100$ mm:

$$I_1 = \frac{BD_1^3}{12} = \frac{0.05 \times 0.1^3}{12}$$

$$\mathbf{I_1 = 4.17 \times 10^{-6} \, m^4}$$

Finding initial radius of gyration about the neutral axis of bending:

$$I_1 = k^2 A$$

$$k = \sqrt{\frac{I}{A}} = \sqrt{\frac{4.17 \times 10^{-6}}{0.05 \times 0.1}}$$

$$\mathbf{k = 0.0289 \, m \quad or \quad 28.9 \, mm}$$

Finding second moment of area, I_2 when stiffness is doubled:

$$I_2 = 2I_1 = 2 \times 4.17 \times 10^{-6}$$

$$\mathbf{I_2 = 8.34 \times 10^{-6} \, m^4}$$

Finding new depth D_2 of the section:

$$I_2 = \frac{BD_2^3}{12}$$

$$D_2 = \sqrt[3]{\frac{12 I_2}{B}} = \sqrt[3]{\frac{12 \times 8.34 \times 10^{-6}}{0.05}}$$

$$\mathbf{D_2 = 0.126 \, m \quad or \quad 126 \, mm}$$

The depth would need to be increased by 26 mm to double the stiffness of the beam, i.e. a 26% increase.

Example 2.4

A rectangular section structural member has a width 75 mm and depth 200 mm. What would be the diameter and radius of gyration of a circular section of the same material and same stiffness?

Finding second moment of area of rectangular section:

$$I = \frac{BD^3}{12} = \frac{0.075 \times 0.2^3}{12}$$
$$I = 50 \times 10^{-6} \, m^4$$

Finding diameter of equivalent circular section:

$$I = \frac{\pi D^4}{64}$$
$$D = \sqrt[4]{\frac{64I}{\pi}} = \sqrt[4]{\frac{64 \times 50 \times 10^{-6}}{\pi}}$$
$$D = 0.179 \, m \quad \text{or} \quad 179 \, mm$$

Finding cross-sectional area of circular section:

$$A = \frac{\pi D^2}{4} = \frac{\pi \times 0.179^2}{4}$$
$$A = 0.0252 \, m^2$$

Finding radius of gyration of circular section:

$$I = k^2 A$$
$$k = \sqrt{\frac{I}{A}} = \sqrt{\frac{50 \times 10^{-6}}{0.0252}}$$
$$k = 0.0445 \, m \quad \text{or} \quad 44.5 \, mm$$

Activity 2.2

A circular section timber beam of diameter 150 mm is to be replaced by a rectangular section of the same material and width 50 mm. What is the required depth of the rectangular section if the two beams are to have the same stiffness? What will be the radius of gyration of the rectangular section taken about its neutral axis of bending?

Problems 2.1

1. A simply supported beam is 10 m long and carries a concentrated load of 150 kN at its centre. If the allowable stress due to bending is 70 MPa, select a suitable section from standard tables which has the least mass.
   [Serial size $762 \times 267 \, mm^2$, $z = 5387 \, cm^3$, $m = 173 \, kg \, m^{-1}$]

2. A simply supported beam of 3.5 m span carries a concentrated load of 60 kN at its centre. If the maximum allowable stress is 165 KPa, select a suitable rolled steel I-section from standard

tables which (a) has the least mass per metre, (b) has the least depth.

[(a) Serial size $305 \times 102 \, \text{mm}^2$, $z = 348 \, \text{cm}^3$, $m = 28.2 \, \text{kg m}^{-1}$]
[(b) Serial size $254 \times 146 \, \text{mm}^2$, $z = 351 \, \text{cm}^3$, $D = 251.4 \, \text{mm}$]

3. A simple cantilever of length 2.6 m carries a concentrated load of 400 kN at its free end. If the maximum allowable stress due to bending is 150 KPa, select a suitable rolled steel I-section from standard tables which (a) has the least mass per metre, (b) has the least depth.

[(a) Serial size $914 \times 305 \, \text{mm}^2$, $z = 7204 \, \text{cm}^3$, $m = 200.9 \, \text{kg m}^{-1}$]
[(b) Serial size $838 \times 292 \, \text{mm}^2$, $z = 7985 \, \text{cm}^3$, $D = 850.9 \, \text{mm}$]

4. A simply supported beam of length 4 m carries a central concentrated load of 250 kN at its centre and a uniformly distributed load of $50 \, \text{kN m}^{-1}$. A standard rolled steel I-section is to be used with a factor of safety of 4. The ultimate tensile strength of the steel is $500 \, \text{MN m}^{-2}$. Select a suitable section from standard tables which (a) has the least weight, (b) has the least possible depth.

[(a) Serial size $406 \times 178 \, \text{mm}^2$, $z = 1063 \, \text{cm}^3$, $m = 60.1 \, \text{kg m}^{-1}$]
[(b) Serial size $356 \times 171 \, \text{mm}^2$, $z = 1071 \, \text{cm}^3$, $D = 363.4 \, \text{mm}$]

5. By how much would the diameter of a solid circular section beam of initial diameter 150 mm need to be increased in order to double its bending stiffness? What would be the percentage increase in diameter?

[28.4 mm, 18.9%]

6. A solid square section beam 100×100 mm is to be replaced by a circular section of the same material and cross-sectional area. Calculate the second moments of area of the two sections and state which has the greater resistance to bending.

[$8.33 \times 10^6 \, \text{mm}^4$, $7.96 \times 10^6 \, \text{mm}^4$, the square section]

Selection of compression members

You will recall that in Chapter 1, we referred to the compression members in framed structures as *struts*. There are, however, other types of compression member encountered in engineering structures which are referred to as *columns, pillars, posts* and *stanchions*. Columns and pillars may be made of any material, e.g. concrete, stone, brick timber. Posts are usually timber and stanchions are generally rolled steel sections. It is the selection of steel stanchions and struts to support given compressive axial loads that we shall be concerned with here. Axial loading means that the load can be considered to act through the centroid of the cross-section.

The rolled steel I-sections used for stanchions have different dimensions to those used for beams. They are generally wider across the flanges and shorter in the web. Their dimensions are specified in BS4: Part 1: 1980 – Universal columns: parallel flanges: dimensions and properties. An extract is shown in Table 2.2. In addition to the load, the length of the stanchion and its type of end fixing must be taken into account when making a selection. The process is not so easy as with beams and involves a little trial and error.

Table 2.2 Universal columns: parallel flanges: dimensions and properties (TO BS4: Part 1: 1993)

Designation	Mass per metre (kg/m)	Depth of section h (mm)	Width of section b (mm)	Thickness of web s (mm)	Thickness of flange t (mm)	Second moment of area Axis x–x (cm⁴)	Second moment of area Axis y–y (cm⁴)	Radius of gyration Axis x–x (cm)	Radius of gyration Axis y–y (cm)	Elastic modulus Axis x–x (cm³)	Elastic modulus Axis y–y (cm³)	Area of section (cm²)	Mass per metre (kg/m)
356 × 406 × 634	633.9	474.6	424.0	47.6	77.0	274800	98130	18.4	11.0	11580	4629	808	633.9
356 × 406 × 551	551.0	455.6	418.5	42.1	67.5	225900	82670	18.0	10.9	9962	3951	702	551.0
356 × 406 × 467	467.0	436.6	412.2	35.8	58.0	183000	67830	17.5	10.7	8383	3291	595	467.0
356 × 406 × 393	393.0	419.0	407.0	30.6	49.2	146600	55370	17.1	10.5	6998	2721	501	393.0
356 × 406 × 340	339.9	406.4	403.0	26.6	42.9	122500	46850	16.8	10.4	6031	2325	433	339.9
356 × 406 × 287	287.1	393.6	399.0	22.6	36.5	99880	38680	16.5	10.3	5075	1939	366	287.1
356 × 406 × 235	235.1	381.0	394.8	18.4	30.2	79080	30990	16.3	10.2	4151	1570	299	235.1
356 × 368 × 202	201.9	374.6	374.7	16.5	27.0	66260	23690	16.1	9.60	3538	1264	257	201.9
356 × 368 × 177	177.0	368.2	372.6	14.4	23.8	57120	20530	15.9	9.54	3103	1102	226	177.0
356 × 368 × 153	152.9	362.0	370.5	12.3	20.7	48590	17550	15.8	9.49	2684	948	195	152.9
356 × 368 × 129	129.0	355.6	368.6	10.4	17.5	40250	14610	15.6	9.43	2264	793	164	129.0
305 × 305 × 283	282.9	365.3	322.2	26.8	44.1	78870	24630	14.8	8.27	4318	1529	360	282.9
305 × 305 × 240	240.0	352.5	318.4	23.0	37.7	64200	20310	14.5	8.15	3643	1276	306	240.0
305 × 305 × 198	198.1	339.9	314.5	19.1	31.4	50900	16300	14.2	8.04	2995	1037	252	198.1
305 × 305 × 158	158.1	327.1	311.2	15.8	25.0	38750	12570	13.9	7.90	2369	808	201	158.1
305 × 305 × 137	136.9	320.5	309.2	13.8	21.7	32810	10700	13.7	7.83	2048	692	174	136.9
305 × 305 × 118	117.9	314.5	307.4	12.0	18.7	27670	9059	13.6	7.77	1760	589	150	117.9
305 × 305 × 97	96.9	307.9	305.3	9.9	15.4	22250	7308	13.4	7.69	1445	479	123	96.9
254 × 254 × 167	167.1	289.1	265.2	19.2	31.7	30000	9870	11.9	6.81	2075	744	213	167.1
254 × 254 × 132	132.0	276.3	261.3	15.3	25.3	22530	7531	11.6	6.69	1631	576	168	132.0
254 × 254 × 107	107.1	266.7	258.8	12.8	20.5	17510	5928	11.3	6.59	1313	458	136	107.1
254 × 254 × 89	88.9	260.3	256.3	10.3	17.3	14270	4857	11.2	6.55	1096	379	113	88.9
254 × 254 × 73	73.1	254.1	254.6	8.6	14.2	11410	3908	11.1	6.48	898	307	93.1	73.1
203 × 203 × 86	86.1	222.2	209.1	12.7	20.5	9449	3127	9.28	5.34	850	299	110	86.1
203 × 203 × 71	71.0	215.8	206.4	10.0	17.3	7618	2537	9.18	5.30	706	246	90.4	71.0
203 × 203 × 60	60.0	209.6	205.8	9.4	14.2	6125	2065	8.96	5.20	584	201	76.4	60.0
203 × 203 × 52	52.0	206.2	204.3	7.9	12.5	5259	1778	8.91	5.18	510	174	66.3	52.0
203 × 203 × 46	46.1	203.2	203.6	7.2	11.0	4568	1548	8.82	5.13	450	152	58.7	46.1
152 × 152 × 37	37.0	161.8	154.4	8.0	11.5	2210	706	6.85	3.87	273	91.5	47.1	37.0
152 × 152 × 30	30.0	157.6	152.9	6.5	9.4	1748	560	6.76	3.83	222	73.3	38.3	30.0
152 × 152 × 23	23.0	152.4	152.2	5.8	6.8	1250	400	6.54	3.70	164	52.6	29.2	23.0

End fixing and effective length

The different combinations of end fixing for compression members are shown in Figure 2.4.

Short columns made from masonry or concrete tend to fail by crushing when the compressive stress reaches a particular value. Comparatively, long struts and stanchions tend to fail by buckling when the load reaches a value known as the *critical load*. The four struts shown in Figure 2.4 all have the same length, L and it may be assumed that they are made from the same material and have the same cross-section. They do however have different end fixings and this affects the critical load that they can carry before buckling. The weakest will be (a), the member with one end direction-fixed and the other one free. The strongest will be (d), with both ends direction-fixed.

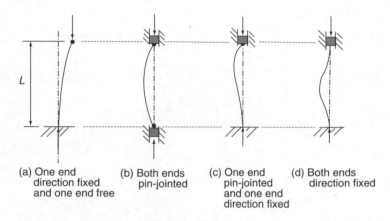

(a) One end direction fixed and one end free   (b) Both ends pin-jointed   (c) One end pin-jointed and one end direction fixed   (d) Both ends direction fixed

Figure 2.4 *Types of end fixing*

The *effective length* of a strut is defined as that part of its actual length over which buckling is likely to occur. The effective length, l of strut (b), with both ends pin-jointed, is the same as its actual length, L, since it will buckle over its full length. An alternative way of defining effective length is to say that it is the length of that strut with pin-jointed ends which will have the same critical load as the strut in question. The effective lengths, l of the struts shown in Figure 2.4, each having an actual length, L are shown in Table 2.3.

Table 2.3 indicates that the weakest strut of (a) has the same critical load as a strut of type (b) which has twice its length. At the other extreme, the strongest strut of type (d) has the same critical load as a strut of type (b) with 0.7 of its length.

Table 2.3 *Effective lengths*

Type of end fixing	Effective length, l
(a) One end direction-fixed, one end free	$l = 2L$
(b) Both ends pin-jointed	$l = L$
(c) One end direction-fixed, one end pin-jointed	$l = 0.85L$
(d) Both ends direction-fixed	$l = 0.7L$

Slenderness ratio (*l*/*k*)

The *slenderness ratio* of a strut or stanchion is one the parameters used in the selection process.

$$\text{slenderness ratio} = \frac{\text{effective length}}{\text{least radius of gyration}}$$

$$\mathbf{S.R.} = \frac{l}{k} \tag{2.7}$$

For the standard rolled steel sections listed in Table 2.2, the least radius of gyration, *k* is the radius of gyration about the *y*–*y* neutral axis of bending. Table 2.4 contains the British Standards recommendations for the allowable compressive stress in struts for increasing values of slenderness ratio. It is taken from BS5950: Part 1: 1985.

Table 2.4 *Allowable compressive stress value for Grade 43 steel*

Slenderness ratio = *l*/*k*	Allowable stress for flange thickness up to 16 mm (N mm^{-2})	Allowable stress for flange thickness 17–40 mm (N mm^{-2})	Allowable stress for flange thickness 41–100 mm (N mm^{-2})
20	272	263	243
30	262	253	234
40	250	241	224
50	237	229	213
55	228	221	206
60	221	214	200
65	211	205	192
70	202	196	185
75	192	187	176
80	181	177	168
85	171	167	159
90	161	157	150
95	150	148	141
100	141	138	133
105	131	129	125
110	123	121	118
115	115	113	110
120	108	107	104
125	101	100	97
130	95	94	92
135	89	88	86
140	84	83	81
145	79	78	77
150	74	74	72
155	70	70	69
160	66	66	65
165	63	63	61
170	60	59	58
175	57	56	56
180	54	54	53
185	51	51	50
190	49	48	48
195	47	46	46
200	44	44	44
250	29	29	29
300	21	21	21
350	15	15	15

Selection process for struts and stanchions

When selecting a standard rolled steel section to carry a specified load, the procedure is as follows:

1. Calculate the effective length of your strut from its actual length and the type of end fixings. State its value in mm.
2. Choose a trial column from Table 2.2 for universal columns. Structural engineers know from experience roughly the size of section required but for a first trial choose one from somewhere near the centre of the table. Make a note of its cross-sectional area, A and radius of gyration, k about the y–y axis of bending. Convert their units to mm^2 and mm. Make a note also of the flange thickness.
3. Calculate the slenderness ratio of your trial column.
4. From Table 2.4 read off the maximum allowable compressive stress, σ for your calculated value of slenderness ratio from the column appropriate to your flange thickness. You should note that the units used for stress are $N\,mm^{-2}$ which is why your cross-sectional area needs to be in mm^2.
5. Calculate the safe axial load that your trial column can carry using the formula,

   safe axial load = allowable stress × cross-sectional area

   $$F = \sigma A \qquad (2.8)$$

6. Compare this with the load which the member has to carry. If it is less or appreciably greater, choose another trial column and repeat the calculations. Eventually, after no more than two or three trials, you should be able to narrow the choice down to a section with the least weight that can carry slightly more than the required load.

Example 2.5

A stanchion of length 10 m with direction-fixed ends is required to carry a load of 500 kN. Select a suitable universal steel column from standard section tables which has the least mass.

Finding effective length:

$l = 0.7L = 0.7 \times 10$

$l = 7\,m$ or $7 \times 10^3\,mm$

1st trial:
Serial size $203 \times 203\,mm^2$ of mass $86\,kg\,m^{-1}$
Cross-sectional area, $A = 110.1\,cm^2 = 110.1 \times 10^2\,mm$
Least radius of gyration, $k = 5.32\,cm = 53.2\,mm$
Flange thickness, $T = 20.5\,mm$

Finding slenderness ratio:

$S.R. = \dfrac{l}{k} = \dfrac{7 \times 10^3}{53.2}$

$S.R. = 132$

Finding allowable compressive stress from Table 2.4:

$\sigma = 89\,\text{N}\,\text{mm}^{-2}$

Finding maximum load which the member can carry:

$F = \sigma A = 89 \times 110.1 \times 10^2$

$F = 980 \times 10^3\,\text{N}$ or   **980 kN**

This is much higher than the 500 kN load and so another trial is necessary.

2nd trial:
Serial size $152 \times 152\,\text{mm}^2$ of mass $37\,\text{kg}\,\text{m}^{-1}$
Cross-sectional area, $A = 47.4\,\text{cm}^2 = 47.4 \times 10^2\,\text{mm}$
Least radius of gyration, $k = 3.87\,\text{cm} = 38.7\,\text{mm}$
Flange thickness, $T = 11.5\,\text{mm}$

Finding slenderness ratio:

$\text{S.R.} = \dfrac{l}{k} = \dfrac{7 \times 10^3}{38.7}$

S.R. = 181

Finding allowable compressive stress from Table 2.4:

$\sigma = 53\,\text{N}\,\text{mm}^{-2}$

Finding maximum load which the member can carry:

$F = \sigma A = 53 \times 47.4 \times 10^2$

$F = 251 \times 10^3\,\text{N}$ or   **251 kN**

This is below the 500 kN load and so another trial is necessary.

3rd trial:
Serial size $203 \times 203\,\text{mm}^2$ of mass $46\,\text{kg}\,\text{m}^{-1}$
Cross-sectional area, $A = 58.8\,\text{cm}^2 = 58.8 \times 10^2\,\text{mm}$
Least radius of gyration, $k = 51.1\,\text{cm} = 51.1\,\text{mm}$

Finding slenderness ratio:

$\text{S.R.} = \dfrac{l}{k} = \dfrac{7 \times 10^3}{51.1}$

S.R. = 137

Finding allowable compressive stress from Table 2.4:

$\sigma = 87\,\text{N}\,\text{mm}^{-2}$

Finding maximum load which the member can carry:

$F = \sigma A = 87 \times 58.8 \times 10^2$

$F = 512 \times 10^3\,\text{N}$ or   **512 kN**

This can carry slightly more than the 500 kN load and is the most suitable section.

Example 2.6

A strut of length 8 m with one end direction-fixed and one end pin-jointed is required to carry a compressive load of 750 kN. Select a suitable universal steel column from standard section tables which has the least mass.

Finding effective length:

$l = 0.85\,L = 0.85 \times 8$

$l = \textbf{6.8 m}$ or $\textbf{6.8} \times \textbf{10}^3$ **mm**

1st trial:
Serial size 254×254 mm^2 of mass 167 kgm^{-1}
Cross-sectional area, $A = 212.4$ cm$^2 = 212.4 \times 10^2$ mm
Least radius of gyration, $K = 6.79$ cm $= 67.9$ mm
Flange thickness, $T = 31.7$ mm

Finding slenderness ratio:

$$\text{S.R.} = \frac{l}{k} = \frac{6.8 \times 10^3}{67.9}$$

$\textbf{S.R.} = \textbf{100}$

Finding allowable compressive stress from Table 2.4:

$\sigma = \textbf{138 N mm}^{-2}$

Finding maximum load which the member can carry:

$F = \sigma A = 138 \times 212.4 \times 10^2$

$F = \textbf{2.93} \times \textbf{10}^6$ **N**   or   **2.93 MN**

This is much higher than the 750 kN load and so another trial is necessary.

2nd trial:
Serial size 203×203 mm^2 of mass 86 kg m^{-1}
Cross-sectional area, $A = 110.1$ cm$^2 = 110.1 \times 10^2$ mm
Least radius of gyration, $k = 5.32$ cm $= 53.2$ mm
Flange thickness, $T = 20.5$ mm

Finding slenderness ratio:

$$\text{S.R.} = \frac{l}{k} = \frac{6.8 \times 10^3}{53.2}$$

$\textbf{S.R.} = \textbf{128}$

Finding allowable compressive stress from Table 2.4:

$\sigma = \textbf{96 N mm}^{-2}$

Finding maximum load which the member can carry:

$F = \sigma A = 96 \times 110.1 \times 10^2$

$F = \textbf{1.06} \times \textbf{10}^6$ **N**   or   **1.06 MN**

This is still greater than the required load of 750 kN and is unsatisfactory. A further trial is needed on a smaller section.

3rd trial:

Serial size $203 \times 203\,mm^2$ of mass $60\,kg\,m^{-1}$
Cross-sectional area, $A = 7.58\,cm^2 = 75.8 \times 10^2\,mm$
Least radius of gyration, $k = 5.19\,cm = 51.9\,mm$
Flange thickness, $T = 11.0\,mm$

Finding slenderness ratio:

$$\text{S.R.} = \frac{l}{k} = \frac{6.8 \times 10^3}{51.9}$$

$$\textbf{S.R.} = \textbf{131}$$

Finding allowable compressive stress from Table 2.4:

$$\sigma = \textbf{93 N mm}^{-2}$$

Finding maximum load which the member can carry:

$$F = \sigma A = 93 \times 575.8 \times 10^2$$

$$\textbf{\textit{F}} = \textbf{705} \times \textbf{10}^3\,\textbf{N} \quad \text{or} \quad \textbf{705 kN}$$

This is slightly below the 750 kN load and so another trial is necessary:

4th trial:

Serial size $203 \times 503\,mm^2$ of mass $71\,kg\,m^{-1}$
Cross-sectional area, $A = 91.1cm^2 = 91.1 \times 10^2\,mm$
Least radius of gyration, $k = 5.28\,cm = 52.8\,mm$
Flange thickness, $T = 17.3\,mm$

Finding slenderness ratio:

$$\text{S.R.} = \frac{l}{k} = \frac{6.8 \times 10^3}{52.8}$$

$$\textbf{S.R.} = \textbf{129}$$

Finding allowable compressive stress from Table 2.4:

$$\sigma = \textbf{95 N mm}^{-2}$$

Finding maximum load which the member can carry:

$$F = \sigma A = 95 \times 91.1 \times 10^2$$

$$\textbf{\textit{F}} = \textbf{865} \times \textbf{10}^3\,\textbf{N} \quad \text{or} \quad \textbf{865 kN}$$

This can carry slightly more than the 750 kN load and is the most suitable section.

Test your knowledge 2.3

1. What does *axial loading* in struts and stanchions mean?
2. What is meant by the *effective length* of a strut?
3. What is the effective length of a strut with one end direction-fixed and one end pin-jointed?
4. What is the *slenderness ratio* of a strut?
5. How do you convert cross-sectional area in cm^2 to mm^2?

Activity 2.3

A strut of length 7.5 m is pin-jointed at both ends and is required to support an axial load of 800 kN. Select a suitable universal column made from 43 grade steel which has the least possible weight. If the ultimate tensile strength of the steel is 500 MPa, what will be the factor of safety in operation for the selected section?

Problems 2.2

1. A stanchion 3.6 m high has one end direction-fixed and one end pin-jointed. The standard section selected for use is Grade 43 steel serial size $202 \times 203 \, \text{mm}^2$ and mass $60 \, \text{kg m}^{-1}$. Determine (a) the slenderness ratio of the stanchion, (b) the maximum compressive load that it can be allowed to carry.

    [59, 1683 kN]

2. A strut of length 4 m is pin-jointed at both ends and is required to carry a compressive load of 2 MN. Select a suitable I-section with the lowest possible weight using standard section tables for Grade 43 steel.

    [Serial size $254 \times 254 \, \text{mm}^2$, mass $73 \, \text{kg m}^{-1}$]

3. A stanchion is 4.3 m long and direction-fixed at both ends. Using standard section tables, select a suitable I-section made from Grade 43 steel which can safely support a compressive load of 2000 kN.

    [Serial size $203 \times 203 \, \text{mm}^2$, mass $71 \, \text{kg m}^{-1}$]

4. Calculate the maximum allowable compressive loads that the following stanchions can carry. (a) Length 6 m with both ends pin-jointed. (b) Length 9 m with one end direction-fixed, one end pin-jointed. (c) Length 12 m with both ends direction-fixed. The material used is Grade 43 steel of serial size $254 \times 254 \, \text{mm}^1$ and mass $89 \, \text{kg m}^{-1}$.

    [1.78 MN, 1.25 MN, 1.09 MN]

5. A strut of length 5 m has one end pin-jointed and one end free. The strut is to carry an axial compressive load of 1500 kN. Select a suitable I-section from standard section tables for Grade 43 steel which has (a) the lowest possible weight, (b) the lowest possible values of depth and breadth.

    [(a) Serial size $305 \times 305 \, \text{mm}^2$, mass $137 \, \text{kg m}^{-1}$]
    [(b) Serial size $254 \times 254 \, \text{mm}^2$, mass $167 \, \text{kg m}^{-1}$]

Frictional resistance

Frictional resistance is experienced when two surfaces in contact move relative to each other with either a sliding or a rolling action. Sometimes, good use is made of frictional resistance such as between the brake pads and discs in the brakes of a car or motor cycle. Motor vehicle clutches also depend on frictional resistance as do the centrifugal clutches that were described in Chapter 1. In a great many other instances however, friction is undesirable and must be kept to a minimum. This is especially the case with bearings, gears, and the other moving parts of engines power transmission systems, and machine tools.

To begin with, we will consider frictional resistance between dry surfaces. Figure 2.5 shows a block of material resting on a dry horizontal surface. Initially, it is in a state of static equilibrium with its weight, W acting downwards balanced by the reaction, R of the horizontal surface acting upwards. When a small horizontal pulling force, P is applied to the block, nothing much happens to begin with. The frictional resistance, F is able to balance the pulling force and the block stays in static equilibrium. As the pulling force is increased, the frictional resistance increases to balance it. Eventually a value of

Key point

The limiting frictional resistance is the resistance when two surfaces just begin to slide over each other.

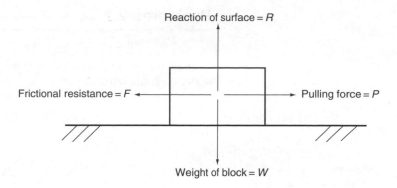

Figure 2.5 *Dry surfaces in sliding contact*

the pulling force is reached above which the frictional resistance cannot rise. This is known as the *limiting frictional resistance* and at this point, the block starts to slide over the surface.

Once an object starts to slide, it is found that the frictional resistance to motion is slightly less than the limiting value, and that the pulling force needed to maintain motion is slightly less than that required to get things moving. The reason for this is thought to be that there are irregularities, such as peaks and valleys, in even the smoothest of surfaces. When at rest, the peaks on the two surfaces are in contact and the weight, W is carried on a very small area, as shown in Figure 2.6.

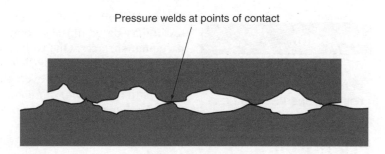

Figure 2.6 *Pressure welds between surfaces*

Key point

The frictional resistance when sliding has begun is slightly less than the limiting frictional resistance.

As a result of the pressure, the surfaces become welded together at the contact points and the bonds have to be broken before sliding can take place. This accounts for the fact that once sliding has begun, the force necessary to maintain the motion, which is called *kinetic frictional resistance*, is less than that to overcome *static friction* or '*stiction*' as it is sometimes called.

There are four statements, sometimes known as *the laws of friction* which, within reasonable limits, apply to dry surfaces in sliding contact. They are associated with the French physicist Charles Augustin de Coulomb who verified them experimentally in 1785.

1. The frictional resistance between dry surfaces in sliding contact is proportional to the normal force between the surfaces.
2. The frictional resistance depends on the nature of the surfaces, i.e. their surface roughness and texture.

3. The frictional resistance is independent of the area of contact between the two surfaces.
4. The frictional resistance is independent of the sliding velocity.

Consider again a block of material sliding on a flat horizontal surface as in Figure 2.7, and the forces which act upon it.

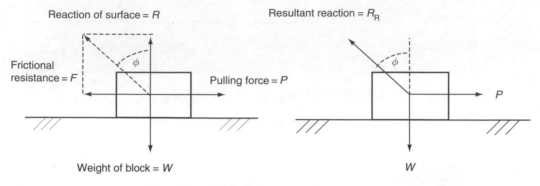

Figure 2.7 *Dry surfaces in sliding contact*

The first law states that frictional resistance is proportional to the force between the surfaces. This is the weight, W. That is,

$$F = \text{constant} \times W$$

The constant is the coefficient of friction, μ for the two surfaces.

$$F = \mu W \tag{2.9}$$

The coefficient of friction depends on the surface roughness and texture, as stated in the second law. It has two values for any two particular surfaces. One is the *coefficient of static friction*, which is also called the *limiting coefficient of friction*. This enables the value of the force required to overcome 'stiction' to be calculated. The other is the *coefficient of kinetic friction* which has a slightly lower value, and enables the value of the frictional resistance in motion to be calculated.

Figure 2.7 shows that the two active forces, P and W, produce a resultant reaction, R_R which is inclined at an angle ϕ to the normal. This is called the *angle of friction*, whose tangent is equal to the coefficient of friction μ. That is,

$$\tan \phi = \frac{F}{W} = \mu \tag{2.10}$$

It should be noted that the laws of friction only apply within reasonable limits. If the force between the surfaces is excessive, or the area of contact is very small or the sliding velocity is high, excessive amounts of heat energy will be generated. This can cause the temperature at the interface to rise to a level where seizure occurs, and the surfaces become welded together. This can easily happen in internal combustion engines if the supply of lubricating oil to the cylinders should fail. In these circumstances, the pistons become welded to the cylinder walls. Figure 2.8 shows graphs of frictional resistance against normal force and coefficient of kinetic friction against sliding velocity, up to the point where seizure occurs.

Key point

The coefficient of friction between dry surfaces in contact depends on surface roughness and texture.

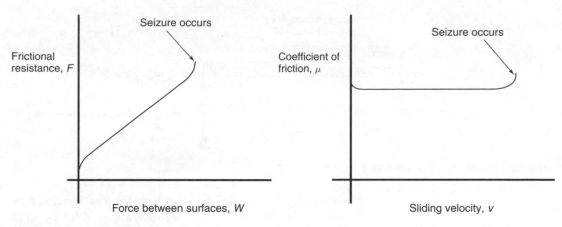

Figure 2.8 *Characteristic graphs*

Typical values of the coefficient of friction between materials commonly used in engineering are shown in Table 2.5. The materials are assumed to be clean and dry but some variation must be expected due to surface finish.

Table 2.5 *Typical coefficient of friction values*

Materials in sliding contact	Coefficient of friction, μ
Steel on cast iron	0.4
Steel on phosphor bronze	0.35
Steel on nylon	0.25
Steel on PTFE	0.05
Brake linings, and disc pads on cast iron	0.35
Rubber tyres skidding on dry asphalt	0.7

Experiments to determine the coefficient of friction

There are two experimental methods which are commonly used to determine the coefficient of friction. The first method lends itself to finding the limiting coefficient of friction whilst the second can, with care, give both the limiting and kinetic coefficients.

Method 1

Apparatus: Horizontal surface and slider faced with the two materials under investigation, free-running pulley, slotted and stackable masses, length of cord and a hanger for the slotted masses.

Procedure:
1. Weigh the slider and make a note of its mass.
2. Assemble the apparatus as shown in Figure 2.9 making sure that the plane surface and the slider are clean and dry.
3. Place a stackable mass of 0.5 kg on the slider and add slotted masses to the hanger until the tension in the cord just overcomes static friction.

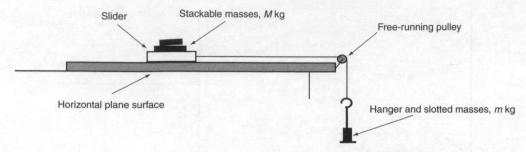

Figure 2.9 *Arrangement of apparatus*

4. Make a note of the total sliding mass and the total hanging mass.
5. Repeat the last two operations for increments of 0.5 kg on the slider until at least six sets of readings have been taken.
6. Plot a graph of hanging mass against sliding mass and calculate the value of its gradient.

The results should give you a straight line graph as shown in Figure 2.10. There may be a little scattering of the points and if so, draw in the line of best fit.

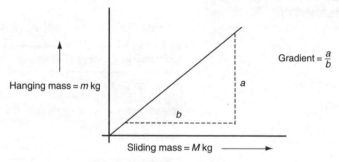

Figure 2.10 *Graph of hanging mass v. sliding mass*

Theory: The total mass of the slider and slotted masses, M is directly proportional to its weight, W. This is the normal force between the surfaces in contact. Similarly, the total hanging mass is directly proportional to its weight. This is the force, F necessary to overcome static friction. The gradient of the graph shown in Figure 2.10 thus gives a mean value of the limiting coefficient of friction. That is,

$$\mu = \frac{F}{W} = \frac{m}{M}$$

$$\mu = \textbf{Gradient of graph} = \frac{a}{b}$$

Method 2

Apparatus: Adjustable inclined plane faced with one of the materials under test and fitted with a protractor to measure the angle of inclination. Slider whose lower surface is faced with the other material and whose upper surface contains a spigot for holding stackable masses (Figure 2.11).

Procedure:
1. Weigh the slider and make a note of its mass. Make sure that the inclined plane surface and the slider are clean and dry.

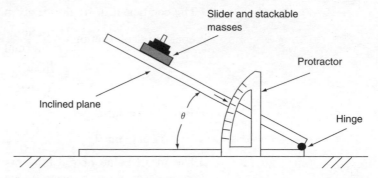

Figure 2.11 *Slider and inclined plane*

2. With the plane horizontal, place the slider at the end opposite the hinge with a mass of 0.5 kg on the slider.
3. Slowly incline the plane and note the angle, θ at which the slider starts to move.
4. Adjust the angle of the plane so that the slider continues to slide slowly down with a steady velocity and again note the value of the angle, θ.
5. Repeat the procedure with increasing values of mass on the slider and tabulate the results.
6. For each set of results, calculate the tangent of the angle at which the slider starts to move and the tangent of the angle at which the slider moves with a slow and steady velocity.

Theory: Figure 2.12 shows the forces acting on the slider. The weight of the slider and masses can be resolved into two components, one which acts normal to the incline and one which acts parallel to the incline. In the limit as motion commences, and when the slider is moving with a steady velocity, the forces normal to the plane and parallel to the plane will be equal and opposite.

Normal reaction, $R = W \cos \theta$
and
Frictional resistance, $F = W \sin \theta$

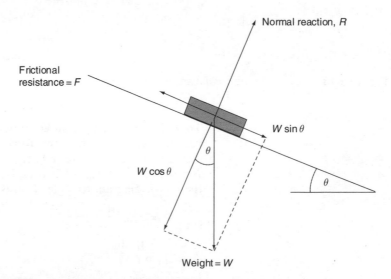

Figure 2.12 *Forces acting on slider*

The coefficient of friction is given by

$$\text{coefficient of friction} = \frac{\text{frictional resistance}}{\text{normal reaction}}$$

$$\mu = \frac{\cancel{W} \sin \theta}{\cancel{W} \cos \theta} = \frac{\sin \theta}{\cos \theta}$$

But $\dfrac{\sin \theta}{\cos \theta} = \tan \theta$

$$\boldsymbol{\mu = \tan \theta}$$

The mean value of the tangents of the angle at which motion commences gives a mean value of the limiting coefficient of friction between the surfaces. Similarly, the mean value of the tangents of the angle at which the slider moves at a steady velocity gives a mean value of the coefficient of kinetic friction.

Motion on a horizontal surface

Consider a body which is being pulled with a force, P inclined at an angle, α to a horizontal surface, and pulling in an upward direction as shown in Figure 2.13. The resultant reaction, R_R will again be inclined at an angle, ϕ to the normal where $\tan \phi = \mu$, the coefficient of friction.

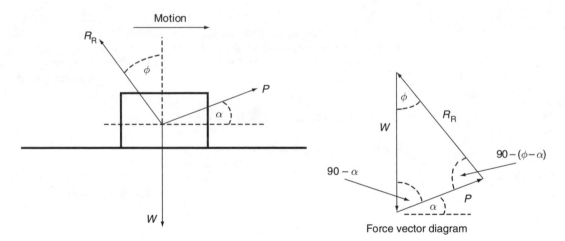

Figure 2.13 *Body on a horizontal plane*

To find the force, P, required to pull the body along with a steady velocity, you can draw the force vector diagram to scale and measure its vector. Alternatively, you may calculate the value of P using equation (2.11). It is derived by applying the sine rule to the vector diagram triangle. That is,

$$\frac{P}{\sin \phi} = \frac{W}{\sin [90 - (\phi - \alpha)]}$$

$$P = \frac{W \sin \phi}{\sin [90 - (\phi - \alpha)]}$$

Now, $\sin [90 - (\phi - \alpha)] = \cos (\phi - \alpha)$

$$P = \frac{W \sin \phi}{\cos (\phi - \alpha)} \tag{2.11}$$

You should note that when $\alpha = 0$, this becomes

$$P = \frac{W \sin \phi}{\cos \phi} = W \tan \phi$$

But, $\tan \phi = \mu$, the coefficient of friction and so $P = \mu W$, as given by equation (2.9).

You should note that the minimum value of the force, P required to pull the body along will be when $\alpha = \phi$ and the denominator of equation (2.11) becomes $\cos 0 = 1$.

Motion on an inclined plane

Consider now a body which is being pulled with a force, P which is inclined at an angle, α to the surface, which is an inclined plane, and pulling in an upward direction as shown in Figure 2.14. The plane is inclined at an angle θ to the horizontal. The resultant reaction, R_R will be inclined at an angle, ϕ to the normal of the plane and as before, $\tan \phi = \mu$, the coefficient of friction.

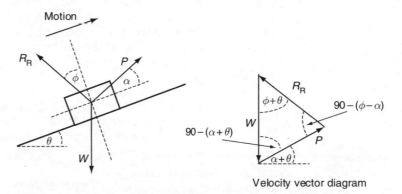

Figure 2.14 *Body on an inclined plane*

You can again find the force, P required to pull the body up the plane with a steady velocity, by drawing the force vector diagram to scale. Alternatively, you can calculate its value using equation (2.12) which is again derived by applying the sine rule to the vector diagram triangle. That is,

$$\frac{P}{\sin (\phi + \theta)} = \frac{W}{\sin [90 - (\phi - \alpha)]}$$

$$P = \frac{W \sin (\phi + \theta)}{\sin [90 - (\phi - \alpha)]}$$

Now, $\sin [90 - (\phi - \alpha)] = \cos (\phi - \alpha)$,

$$P = \frac{W \sin (\phi + \theta)}{\cos (\phi - \alpha)} \tag{2.12}$$

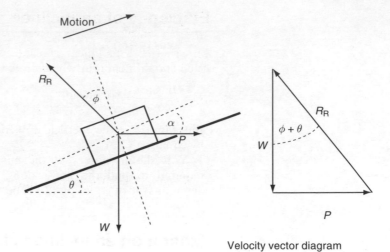

Figure 2.15 *Body on an inclined plane with horizontal pulling force*

You should again note that the minimum value of the force, P required to pull the body up the plane will be that when $\alpha = \phi$ and the denominator becomes $\cos 0 = 1$.

A special case, where the pulling force, P is horizontal, is shown in Figure 2.15. It is the condition which we shall encounter shortly when we consider frictional resistance in screw threads.

The force, P required to pull the body up the incline with a steady velocity can again be found by drawing the force vector diagram to scale. It will be noted that this is now a right-angle triangle which makes the value of P easy to calculate. That is,

$$P = W \tan (\phi + \theta) \tag{2.13}$$

Figure 2.16 shows a body being pushed or pulled down an inclined plane by a horizontal force.

The horizontal force, P required to push the body down the incline is given by

$$P = W \tan (\phi - \theta) \tag{2.14}$$

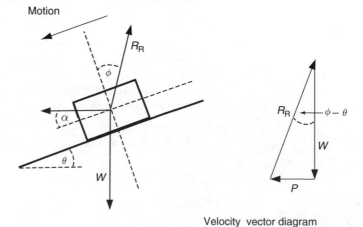

Velocity vector diagram

Figure 2.16 *Body moving down an inclined plane due to a horizontal force*

Efficiency of an inclined plane

The efficiency of an inclined plane, when used as a lifting device, is given by the formula,

$$\text{efficiency} = \frac{\text{work output}}{\text{work input}}$$

Consider the general case where the pulling force, P is inclined at some angle, α to the plane, as shown in Figure 2.17. Suppose that the body of weight, W moves through a distance, x up the incline. In so doing it will be raised through a vertical distance, h. The effective pulling force is the component $P\cos\alpha$ acting parallel to the plane.

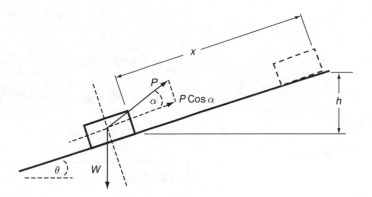

Figure 2.17 *Movement up an inclined plane*

$$\text{work input} = P\cos\alpha\, x$$

$$\text{work output} = W\, h = W\, x\sin\theta$$

The efficiency will thus be given by

$$\text{efficiency} = \frac{W\, x\sin\theta}{P\cos\alpha\, x}$$

$$\text{efficiency} = \frac{W\sin\theta}{P\cos\alpha}$$

Substituting for P from equation (2.12)

$$\text{efficiency} = \frac{W\sin\theta\cos(\phi - \alpha)}{W\sin(\phi + \theta)\cos\alpha}$$

$$\textbf{efficiency} = \frac{\sin\theta\cos(\phi - \alpha)}{\cos\alpha\sin(\phi + \theta)} \qquad (2.15)$$

When the force, P is parallel to the plane, $\alpha = 0$. Substituting for α in equation (2.15) gives

$$\textbf{efficiency} = \frac{\sin\theta\cos\phi}{\sin(\phi + \theta)} \qquad (2.16)$$

When the force, P is horizontal, $\alpha = -\theta$. Substituting for α in equation (2.15) gives

$$\text{efficiency} = \frac{\sin\theta\cos[\phi - (-\theta)]}{\cos(-\theta)\sin(\phi + \theta)}$$

Now $\cos(-\theta) = \cos\theta$, because cosine is positive in the 1st and 4th quadrants

$$\text{efficiency} = \frac{\sin\theta\cos(\phi+\theta)}{\cos\theta\sin(\phi+\theta)}$$

But

$$\frac{\sin\theta}{\cos\theta} = \tan\theta \quad \text{and} \quad \frac{\cos(\phi+\theta)}{\sin(\phi+\theta)} = \frac{1}{\tan(\phi+\theta)}$$

$$\textbf{efficiency} = \frac{\tan\theta}{\tan(\phi+\theta)} \tag{2.17}$$

The other special case is when $\alpha = \phi$ where the force, P is a minimum. This is the condition which will give maximum efficiency for a given angle of inclination, θ and a given coefficient of friction, μ. Remember that throughout these derivations, ϕ is the angle of friction and $\tan\phi = \mu$. Substituting for α in equation (2.15) gives

$$\text{maximum efficiency} = \frac{\sin\theta\cos(\phi-\phi)}{\cos\phi\sin(\phi+\theta)}$$

$$\textbf{maximum efficiency} = \frac{\sin\theta}{\cos\phi\sin(\phi+\theta)} \tag{2.18}$$

Example 2.7

A casting of mass 50 kg is pulled along a horizontal floor at a steady speed by a horizontal force of 175 N. It is then required to be pulled up a ramp of slope 1 in 5 (sine) whose surface has the same coefficient of friction. Determine (a) the coefficient of friction, (b) the horizontal force required to pull the casting up the ramp, (c) the minimum force required and its direction, (d) the maximum efficiency of the lifting operation.

(a) Finding coefficient of kinetic friction:

$$\mu = \frac{F}{W}$$

Now $W = mg$ and when the casting is being pulled at a steady speed, $F = P$.

$$\mu = \frac{P}{mg} = \frac{175}{50 \times 9.81}$$

$$\mu = 0.357$$

(b) Finding angle of friction:

$$\phi = \tan^{-1}\mu = \tan^{-1}0.375$$

$$\phi = 19.6°$$

Finding angle of ramp:

$$\sin\theta = \frac{1}{5} = 0.2$$

$$\theta = 11.5°$$

Finding horizontal force required to pull the casting up the ramp using equation (2.13):

$$P = W\tan(\phi + \theta)$$
$$P = 50 \times 9.81 \times \tan(19.6 + 11.5)$$
$$\mathbf{P = 296\,N}$$

(c) Finding minimum force required to pull the casting up the ramp using equation (2.12):

$$P = \frac{W\sin(\phi + \theta)}{\cos(\phi - \alpha)}$$

This is a minimum when $\alpha = \phi$

$$P_{min} = \frac{W\sin(\phi + \theta)}{\cos(\phi - \phi)} = \frac{W\sin(\phi + \theta)}{\cos(0)}$$

Now, $\cos 0 = 1$

$$P_{min} = W\sin(\phi + \theta)$$
$$P_{min} = 50 \times 9.81 \times \sin(19.6 + 11.5)$$
$$\mathbf{P_{min} = 253\,N}$$

(d) Finding maximum efficiency of ramp as a lifting device using equation (2.17):

$$\text{maximum efficiency} = \frac{\sin\theta}{\cos\phi\sin(\phi + \theta)}$$
$$\text{maximum efficiency} = \frac{\sin 11.5}{\cos 19.6 \times \sin(19.6 + 11.5)}$$
$$\textbf{maximum efficiency} = \mathbf{0.410} \quad \text{or} \quad \mathbf{41\%}$$

Activity 2.4

A load of mass 1 tonne is hauled up and down a ramp of slope 1 in 4 (sine). The coefficient of friction between the contact surfaces is 0.4. Determine (a) the horizontal force required to move the load up the ramp, (b) the horizontal force required to move the load down the ramp, (c) the minimum force which is required in both directions, (d) the maximum efficiency of the system as a lifting device.

Problems 2.3

1. A crate of mass 40 kg rests on a horizontal surface with which the coefficient friction is 0.4. Determine (a) the horizontal force required to move the crate along at a steady speed, (b) the horizontal force required to pull it up a 15° ramp which has the same coefficient of friction.

[157 N, 294 N]

2. A body of mass 35 kg is pulled along a horizontal surface at a steady speed by a horizontal force of 120 N. It is then pulled up an incline

of 1 in 4 which has the same surface material. Determine (a) the coefficient of friction, (b) the pulling force required parallel to the incline, (c) the efficiency of the system for raising the mass.

[0.35, 2.2 N, 42.5%]

3. An object of mass 100 kg rests on an incline of 1 in 5. If the coefficient of friction is 0.45, determine the force required to pull the object up the slope at a steady speed when, (a) the force is acting at an angle of 30° to the incline, (b) the force is horizontal, (c) the force is parallel to the incline, (d) the force is the minimum required.

[576 N, 707 N, 629 N, 573 N]

4. A body of mass 15 kg rests on an adjustable inclined plane. The angle of inclination is increased until the body slides down the plane with a uniform velocity and this is found to occur at an angle of 22°. Determine (a) the coefficient of friction, (b) the horizontal force required to pull the body up the incline, (c) minimum force required to pull the body up the incline, (d) the maximum efficiency of the lifting operation.

[0.404, 91.9 N, 80 N, 35.3%]

5. A casting of mass 40 kg rests on a surface which is inclined at an angle of 25° to the horizontal. If a horizontal force of 65 N is required to push it down the incline at a steady speed, determine (a) the coefficient of friction, (b) the horizontal force required to pull the casting up the incline, (c) the minimum force required to pull the casting up the incline, (d) the maximum efficiency of the inclined plane as a lifting device.

[0.685, 664 N, 338 N, 59.5%]

Friction in screw threads

Single and multi-start screw threads are used for power transmission in screw jacks, presses and machine tools. A square section screw thread can be regarded as an inclined plane which ascends in a spiralling direction (Figure 2.18).

Turning the screw has the effect of moving the load up or down the incline whose angle is the helix angle, θ of the thread. The distance moved by the load for one revolution of the thread is called the *lead*. For a single-start thread the lead, l is equal to the pitch, p and for a two-start thread, $l = 2p$ etc. If d, is the thread diameter, the helix angle can be found from

$$\tan \theta = \frac{\text{lead of thread}}{\text{circumference of thread}}$$

$$\boldsymbol{\tan \theta = \frac{l}{\pi d}} \qquad (2.19)$$

For a single-start thread, $l = p$ and the helix angle is given by

$$\boldsymbol{\tan \theta = \frac{p}{\pi d}} \qquad (2.20)$$

The tangential force, P required to raise the load is the same as that given by equation (2.13) for a load raised on an inclined plane by a horizontal force.

$$P = W \tan(\phi + \theta)$$

Key point

The lead of a screw thread is the axial distance that it moves through a stationary nut during one complete revolution.

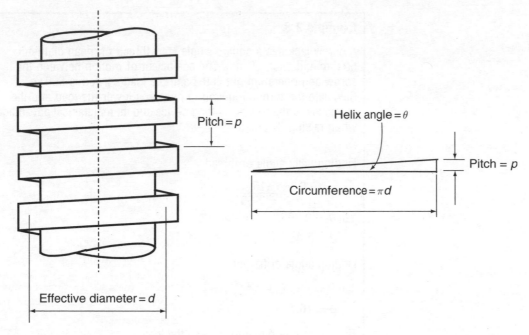

Figure 2.18 *Square screw thread*

The torque required to raise the load is given by

torque = tangential force × effective radius

$$T = P\frac{d}{2} = W\tan(\phi + \theta) \times \frac{d}{2}$$

$$T = \frac{Wd}{2}\tan(\phi + \theta) \tag{2.21}$$

The efficiency of the system when the load is being raised is given by equation (2.17)

$$\textbf{efficiency} = \frac{\tan\theta}{\tan(\phi + \theta)} \tag{2.22}$$

The tangential force, P required to lower the load is the same as that given by equation (2.14) for the inclined plane. That is,

$$P = W\tan(\phi - \theta)$$

The torque required to lower the load is given by

$$T = \frac{Wd}{2}\tan(\phi - \theta) \tag{2.23}$$

When the angle of friction, ϕ is equal to the helix angle of the thread, θ, then $\tan(\phi - \theta) = 0$. The torque required to lower the load is then zero, and the screw will tend to overhaul. In this condition the efficiency of the screw as a lifting device will be

$$\text{efficiency} = \frac{\tan\theta}{\tan(\theta + \theta)}$$

$$\textbf{efficiency} = \frac{\tan\theta}{\tan 2\theta} \tag{2.24}$$

Key point

A screw jack will overhaul when the coefficient of friction, μ is less than the tangent of the helix angle of the thread, i.e. when $\mu < \tan\theta$.

Example 2.8

A screw jack has a square single start thread of mean diameter 50 mm and pitch 15 mm. The coefficient of friction between the screw and operating nut is 0.3 and the load carried is 500 kg. Calculate the torque required to raise and lower the load and the efficiency of the jack as a lifting device and its mechanical advantage when raising this load.

Finding helix angle of thread:

$$\tan \theta = \frac{p}{\pi d} = \frac{15}{\pi \times 50}$$
$$\tan \theta = 0.095$$
$$\boldsymbol{\theta = 5.45°}$$

Finding angle of friction:

$$\tan \theta = \mu = 0.3$$
$$\boldsymbol{\phi = 16.7°}$$

Finding torque required to raise the load:

$$T = \frac{Wd}{2}\tan(\phi + \theta)$$
$$T = \frac{500 \times 9.81 \times 50 \times 10^{-3}}{2} \times \tan(16.7 + 5.45)$$
$$\boldsymbol{T = 49.9\,Nm}$$

Finding torque required to lower the load:

$$T = \frac{Wd}{2}\tan(\phi - \theta)$$
$$T = \frac{500 \times 9.81 \times 50 \times 10^{-3}}{2} \times \tan(16.7 - 5.45)$$
$$\boldsymbol{T = 24.4\,N\,m}$$

Finding efficiency of the system as a lifting device:

$$\text{efficiency} = \frac{\tan \theta}{\tan(\phi + \theta)}$$
$$\text{efficiency} = \frac{\tan 5.45}{\tan(16.7 + 5.45)}$$
$$\boldsymbol{\text{efficiency} = 0.234} \quad \text{or} \quad \boldsymbol{23.4\%}$$

Finding velocity ratio of the jack (see Chapter 1):

$$\text{VR} = \frac{\pi d}{p} = \frac{\pi \times 50}{15}$$
$$\boldsymbol{\text{VR} = 10.47}$$

Finding mechanical advantage (see Chapter 1):

$$\text{efficiency} = \frac{\text{mechanical advantage}}{\text{velocity ratio}} = \frac{\text{MA}}{\text{VR}}$$
$$\text{MA} = \text{efficiency} \times \text{VR}$$
$$\text{MA} = 0.234 \times 10.47$$
$$\boldsymbol{\text{MA} = 2.45}$$

Rolling resistance

The frictional resistance between surfaces which roll over each other is generally much less than that which occurs due to sliding contact. The wheels of vehicles experience resistance to motion because as they rotate, the tyre is deformed at the point of contact with the road surface. With heavy vehicles, it is also possible that the road surface becomes slightly deformed. Both have the effect of opposing the rolling motion of the wheels. The same kind of rolling resistance is encountered in ball, roller and needle bearings when operating under load (Figure 2.19).

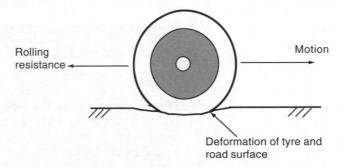

Figure 2.19 *Rolling resistance*

Effects of lubrication

When a lubricant such as oil or grease is introduced between surfaces in sliding contact, the effect is to separate them. The interlocking and pressure welding of the irregularities in the surfaces is then prevented (Figure 2.20).

As the surfaces slide over each other, a shearing action takes place in the lubricant and the resistance to motion is now termed *viscous drag*. Its value is much smaller than when the surfaces are in dry contact. Typically, the coefficient of friction between metal

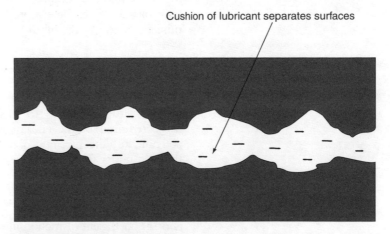

Figure 2.20 *Lubricated surfaces*

surfaces is reduced from $\mu = 0.3-0.4$ to $\mu = 0.05-0.15$ depending on the viscosity of the lubricant and the thickness of the lubricant film. The laws of friction which apply to dry surfaces no longer apply and it is found that,

1. The viscous resistance varies directly with the area of contact between the two surfaces.
2. The viscous resistance varies with the sliding velocity.
3. The viscous resistance varies inversely with the thickness of the lubricant film.
4. The viscous resistance varies directly with the viscosity of the lubricant.

The calculation of viscous resistance in slideways and bearings and the associated power loss will be explained in Chapter 5.

Activity 2.5

Whilst building repairs are being carried out, a screw jack is used to support a roof beam. The beam is raised off its support by an input torque of 10 N m. The jack has a square single-start thread of pitch 6 mm and effective diameter 30 mm. The coefficient of friction between the screw and the nut is 0.25. Calculate the load supported by the jack and the torque required to lower the beam back into position. If lubricating the screw thread reduces its coefficient of friction to 0.1, determine whether it is likely to overhaul.

Problems 2.4

1. Calculate the torque required to raise a load of 1 kN by means of a screw jack. The screw has a single start square thread of pitch 5 mm and a mean diameter of 20 mm. The coefficient of friction between the nut and screw is 0.15. Calculate also the efficiency of the jack when raising this load.

   [2.32 N m, 34.4%]

2. A single start screw jack has a square section thread of pitch 6 mm and effective diameter 35 mm. The coefficient of friction between the screw and the stationary nut is 0.2 and the jack supports a load of 0.5 tonne. Determine (a) the input torque required to raise the load, (b) the input torque required to lower the load.

   [22.1 N m, 12.4 N m]

3. A load of 1 tonne is to be raised by a screw jack which has a single start thread of pitch 3 mm and mean diameter 25 mm. The coefficient of friction in operation is 0.15. Determine (a) the torque required to raise the load, (b) the tangential force required at the end of an operating handle of length 600 mm, (c) the mechanical advantage, (d) the efficiency of the jack under these operating conditions.

   [23.2 N m, 38.7 N, 253, 20.2%]

4. A fly-press has a multi-start square section thread of lead 20 mm and mean diameter 30 mm. The mass of the rotating parts is 35 kg and the coefficient of friction in the thread is 0.15. Determine the tangential force that must be applied to the operating arm of radius

500 mm in order to raise the screw and show that the rotating parts will fall under the effects of gravity when the arm is released.

[128 N, $\phi < \theta$, i.e. machine will overhaul]

5. A sluice gate of weight 30 kN has water to one side which exerts a normal thrust of 1.24 MN. The gate can be raised in guides by means of a single start lead-screw operated by a motor through a 3:1 reduction gearbox. The coefficient of friction between the gate and its guides and in the screw thread is 0.1. The lead-screw has a square thread of pitch 25 mm and mean diameter 75 mm. Assuming no losses on the gearbox, determine the motor torque required to raise the gate.

[399 N m]

Rotational kinetics

You should already have some knowledge of angular motion. From the work which you have done to-date you will be aware of the equations for uniform angular motion. As a reminder, here they are again.

$$\omega_2 = \omega_1 + \alpha t \tag{2.25}$$

$$\theta = \omega_1 t + \frac{1}{2}\alpha t^2 \tag{2.26}$$

$$\theta = \frac{1}{2}(\omega_1 + \omega_2)t \tag{2.27}$$

$$\omega_2^2 = \omega_1^2 + 2\alpha\theta \tag{2.28}$$

where ω_1 and ω_2 are initial and final angular velocities, α is the uniform angular acceleration, θ is the angle turned in radians and t is the time taken. You should also know how to convert tangential velocity, v to angular velocity, ω and rotational speed, N in revolutions per minute into angular velocity, $\omega\,\mathrm{rad\,s}^{-1}$. That is,

$$\omega = \frac{2\pi N}{60} \tag{2.29}$$

$$\omega = \frac{v}{r} \tag{2.30}$$

> **Key point**
>
> To convert tangential velocity and tangential and tangential acceleration to angular velocity and acceleration, divide by the radius of rotation.

The work done by the torque, T which turns a body through some angle, θ rad is given by

$$W = T\theta \text{ Joules} \tag{2.31}$$

The average power developed over the period during which the torque acts is the average amount of work done per second.

$$\text{average power} = \frac{W}{t}$$

$$\textbf{average power} = \frac{T\theta}{t} \textbf{ Watts} \tag{2.32}$$

But θ/t is the average angular velocity and so the average power can also be found from

average power = torque × average angular velocity

The power which is being produced at any instant when the angular velocity is ω rad s^{-1} is given by

instantaneous power = torque × instantaneous angular velocity

$$\textbf{instantaneous power} = T\omega \textbf{ Watts} \tag{2.33}$$

Moment of inertia

Consider now the torque, T required to overcome the inertia of a rigid body of mass m and produce angular acceleration, α (Figure 2.21).

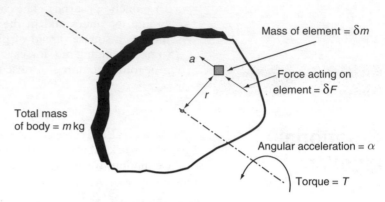

Figure 2.21 *Rotation of a rigid body*

The body can be considered to be made up of a large number of elements of mass δm, one of which is shown. The radius of rotation of the element is r, and its linear tangential acceleration at the instant shown is a. The force, δF acting on the element is given by Newton's second law of motion. That is,

$\delta F = \delta m\, a$

Now the linear acceleration of the element is given by, $a = \alpha r$

$\delta F = \alpha r\, \delta m$

The torque, δT acting on the element is given by the product of the force, δF and the radius, r. That is,

$\delta T = \delta F\, r$

$\delta T = \alpha r\, \delta m\, r$

$\delta T = \alpha r^2\, \delta m$

The total accelerating torque is the sum of all such elements:

$T = \Sigma(\alpha r^2\, \delta m)$

All of the elements have the same angular acceleration, α and so

$T = \alpha\Sigma(r^2\, \delta m)$

As $\delta m \longrightarrow 0$ this becomes

$$T = \alpha \int r^2\, dm \qquad (2.34)$$

The integral term is known as the *second moment of mass* or *mass moment of inertia* of the body which is given the symbol I. Its units are $kg\,m^2$. You will note that it is the sum of the products of the mass elements and the square of their radii. It is very closely related to second moment of area which we discussed when dealing with beams and struts. That, you will recall is the sum of the products of the elements of area and the square of their radii. The above expression for total accelerating torque can thus be written as

$$T = I\alpha \qquad (2.35)$$

This is in fact an expression of Newton's second law of motion which applies to rotating rigid bodies. You might also recall the concept of *radius of gyration*. The radius of gyration, k of a rotating body is the root-mean square radius of the elements. It can be thought of as the radius at which the mass of a rotating body is concentrated. The mass moment of inertia can thus be written as

$$I = mk^2 \qquad (2.36)$$

The values of moment of inertia and radius of gyration of engineering components such as vehicle wheels, gears, pulleys, motor armatures and turbine rotors can often only be found by experiment. It is however possible to derive expressions for some basic objects.

Key point

The radius of gyration is the radius at which the mass of a rotating body can be considered to be concentrated.

1. Concentrated mass, m with radius of rotation, r (Figure 2.22).

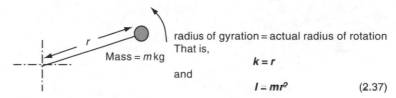

radius of gyration = actual radius of rotation
That is,

$$k = r$$

and

$$I = mr^2 \qquad (2.37)$$

Figure 2.22 *Rotating mass*

2. Thin rotating ring of mass, m and mean radius, r rotating about a polar axis (Figure 2.23).

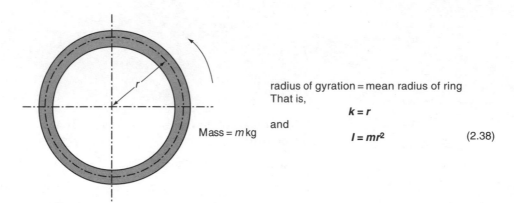

radius of gyration = mean radius of ring
That is,

$$k = r$$

and

$$I = mr^2 \qquad (2.38)$$

Figure 2.23 *Rotating ring*

3. Disc of mass, m radius, r and thickness, t rotating about a polar axis.

Consider the disc to be made up of an infinite number of concentric elemental rings. The one shown is at radius, x and has a thickness, dx (Figure 2.24). Let the density of the material be $\rho\,\mathrm{kg\,m^{-3}}$.

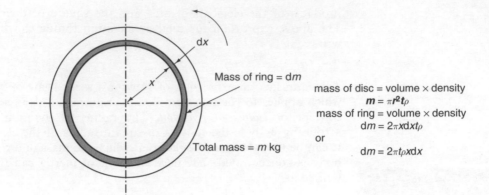

Figure 2.24 *Rotating disc*

The moment of inertia, dI of the elemental ring is given by equation (2.38). That is,

$$dI = x^2 dm$$

$$dI = x^2 2\pi t \,\rho x \,dx$$

$$\mathbf{dI = 2\pi t\, \rho\, x^3\, dx}$$

The moment of inertia of the disc is the sum of all such elemental moments.

$$I = \int_0^r 2\pi t\, \rho\, x^3 \,dx$$

$$I = 2\pi t\, \rho \int_0^r x^3 dx$$

$$I = 2\pi t\, \rho \left(\frac{x^4}{4}\right)_0^r$$

$$I = 2\pi t\, \rho\, \frac{r^4}{4}$$

$$\mathbf{I = \pi t\, \rho\, \frac{r^4}{2}}$$

Now the mass of the disc, $m = \pi r^2\, t\, \rho$ and so the moment of inertia can be written as

$$\mathbf{I = \frac{mr^2}{2}} \tag{2.39}$$

Since, $I = mk^2$, the radius of gyration of the disc will be given by

$$\mathbf{k = \frac{r}{\sqrt{2}} = 0.7071r} \tag{2.40}$$

Key point

The radius of gyration of a rotating body is a root-mean-square radius.

This is the root-mean-square radius. You may recall that the same multiplying factor of 0.7071 is used to find the root-mean-square value of alternating current and voltage.

4. Rod of mass, m length, l, cross-sectional area, A and density, ρ rotating about an axis through its centre (Figure 2.25).

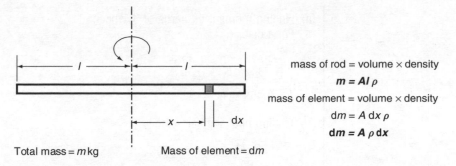

Figure 2.25 *Rod rotating about a central axis*

The moment of inertia, dI of the element is given by

$$dI = x^2 \, dm$$

$$dI = x^2 \, A\rho \, dx$$

$$\mathbf{dI = A\rho \, x^2 dx}$$

The moment of inertia of the rod is the sum of all such elemental moments.

$$I = \int_{-l/2}^{+l/2} A\,\rho\,x^2\,dx$$

$$I = A\,\rho \int_{-l/2}^{+l/2} x^2\,dx$$

$$I = A\,\rho\left(\frac{x^3}{3}\right)_{-l/2}^{+l/2}$$

$$I = A\,\rho\left[\frac{l^3}{24} - \left(\frac{-l^3}{24}\right)\right]$$

$$\mathbf{I = A\rho\frac{l^3}{12}}$$

Now the mass of the rod, $m = Al\rho$ and so the moment of inertia can be written as

$$\mathbf{I = m\frac{l^2}{12}} \tag{2.41}$$

Since, $I = mk^2$ the radius of gyration of the rod will be given by

$$\mathbf{k = \frac{l}{\sqrt{12}} = 0.289\,l} \tag{2.42}$$

Example 2.9

A flywheel, which may be regarded as a disc of diameter 350 mm and mass 30 kg is accelerated uniformly from rest by an electric motor to a speed of 500 rpm in a time of 30 s. The drive from the motor is then disconnected and a brake is applied which brings it uniformly to rest whilst rotating through 100 complete revolutions. Neglecting bearing friction and air resistance, determine (a) the moment of inertia of the disc, (b) the accelerating torque, (c) the breaking torque.

(a) Finding moment of inertia of flywheel:
For a disc,

$$I = \frac{mr^2}{2} = \frac{30 \times 0.175^2}{2}$$

$$I = 0.459 \, \text{kg m}^2$$

(b) Finding final angular velocity:

$$\omega_2 = \frac{2\pi N_2}{60} = \frac{2 \times \pi \times 500}{60}$$

$$\omega_2 = 52.4 \, \text{rad s}^{-1}$$

Finding final angular acceleration, α_1:

$$\omega_2 = \omega_1 + \alpha_1 t$$

$$\alpha_1 = \frac{\omega_2 - \omega_1}{t} = \frac{52.4 - 0}{30}$$

$$\alpha_1 = 1.75 \, \text{rad s}^{-1}$$

Finding accelerating torque, T_1:

$$T_1 = I\alpha_1 = 0.459 \times 1.75$$

$$T_1 = 0.803 \, \text{N m}$$

(c) Finding angle turned, θ rad whilst braking:

$$\theta = \text{number of revolutions} \times 2\pi = 100 \times 2\pi$$

$$\theta = 628 \, \text{rad}$$

Finding angular retardation, α_2 (final angular velocity, $\omega_3 = 0$):

$$\omega_3^2 = \omega_2^2 + 2\alpha_2\theta$$

$$\alpha_2 = \frac{\omega_3^2 - \omega_2^2}{\theta} = \frac{0 - 52.4^2}{628}$$

$$\alpha_2 = -4.37 \, \text{rad s}^{-1} \text{ (negative sign denotes retardation)}$$

Finding braking torque:

$$T_2 = I\alpha_2 = 0.459 \times 4.37$$

$$T_2 = 2.01 \, \text{N m}$$

Test your knowledge 2.6

1. How may the work done in rotational motion be found from the applied torque and angle turned?
2. What is the *radius of gyration* of a rotating body?
3. How may the moment of inertia of a rotating body be found from its mass and radius of gyration?
4. What is the expression for the radius of gyration of a disc rotating about a polar axis?

Activity 2.6

A flywheel of mass 50 kg has a radius of gyration of 120 mm is accelerated from an initial speed of 200 rpm to a final speed of 1000 rpm in a time of 25 s. If the torque needed to overcome bearing friction is 1.5 N m, determine (a) the total applied torque, (b) the work done, (c) the average input power from the driving motor, (d) the input power when reaching the final speed.

Problems 2.5

1. A winding drum of mass 1.5 tonnes has a radius of gyration of 2 m. What will be the torque required (a) to accelerate it uniformly from rest to a speed of 150 rpm in a time of 1.5 min, (b) to bring it to rest over a period of 3 min? Ignore frictional resistance.
[1.04 kN m, 522 N m]

2. A flywheel of mass 2 tonne, diameter 2 m and radius of gyration 950 mm runs at a speed of 1000 rpm. What will be the tangential braking force applied to its outer diameter that is required to bring it uniformly to rest whilst turning through 200 revolutions?
[7.87 kN]

3. A flywheel, which can be regarded as a solid disc of diameter 300 mm and mass 25 kg, is accelerated from 500 rpm to 3000 rpm in 20 s. If the friction torque which must be overcome is a constant 5 N m, determine the magnitude of the total applied torque.
[7.95 N m]

4. A flywheel of mass 850 kg is retarded uniformly from a speed of 1000 rpm to 500 rpm in a time of 5 min under the action of a 124 N m braking torque. Determine (a) the angular retardation, (b) the number of revolutions made in coming to rest, (c) the moment of inertia and radius of gyration of the flywheel.
[0.175 rad s^{-1}, 3750 revs, 708 kg m^2, 913 mm]

5. A rotating drum is a thin cylinder 1.3 m in diameter and 600 mm long. The material thickness is 5 mm and it is made from mild steel of density 7800 kg m^{-3}. Calculate (a) the moment of inertia of the drum when rotating about its polar axis, (b) the time taken for it to reach a speed of 3600 rpm from rest if the driving torque is 55 N m and there is a friction torque of 5 N m.
[31.8 kg m^2, 4 min]

Parallel axis theorem

Provided that the moment of inertia of a body about an axis through its centre of gravity is known, its value about any parallel axis can be found by using the parallel axis theorem (Figure 2.26).

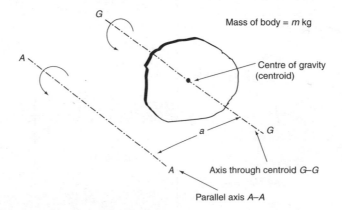

Figure 2.26 *Parallel axes of rotation*

The parallel axis theorem states that if the moment of inertia about an axis through the centroid of a body is I_G, its value I_A about a parallel axis distance a, away is given by

$$I_A = I_G + ma^2 \tag{2.43}$$

Perpendicular axis theorem

The perpendicular axis theorem relates the polar moment of inertia of a lamina about an axis through its centroid to two mutually perpendicular axes in the plane of the body (Figure 2.27).

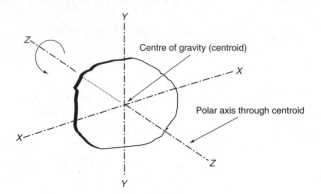

Figure 2.27 *Perpendicular axes of rotation*

The perpendicular axis theorem states that if I_Z is the moment of inertia about a polar axis through the centroid of a lamina, the moments of inertia, I_X and I_Y about two perpendicular axes in the plane of the body are related by the formula,

$$I_Z = I_X + I_Y \tag{2.44}$$

Example 2.10

A rectangular plate measuring 300 mm × 900 mm has a mass of 7.5 kg (Figure 2.28). What will be its moment of inertia and radius of gyration when (a) rotating about a polar axis through its centroid, (b) rotating about a parallel axis through one of its corners?

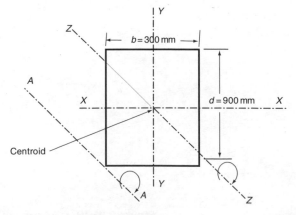

Figure 2.28 *Rectangular plate*

(a) Finding moment of inertia when rotating about the axis *X–X*. The plate can be considered to be a wide, thin rod of length *d* rotating about an axis through its centroid.

$$I_X = \frac{md^2}{12} = \frac{7.5 \times 0.9^2}{12}$$

$$I_X = 0.506 \text{ kg m}^2$$

Finding moment of inertia when rotating about the axis, *Y–Y* through its centroid:

$$I_Y = \frac{mb^2}{12} = \frac{7.5 \times 0.3^2}{12}$$

$$I_Y = 0.056 \text{ kg m}^2$$

Finding moment of inertia when rotating about the polar axis *Z–Z* through the centroid. The perpendicular axis theorem gives

$$I_Z = I_X + I_Y = 0.506 + 0.056$$

$$I_Z = 0.562 \text{ kg m}^2$$

Finding radius of gyration, k_z about axis *Z–Z*:

$$I_Z = m k_z^2$$

$$k_z = \sqrt{\frac{I_Z}{m}} = \sqrt{\frac{0.562}{7.5}}$$

$$k_z = 0.274 \text{ m} \quad \text{or} \quad 0.274 \text{ mm}$$

(b) Finding distance, *a* from centroid to the corner using Pythagoras' theorem:

$$a = \sqrt{\frac{b^2}{2} + \frac{d^2}{2}} = \sqrt{450^2 + 150^2}$$

$$a = 474 \text{ mm} \quad \text{or} \quad 0.474 \text{ m}$$

Finding moment of inertia when rotating about the axis *A–A* which is parallel to the polar axis *Z–Z*. The parallel axis theorem gives

$$I_A = I_Z + m a^2 = 0.562 + (7.5 \times 0.474^2)$$

$$I_A = 2.25 \text{ kg m}^2$$

Finding radius of gyration, k_a about axis *A–A*:

$$I_A = m k_a^2$$

$$k_a = \sqrt{\frac{I_A}{m}} = \sqrt{\frac{2.25}{7.5}}$$

$$k_a = 0.548 \text{ m} \quad \text{or} \quad 548 \text{ mm}$$

Rotational kinetic energy

The kinetic energy stored in a moving body is the work done by the force which has accelerated it from rest up to the particular speed at which it is travelling. You will recall that for a body of mass, m kg which is moving with linear velocity, v m s^{-1} its kinetic energy is given by the formula

$$KE = \frac{1}{2}mv^2 \text{ Joules} \tag{2.45}$$

Kinetic energy is also stored in a rotating body. It is the work done by the torque which has accelerated it from rest up to its particular angular velocity. Consider now a rotating rigid body of mass m kg rotating with angular velocity ω rad s^{-1} as shown in Figure 2.29.

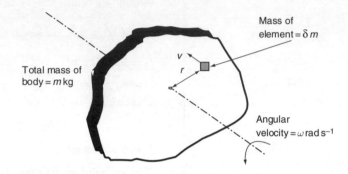

Figure 2.29 *Rotation of a rigid body*

The small element of mass δm at radius r from the axis of rotation has a tangential velocity, v m s^{-1} at the instant shown. If the angular velocity is ω rad s^{-1}, its tangential velocity can be written as

$$v = \omega r$$

The small amount of kinetic energy, dE stored in the element will be

$$\mathrm{d}E = \frac{1}{2}\delta m v^2$$

Substituting for v gives

$$\mathrm{d}E = \frac{1}{2}\delta m (\omega r)^2 = \frac{1}{2}\delta m \,\omega^2\, r^2$$

$$\mathrm{d}E = \frac{1}{2}\omega^2 \delta m\, r^2$$

The total kinetic energy stored in the body is the sum of all such amounts:

$$KE = \sum \left(\frac{1}{2}\omega^2 \delta m\, r^2 \right)$$

Now the $\frac{1}{2}\omega^2$ term is a constant for all elements and can be taken outside the summation sign:

$$KE = \frac{1}{2}\omega^2 \sum (\delta m\, r^2)$$

But the $\Sigma(\delta m\, r^2)$ term is the moment of inertia, I of the body giving

$$KE = \frac{1}{2}\omega^2 I$$

or

$$KE = \frac{1}{2}I\omega^2 \text{ Joules} \tag{2.46}$$

If a body has an initial angular velocity, ω_1 and is accelerated by an applied torque, T to a final angular velocity, ω_2 whilst rotating through an angle of θ rad, the work done, W by the accelerating torque is equal to the change of kinetic energy. That is,

work done = change of kinetic energy

$$W = \frac{1}{2}I\omega_2^2 - \frac{1}{2}I\omega_1^2$$

$$W = \frac{1}{2}I(\omega_2^2 - \omega_1^2) \text{ Joules} \qquad (2.47)$$

The work done is also equal to the product of the accelerating torque and the angle turned. That is,

work done = torque × angle turned

$$W = T\theta \text{ Joules} \qquad (2.48)$$

Equating these gives a method of calculating the accelerating torque.

Example 2.11

A flywheel of mass 15 kg and radius of gyration 120 mm is accelerated from an initial speed of 500 rpm to a final speed of 1200 rpm whilst turning through 100 complete revolutions. Neglecting the effects of friction, determine (a) the work done, (b) the accelerating torque, (c) the maximum power developed.

(a) Finding moment of inertia of flywheel:

$$I = mk^2 = 15 \times 0.12^2$$

$$I = 0.216 \text{ kg m}^2$$

Finding initial angular velocity:

$$\omega_1 = N_1 \times \frac{2\pi}{60} = \frac{500 \times 2\pi}{60}$$

$$\omega_1 = 52.4 \text{ rad s}^{-1}$$

Finding final angular velocity:

$$\omega_2 = N_2 \times \frac{2\pi}{60} = \frac{1200 \times 2\pi}{60}$$

$$\omega_2 = 125.7 \text{ rad s}^{-1}$$

Finding work done by accelerating torque:

$$W = \frac{1}{2}I(\omega_2^2 - \omega_1^2)$$

$$W = \frac{1}{2} \times 0.216 \times (125.7^2 - 52.4^2)$$

$$W = 1410 \text{ J} \quad \text{or} \quad 1.41 \text{ kJ}$$

(b) Finding angle turned in radians:

$$\theta = \text{number of revolutions} \times 2\pi$$

$$\theta = 100 \times 2\pi$$

$$\theta = 628 \text{ rad}$$

Finding accelerating torque:

$$W = T\theta$$

$$T = \frac{W}{\theta} = \frac{1410}{628}$$

$$T = 2.25 \, \text{N m}$$

(c) Finding maximum power developed:

maximum power = torque × maximum angular velocity

maximum power = $T\omega_2$ = 2.25 × 125.7

maximum power = 283 W

Activity 2.7

When rotating about an axis through its centre of gravity, which is perpendicular to its polar axis, the moment of inertia of a rod of mass m kg and length l m is given by the expression

$$I = \frac{ml^2}{12} \, \text{kg m}^2$$

Show, using the parallel axis theorem, that its moment of inertia when rotating about a parallel axis through one of its ends is given by the expression

$$I = \frac{ml^2}{3} \, \text{kg m}^2$$

A bar of length 1 m and mass of 2.5 kg rotates about an axis through one of its ends. If its speed is increased from 50 rpm to 120 rpm in a time of 20 s, determine (a) the change of kinetic energy, (b) the accelerating torque, (c) the maximum power developed.

Problems 2.6

1. A thin rotating ring of mass 2 kg has an outer diameter of 500 mm and an inner diameter of 450 mm. Determine the moment of inertia of the ring when rotating about (a) a polar axis through its centre of gravity, (b) an axis along a diameter through the centre of gravity, (c) a tangential axis parallel to a diameter.

   [0.113 kg m^2, 0.0564 kg m^2, 0.181 kg m^2]

2. A square lamina has sides of length 250 mm and a mass of 1.5 kg. Determine the moment of inertia of the lamina when rotating about an axis through its centroid and perpendicular to its plane.

   [15.6 × 10^{-3} kg m^2, 102 mm]

3. A flywheel of mass 15 kg, which can be treated as a disc of diameter 400 mm, is accelerated uniformly from a speed of 100 rpm to 1000 rpm in a time of 20 s. Neglecting friction and using an energy method, determine (a) the work done, (b) the external torque applied, (c) the maximum power developed.

   [1.63 kJ, 1.41 Nm, 148 W]

4. A flywheel of mass 50 kg, diameter 600 mm and radius of gyration 200 mm is rotating freely at a speed of 1200 rpm when a brake is applied to its perimeter. If the flywheel is retarded uniformly to rest whilst turning through 300 complete revolutions, determine using an energy method, (a) the work done during braking, (b) the braking torque, (c) the tangential braking force.

[15.9 kJ, 8.42 N m, 28.1 N]

5. A winding drum of mass 0.5 tonnes and radius of gyration 750 mm is accelerated uniformly from rest to a speed of 200 rpm in a time of 30 s after which it is retarded uniformly to rest whilst rotating through 60 revolutions. Determine using an energy method, (a) the work done by the driving motor, (b) the accelerating torque, (c) the braking force applied tangentially at a radius of 1200 mm. Neglect the effects bearing friction and air resistance.

[61.4 kJ, 195 N m, 136 N]

Combined linear and angular motion

Linear and angular motion very often occur together in dynamic engineering systems. In fact, it is the purpose of some systems to convert angular motion into linear motion and vice versa. A common example is a hoist in which a load is raised by a cable which passes around a winding drum. Another example is a vehicle in which the angular motion of the driving wheels propels it in a linear direction. It is often required to find the driving torque, work done and power requirements for these systems and there are two alternative methods of finding them.

1. By the application of D'Alembert's principle and Newton's second law of motion.
2. By the application of the principle of conservation of energy.

Application of D'Alembert's principle and Newton's second law of motion

In the study of applied mechanics, which is often called *Newtonian mechanics*, the resistance to motion is usually due to one or more of three things.

1. Frictional resistance, which may also include air resistance.
2. Gravitational force, if the motion is in an upward direction.
3. Inertia resistance, if a body is being accelerated or retarded.

Gravitational force is the weight, or a component of the weight, of a body and we have already investigated frictional resistance. Inertia is the resistance of a body to being speeded up or slowed down, i.e. accelerated or retarded. It is described by Newton's first law of motion which states that:

A body continues in a state of rest or travels at a uniform velocity unless acted upon by some external force.

Newton's second law of motion relates the accelerating force to the rate of change of momentum. It states that:

The rate of change of momentum of a body is proportional to the force which is producing the change.

You will recall that it is from this law that we obtain the equation, $F = ma$. Inertia resistance is further described by Newton's third law of motion which states that:

To every action there is an equal and opposite reaction.

Inertia resistance is the reaction to the external force which is causing the change of velocity, i.e. it is equal and opposite to the force given by the equation $F = ma$.

D'Alembert's principle states that:

If the inertial reaction to motion is treated as an external force, a body in uniform motion can be treated as if it were in static equilibrium.

Consider now a simple hoist where a load of mass m kg is raised with acceleration a m s^{-2} by a light cable. This is wound around a winding drum of radius, r m and moment of inertia I kg m^2. Let the angular acceleration of the winding drum be α rad s^{-2}, and the angle turned be θ rad in a time of t s (Figure 2.30).

Applying D'Alembert's principle to the load, the forces acting on it are as shown in Figure 2.31.

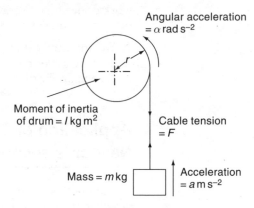

Figure 2.30 *Simple hoist*

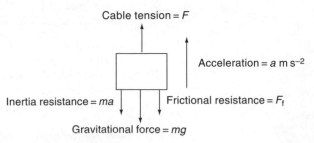

Figure 2.31 *Forces acting on the load*

Equating the upward and downward forces on the load gives

$$\frac{\text{applied}}{\text{force}} = \frac{\text{inertia}}{\text{resistance}} + \frac{\text{gravitational}}{\text{force}} + \frac{\text{frictional}}{\text{resistance}}$$

$$F = ma + mg + F_f \qquad (2.49)$$

Applying D'Alembert's principle to the torques acting on the winding drum, these are as shown in Figure 2.32.

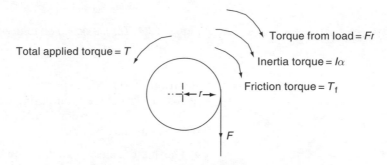

Figure 2.32 *Torques acting on winding drum*

Equating clockwise and anticlockwise torques gives

$$\frac{\text{applied}}{\text{torque}} = \frac{\text{inertia}}{\text{torque}} + \frac{\text{torque from}}{\text{load}} + \frac{\text{friction}}{\text{torque}}$$

$$T = I\alpha + Fr + T_f \qquad (2.50)$$

Consider now a vehicle which is accelerating whilst travelling up an incline (Figure 2.33).

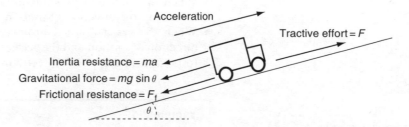

Figure 2.33 *Vehicle on an incline*

Applying D'Alembert's principle and equating the forces acting parallel to the incline

$$\frac{\text{tractive}}{\text{effort}} = \frac{\text{inertia}}{\text{force}} + \frac{\text{gravitational}}{\text{force}} + \frac{\text{frictional}}{\text{resistance}}$$

$$F = ma + mg\sin\theta + F_f \qquad (2.51)$$

Applying D'Alembert's principle to the torque acting on the wheels (Figure 2.34).

$$\frac{\text{total torque}}{\text{applied}} = \frac{\text{inertia}}{\text{torque}} + \frac{\text{torque to}}{\text{tractive effort}} + \frac{\text{friction torque}}{\text{in bearings}}$$

$$T = I\alpha + Fr + T_f \qquad (2.52)$$

Having calculated the value of the force, F and the value of the total applied torque, T, the work done, W can be found using

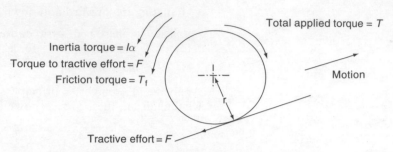

Figure 2.34 *Torque acting on road wheels*

equation (2.31). Also, the average power developed and instantaneous power can be found using equations (2.32) and (2.33).

There are of course a number of other possible ways in which the systems may be operating.

- The load/vehicle travelling upwards or downwards with uniform velocity. Here the inertia resistance and inertia torque are zero since the acceleration is zero.
- The load/vehicle travelling upwards but retarding.
- The load/vehicle travelling downwards and accelerating.
- The load/vehicle travelling downwards and retarding.

The force and torque equations can be obtained for each of these modes of operation from the appropriate system diagram. Points to remember are:

- The friction forces always act in the opposite direction to the motion.
- When the load/vehicle is moving downwards the gravitational force is in the direction of motion.
- When the systems are retarding, the inertia forces are in the direction of motion and the acceleration arrow should be drawn to point in the opposite direction.

Example 2.12

The winding drum of a hoist has a mass of 150 kg, an outer diameter of 600 mm and a radius of gyration 200 mm. A light cable wound around the drum carries a load of 100 kg. This is accelerated from an initial speed of $1.5\,\mathrm{m\,s^{-1}}$ to a final speed of $3\,\mathrm{m\,s^{-1}}$ whilst travelling upwards through a distance of 5 m. The frictional resistance to the linear motion of the load is 150 N and the friction torque in the bearings of the drum is 5 N m. Determine the work done, the input torque applied and the maximum input power delivered by the driving motor. (See Figures 2.30, 2.31 and 2.32 for the system diagrams.)

Finding the linear acceleration of the load:

$$v^2 = u^2 + 2\,a\,s$$

$$a = \frac{v^2 - u^2}{2s} = \frac{3^2 - 1.5^2}{2 \times 5}$$

$$a = 0.675\ \mathrm{m\,s^{-2}}$$

Finding angular acceleration of the winding drum:

$$\alpha = \frac{a}{r} = \frac{0.675}{0.3}$$

$\alpha = \mathbf{2.25\ rad\ s^{-2}}$

Finding moment of inertia of winding drum:

$$I = mk^2 = 150 \times 0.2^2$$

$I = \mathbf{6.0\ kg\ m^2}$

Finding the cable tension using equation (2.49):

$$F = ma + mg + F_f$$
$$F = (100 \times 0.675) + (100 \times 9.81) + 150$$

$F = \mathbf{1199\ N}$

Finding input torque from driving motor using equation (2.50):

$$T = I\alpha + Fr + T_f$$
$$T = (6.0 \times 2.25) + (1199 \times 0.3) + (5)$$

$T = \mathbf{378\ Nm}$

Finding angle turned by drum:

$$\theta = \frac{s}{r} = \frac{5}{0.3}$$

$\theta = \mathbf{16.7\ rad}$

Finding work done during period of acceleration:

$$W = T\theta = 378 \times 16.7$$

$W = \mathbf{6300\ J}$ or   **6.3 kJ**

Finding maximum angular velocity:

$$\omega_2 = \frac{v}{r} = \frac{3}{0.3}$$

$\omega_2 = \mathbf{10\ rad\ s^{-1}}$

Finding maximum power delivered by motor:

$$\text{maximum power} = T\omega_2 = 378 \times 10$$

maximum power $= $ 3780 W   or   **3.78 kW**

Example 2.13

A vehicle of mass 750 kg accelerates up an incline of 1 in 10 (sine) increasing its speed from $20\ km\ h^{-1}$ to $60\ km\ h^{-1}$ whilst travelling through a distance of 100 m up the slope. The four wheels each have a mass of 10 kg, a diameter of 0.5 m and a radius of gyration of 200 mm. The rolling friction, air resistance and bearing friction amount to a total resistance to motion of 0.5 kN. Calculate the tractive effort between the driving wheels and the road surface, the work done during the period of acceleration and the average power developed. (See Figures 2.33 and 2.34 for system diagrams.)

Finding velocities in $m\,s^{-1}$:

$$u = \frac{20 \times 10^3}{60^2}, \qquad v = \frac{60 \times 10^3}{60^2}$$

$$u = 5.56\,m\,s^{-1}, \qquad v = 16.7\,m\,s^{-1}$$

Finding linear acceleration of vehicle:

$$v^2 = u^2 + 2a\,s$$

$$a = \frac{v^2 - u^2}{2s} = \frac{16.7^2 - 5.56^2}{2 \times 100}$$

$$a = 1.24\,m\,s^{-2}$$

Finding angular acceleration of the wheels:

$$\alpha = \frac{a}{r} = \frac{1.24}{0.25}$$

$$\alpha = 4.96\,m\,s^{-2}$$

Finding total moment of inertia of wheels:

$$I = 4 \times mk^2 = 4 \times 10 \times 0.2^2$$

$$I = 1.6\,kg\,m^2$$

Finding tractive effort using equation (2.51):

$$F = ma + mg\sin\theta + F_f$$

$$F = (750 \times 1.24) + (750 \times 9.81 \times 0.1) + (0.5 \times 10^3)$$

$$F = 2166\,N$$

Finding torque delivered to driving wheels using equation (2.52):

$$T = I\alpha + Fr + T_f$$

(Note that the friction torque in the bearings is included in the overall frictional resistance to motion and can be neglected in this calculation.)

$$T = (1.6 \times 4.96) + (2166 \times 0.25) + 0$$

$$T = 549\,N\,m$$

Finding angle turned by wheels during period of acceleration:

$$\theta = \frac{s}{r} = \frac{100}{0.25}$$

$$\theta = 400\,rad$$

Finding work done during period of acceleration:

$$W = T\theta = 549 \times 400$$

$$W = 220 \times 10^3\,J \quad or \quad 220\,kJ$$

Finding time taken:

$$v = u + at$$

$$t = \frac{v - u}{a} = \frac{16.7 - 5.56}{1.24}$$

$$t = 8.98\,s$$

Finding average power developed:

$$\text{average power} = \frac{\text{work done}}{\text{time taken}} = \frac{W}{t}$$

$$\text{average power} = \frac{220 \times 10^3}{8.98}$$

average power $= 24.5 \times 10^3$ **W** or **24.5 kW**

Application of the principle of conservation of energy

The principle of conservation of energy, as applied to mechanical systems such as a hoist or a motor vehicle, can be stated as follows:

The total energy contained in a system and its surroundings is constant although changes may take place from one energy form to another during the operation of the system.

Figure 2.35 shows energy in the form of work entering a system from a driving motor or engine. Some of this changes the potential energy of the system, some of it changes the kinetic energy and the remainder is wasted in overcoming friction. The system input and outputs can be equated as follows:

$$\textbf{work} \atop \textbf{input} = {\textbf{change of} \atop \textbf{potential energy}} + {\textbf{change of} \atop \textbf{kinetic energy}} + {\textbf{work done} \atop \textbf{against friction}}$$

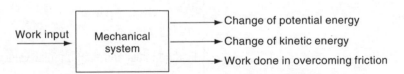

Figure 2.35 *Energy changes in a mechanical system*

If the height of a body of mass m kg, measured above some particular level, changes from an initial value h_1 to a final value h_2, the change of potential energy will be

change of $\textbf{\textit{PE}} = mg(h_2 - h_1)$

If the linear velocity of the body changes from v_1 to v_2, the change of linear kinetic energy will be

change of linear $\textbf{\textit{KE}} = \frac{1}{2}m(v_2^2 - v_1^2)$

If the angular velocity of the body changes from ω_1 to ω_2, the change of angular kinetic energy will be

change of angular $\textbf{\textit{KE}} = \frac{1}{2}m(\omega_2^2 - \omega_1^2)$

If the frictional resistance to linear motion is F_f, the work done in overcoming friction whilst moving through a distance, s will be

work done against friction $= F_f s$

If the friction torque resisting angular motion is T_f, the work done in overcoming friction whilst turning through an angle, θ will be

work done against friction $= T_f \theta$

If these terms are substituted in the energy equation, the work input, W is given by

$$W = mg(h_2 - h_1) + \frac{1}{2}m(v_2^2 - v_1^2) + \frac{1}{2}m(\omega_2^2 - \omega_1^2)$$
$$+ F_f s + T_f \theta \tag{2.53}$$

This is a rather long expression but it sometimes provides a quicker alternative method for solving problems such as Examples 2.12 and 2.13. It is now proposed to do these again and you may decide for yourself which method you prefer.

Example 2.14

The winding drum of a hoist has a mass of 150 kg, an outer diameter of 600 mm and a radius of gyration 200 mm. A light cable wound around the drum carries a load of 100 kg. This is accelerated from an initial speed of $1.5\,\mathrm{m\,s^{-1}}$ to a final speed of $3\,\mathrm{m\,s^{-1}}$ whilst travelling upwards through a distance of 5 m. The frictional resistance to the linear motion of the load is 150 N and the friction torque in the bearings of the drum is 5 N m. Determine the work done, the input torque applied and the maximum input power delivered by the driving motor.

Finding angle turned by winding drum:

$$\theta = \frac{h_2 - h_1}{r} = \frac{5}{0.3}$$

$$\theta = 16.7\,\mathrm{rad}$$

Finding initial and final angular velocities of the winding drum:

$$\omega_1 = \frac{v_1}{r} = \frac{1.5}{0.3}, \qquad \omega_2 = \frac{v_2}{r} = \frac{3}{0.3}$$

$$\omega_1 = 5\,\mathrm{rad\,s^{-1}}, \qquad \omega_2 = 10\,\mathrm{rad\,s^{-1}}$$

Finding moment of inertia of winding drum:

$$I = mk^2 = 150 \times 0.2^2$$

$$I = 6.0\ \mathrm{kg\,m^2}$$

Finding work done by driving motor:

$$W = mg(h_2 - h_1) + \frac{1}{2}m(v_2^2 - v_1^2) + \frac{1}{2}m(\omega_2^2 - \omega_1^2) + F_f s + T_f \theta$$

$$W = (100 \times 9.81 \times 5) + \left[\frac{1}{2} \times 100 \times (3^2 - 1.5^2)\right]$$
$$+ \left[\frac{1}{2} \times 6.0 \times (10^2 - 5^2)\right] + (150 \times 5) + (5 \times 16.7)$$

$$W = 6301\,\mathrm{J} \quad \text{or} \quad 6.3\,\mathrm{kJ}$$

Finding input torque required:

$$W = T\theta$$

$$T = \frac{W}{\theta} = \frac{6301}{16.7}$$

$$T = 377\,\text{N m}$$

Finding maximum power delivered by motor:

$$\text{maximum power} = T\omega_2 = 377 \times 10$$

maximum power = 3770 W   or   3.77 kW

You will note that the answers obtained are the same as for Example 2.12 except for the last decimal place. This can be attributed to the rounding off of values.

Example 2.15

A vehicle of mass 750 kg accelerates up an incline of 1 in 10 (sine) increasing its speed from 20 km h^{-1} to 60 km h^{-1} whilst travelling through a distance of 100 m up the slope. Each of the four wheels has a mass of 10 kg, a diameter of 0.5 m and a radius of gyration of 200 mm. The rolling friction, air resistance and bearing friction amount to a total resistance to motion of 0.5 kN. Calculate the tractive effort between the driving wheels and the road surface, the work done during the period of acceleration and the average power developed. (See Figures 2.31 and 2.32 for system diagrams.)

Finding vertical height, $h_2 - h_1$, gained by the vehicle (Figure 2.36):

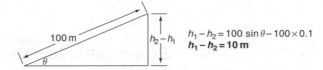

$$h_1 - h_2 = 100 \sin\theta - 100 \times 0.1$$
$$h_1 - h_2 = 10\,\text{m}$$

Figure 2.36 *Vertical height gained*

Finding initial and final velocities in m s^{-1}:

$$v_1 = \frac{20 \times 10^3}{60^2}, \qquad v_2 = \frac{60 \times 10^3}{60^2}$$

$$v_1 = 5.56\,\text{m s}^{-1}, \qquad v_2 = 16.7\,\text{m s}^{-1}$$

Finding initial and final angular velocities of wheels in rad s^{-1}:

$$\omega_1 = \frac{v_1}{r} = \frac{5.56}{0.25}, \qquad \omega_2 = \frac{v_2}{r} = \frac{16.67}{0.25}$$

$$\omega_1 = 22.2\,\text{rad s}^{-1}, \qquad \omega_2 = 66.7\,\text{rad s}^{-1}$$

Finding total moment of inertia of wheels:

$$I = 4 \times mk^2 = 4 \times 10 \times 0.2^2$$

$$I = 1.6\,\text{kg m}^2$$

Finding work done in accelerating the vehicle:

$$W = mg(h_2 - h_1) + \left[\tfrac{1}{2}m(v_2^2 - v_1^2)\right] + \left[\tfrac{1}{2}I(\omega_2^2 - \omega_1^2)\right] + F_1 s + T_f \theta$$

$$W = (750 \times 9.81 \times 10) + \left[\tfrac{1}{2} \times 750 \times (16.7^2 - 5.56^2)\right]$$
$$+ \left[\tfrac{1}{2} \times 1.6 \times (66.7^2 - 22.2^2)\right] + (500 \times 100)$$

$W = 220 \times 10^3 \, \text{J}$ or $220 \, \text{kJ}$

Finding angle turned by wheels during period of acceleration:

$$\theta = \frac{s}{r} = \frac{100}{0.25}$$

$\theta = 400 \, \text{rad}$

Finding torque applied to road wheels:

$$W = T\theta$$

$$T = \frac{W}{\theta} = \frac{220 \times 10^3}{400}$$

$T = 550 \, \text{N m}$

Finding average power developed:

average power = torque $\times$ average angular velocity

$$\text{average power} = T\frac{(\omega_2 + \omega_1)}{2} = 550 \times \frac{(66.7 + 22.2)}{2}$$

average power $= 24.4 \times 10^3 \, \text{W}$ or   **24.4 kW**

You will note that the answers obtained are the same as for Example 2.13 except for the last decimal place. This can again can be attributed to the rounding off of values.

Experiments to determine moment of inertia and radius of gyration

There are a number of experimental methods which may be used to determine the moment of inertia and radius of gyration of a rotating body. Two, which are within the scope of this unit, will be described. The first may be applied to bodies which rotate on an axle which is supported in bearings. The second method makes use of an inclined plane or track.

Method 1

Apparatus: Rotating body under investigation which is mounted on an axle. The axle may be supported in bearings or between free-running centres at some distance above ground level as shown in Figure 2.37. Length of cord, hanger and slotted masses, metre rule, stop watch.

Procedure:
1. Measure the diameter of the axle and record the value of its radius, r.
2. Wind the cord around the axle and attach the hanger to its free end.

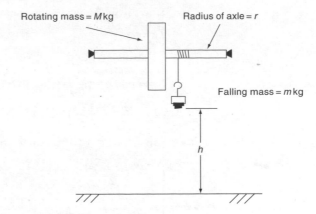

Figure 2.37 *Arrangement of apparatus*

3. Place small slotted masses on the hanger until bearing friction is just overcome. The hanger and masses should then fall with a slow uniform velocity. Do not record these masses.
4. Rewind the cord and place an additional mass of m kg on the hanger.
5. Measure and record the distance, h from the base of the hanger to the ground.
6. Release the hanger and record the time for it to fall to the ground. Repeat this operation a number of times and obtain the average time of fall.
7. Calculate the moment of inertia and radius of gyration of the rotating body using the following derived formulae.

Theory: Finding acceleration of falling mass:

$$h = ut + \frac{1}{2}a\,t^2$$

But $u = 0$,

$$h = \frac{1}{2}a\,t^2$$

$$a = \frac{2h}{t^2} \tag{i}$$

Finding angular acceleration of rotating body:

$$\alpha = \frac{a}{r}$$

$$\alpha = \frac{2h}{rt^2} \tag{ii}$$

Finding accelerating force in the cord. *Note*: The inertia force is in the upward direction, opposite to the direction of the acceleration.

$$F = mg - ma = m(g - a)$$

$$F = m\left(g - \frac{2h}{t^2}\right) \tag{iii}$$

Finding accelerating torque acting on the rotating body:

$$T = Fr$$

$$T = m\left(g - \frac{2h}{t^2}\right)r$$

$$T = mr\left(g - \frac{2h}{t^2}\right) \tag{iv}$$

Now, from Newton's second law of motion,

$$T = I\alpha \tag{v}$$

Equating (iv) and (v) gives

$$I\alpha = mr\left(g - \frac{2h}{t^2}\right)$$

$$\boldsymbol{I = \frac{mr}{\alpha}\left(g - \frac{2h}{t^2}\right)} \tag{vi}$$

Substituting α from equation (ii) gives

$$I = \frac{mr\left(g - \frac{2h}{t^2}\right)}{\frac{2h}{rt^2}}$$

$$\boldsymbol{I = \frac{mr^2t^2}{2h}\left(g - \frac{2h}{t^2}\right)} \tag{vii}$$

Finding radius of gyration:

$$I = mk^2$$

$$\boldsymbol{k = \sqrt{\frac{I}{m}}} \tag{ix}$$

Note: Equation (vii) may also be obtained by applying the energy equation to the system.

Method 2

Apparatus: Rotating body under investigation, inclined plane or track, metre rule, stopwatch, clinometer (if available).

If the body under investigation is mounted on an axle or shaft it may be positioned to roll down inclined parallel rails, as shown in Figure 2.38(a). If the body is cylindrical, it may be positioned to roll down an inclined plane as shown in Figure 2.38(b).

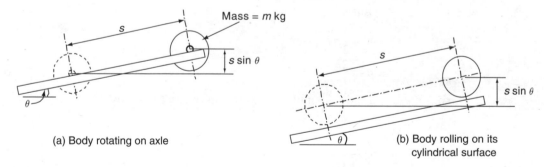

(a) Body rotating on axle

(b) Body rolling on its cylindrical surface

Figure 2.38 *Arrangement of apparatus*

Procedure:
1. Record the mass of the body under investigation. If it is to roll on an axle, record the axle diameter. If it is to roll on its cylindrical surface, record its outer diameter.
2. Set the track or inclined plane to a small angle, i.e. less than 10°. It is best to use a clinometer for this if one is available, even if the track is fitted with a protractor.
3. Mark the start and finish positions on the track and record the distance, *s* between them.
4. Place the body in the start position, release it and at the same time start the stopwatch.

5. Record the time taken for the body to roll through the distance, s. Repeat the operation several times and obtain an average value of the time, t.
6. Use the collected data to calculate the values of moment of inertia and radius of gyration of the body.

Theory: Finding vertical distance travelled by body (this will be negative since the body loses height):

$$h_1 - h_2 = -s\sin\theta \tag{i}$$

Finding the final linear velocity of the body:

$$v_2 = 2 \times \text{average velocity}$$

$$v_2 = \frac{2s}{t} \tag{ii}$$

The final angular velocity is given by

$$\omega_2 = \frac{v_2}{r}$$

$$\omega_2 = \frac{2s}{rt} \tag{iii}$$

The energy equation for the system is

$$W = mg(h_2 - h_1) + \frac{1}{2}m(v_2^2 - v_1^2) + \frac{1}{2}I(\omega_2^2 - \omega_1^2) + F_f s + T_f\theta$$

There is no work input and so W is zero. Also, v_1 and ω_1 are zero. The effects of rolling friction will be small and the friction work terms can also be neglected. This leaves

$$0 = mg(h_2 - h_1) + \frac{1}{2}mv_2^2 + \frac{1}{2}I\omega_2^2$$

$$\frac{1}{2}I\omega_2^2 = -mg(h_2 - h_1) - \frac{1}{2}mv_2^2 \tag{iv}$$

Substituting for $h_1 - h_2$ and v_2 and ω_2 gives

$$\frac{1}{2}I\omega_2^2 = -mg(h_2 - h_1) - \frac{1}{2}mv_2^2$$

$$\frac{1}{2}I\left(\frac{2s}{rt}\right)^2 = -mg(-s\sin\theta) - \frac{1}{2}m\left(\frac{2s}{t}\right)^2$$

$$\frac{1}{2}I\left(\frac{4s^2}{r^2t^2}\right) = mg\,s\sin\theta - \frac{1}{2}m\left(\frac{4s^2}{t^2}\right)$$

$$I\left(\frac{2s^2}{r^2t^2}\right) = ms\left(g\sin\theta - \frac{2s}{t^2}\right)$$

$$I = \frac{msr^2t^2}{2s^2}\left(g\sin\theta - \frac{2s}{t^2}\right)$$

$$I = \frac{mr^2t^2}{2s}\left(g\sin\theta - \frac{2s}{t^2}\right) \tag{v}$$

Finding radius of gyration:

$$I = mk^2$$

$$k = \sqrt{\frac{I}{m}} \tag{vi}$$

Note: Equation (v) may also be obtained by applying the D'Alembert's principle and Newton's second law of motion to the system.

Test your knowledge 2.8

1. What are the three main kinds of resistance to motion?
2. What is D'Alembert's principle?
3. What is meant by the *tractive effort* of a vehicle?
4. What is the component of the weight of a body which acts parallel to an incline?
5. If the load on a hoist is travelling upwards and slowing down, in which direction will the inertia force be acting?

Activity 2.8

The trolley shown in Figure 2.39 is accelerated up the incline from rest to a speed of $3.5\,\mathrm{m\,s^{-2}}$ in a time of 15 s. The winding drum has a diameter of 1 m and its radius of gyration is 350 mm. The mass of the winding rope and inertia of the trolley wheels can be neglected. Rolling friction amounts to 150 N and the friction torque in the bearings of the winding drum is 15 N m. Determine the torque required at the winding drum, the work done and the maximum input power from the driving motor. Verify your answers using an alternative method of calculation.

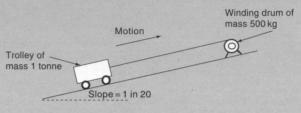

Winding drum of mass 500 kg

Motion

Trolley of mass 1 tonne

Slope = 1 in 20

Figure 2.39 *Trolley on an incline*

Problems 2.7

1. The winding drum of a hoist has an effective diameter of 1 m, a mass of 220 kg and a radius of gyration of 300 mm. A light cable wrapped around the winding drum carries a load of 100 kg which is accelerated uniformly upwards from rest to a speed of $1.75\,\mathrm{m\,s^{-1}}$ in a time of 5 s. Neglecting friction, determine (a) the distance travelled by the load in this time, (b) the angle turned by the winding drum, (c) the work done, (d) the input torque.

   [4.38 m, 8.75 rad, 4.57 kJ, 522 N m]

2. A truck of mass 1.35 tonnes accelerates from rest up an incline of gradient 1 in 12 through a distance of 120 m. Each wheel has a mass of 34 kg and a radius of gyration of 375 mm. The diameter of the wheels is 900 mm and the total resistance to motion is 350 N. Determine using an energy method or otherwise, the tractive effort between the wheels and the road.

   [3.0 kN]

3. A lift cage of mass 4 tonnes is to be raised through a distance of 10 m with an acceleration of $1.5\,\mathrm{m\,s^{-2}}$. The winding drum has a mass of 750 kg, an outer diameter of 1.5 m diameter and a radius of gyration is 600 mm. A torque of 3 kN m is required to overcome friction. Determine (a) the total input torque required, (b) the average power developed.

   [39.5 kN m, 144 kW]

4. A rotor of mass 11 kg test is mounted with its shaft supported horizontally in free-running bearings. A light cord is wrapped around the shaft, which is 25 mm in diameter, and a mass of 2 kg is attached to its free end. When released, the mass falls through a distance of 900 mm to the floor in a time of 20 s. Calculate the moment of inertia and radius of gyration of the rotor.

   [0.681 kg m^2, 249 mm]

Natural vibrations

There are certain mechanical systems which will vibrate if they are disturbed from their equilibrium position. A mass suspended or supported on a spring will vibrate if disturbed before finally settling back to its equilibrium position. This is a system with elasticity. Practical examples are the suspension systems in motor vehicles, motor cycles and some of the modern designs of bicycle. Another type of mechanical system which will vibrate is a mass suspended by a cord. This will swing from side to side if disturbed from its equilibrium position. The pendulum in a clock is an example of such a system.

A *naturally*, or *freely vibrating* system is one which would not experience any frictional resistance, air resistance or any other kind of resistance to its motion. If such a system could exist, the *natural vibrations* would never die away and there would be perpetual motion. This of course does not happen in practice, and all vibrations eventually die away due to *damping forces*. They are said to *attenuate*.

Simple harmonic motion

Simple harmonic motion (SHM) is a form of oscillating motion which is often associated with freely vibrating mechanical systems. It is defined as follows:

> *When a body moves in such a manner that its acceleration is directed towards, and is proportional to its distance from a fixed point in its path, it is said to move with simple harmonic motion.*

The fixed point referred to is the point at the centre of the motion path. It is the equilibrium position to which all practical systems eventually return. The definition states that for a system which describes SHM,

acceleration = constant × displacement from centre of path

From this equation, it can be seen that the acceleration is continually changing. As a result, the equations that we have used for uniform acceleration cannot be applied. Some new expressions are required to find the displacement, velocity and acceleration at any instant in time for a body which moves with SHM. These may be derived by considering a body which moves in a circular path of radius, r with uniform angular velocity, $\omega \, \mathrm{rad \, s^{-1}}$ as shown in Figure 2.40.

As the body P travels with uniform angular velocity, its projection or shadow P′ on the diameter oscillates between A and B. At time t s after passing through the point A, the body P has travelled through an angle of θ rad and the projection P′ is distance x from the centre O. The angle θ rad is given by

$$\theta = \omega t$$

The distance x of P′ from the centre, O will be

$$x = r \cos \theta$$

or

$$x = r \cos \omega t \qquad (2.54)$$

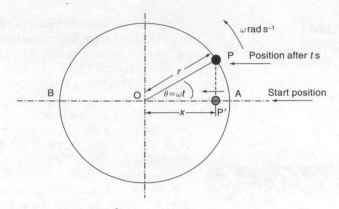

Figure 2.40 *Uniform circular motion*

The velocity, v of P′ towards the centre, O is the rate of change of distance x with time. This can be obtained by differentiating equation (2.54) with respect to the time t.

$$v = \frac{dx}{dt} = \frac{d}{dt}(r\cos\omega t)$$

$$\boldsymbol{v = -\omega r \sin \omega t} \qquad (2.55)$$

Considering the triangle OPP′ (Figure 2.41),

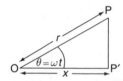

Figure 2.41 *Triangle OPP′*

By Pythagoras' theorem,

$$PP' = \sqrt{r^2 - x^2}$$

and

$$\sin \omega t = \frac{\sqrt{r^2 - x^2}}{r}$$

Substituting for $\sin \omega t$ in equation (2.55) gives

$$v = -\omega r \frac{\sqrt{r^2 - x^2}}{r}$$

$$\boldsymbol{v = -\omega \sqrt{r^2 - x^2}} \qquad (2.56)$$

Equation (2.56) enables the velocity of P′ to be calculated at any distance x from the centre O. The negative sign denotes that when x is increasing, the velocity is decreasing. In many applications it may be disregarded. The acceleration, a of P′ is the rate of change of velocity with time. This can be obtained by differentiating equation (2.55) with respect to the time t.

$$a = \frac{dv}{dt} = \frac{d}{dt}(-\omega r \sin \omega t)$$

$$a = -\omega^2 r \cos \omega t$$

But from equation (2.54), $x = r \cos \omega t$,

$$\boldsymbol{a = -\omega^2 x} \qquad (2.57)$$

This is of the form

acceleration = constant × displacement from centre of path

By definition, the projection P′ thus describes SHM along the diameter AB of the circle. The negative sign in equation (2.57) denotes that when x is increasing, the projection P′ is retarding. In many applications it may be disregarded.

The time taken for one complete oscillation of P′ is known as the *periodic time* of the motion. It is also the time taken for the body P to complete one revolution of its circular path. The periodic time, T is thus given by

$$T = \frac{2\pi}{\omega} \text{ s} \tag{2.58}$$

For very rapid oscillations it is more convenient to consider the frequency measured in Hertz. This is the reciprocal of the periodic time.

$$f = \frac{\omega}{2\pi} \text{ Hz} \tag{2.59}$$

In all of the above equations the term $\omega \text{ rad s}^{-1}$ is the angular velocity of the circular motion which is generating the SHM. It is sometimes called the *circular frequency* of the motion and should not be confused with the frequency of vibration, given by equation (2.59). The projection P′ has maximum displacement from the central position at A and B where $x = r$. The distance, r is called the *amplitude* of the motion.

Examination of equation (2.56) shows that the projection P′ is travelling fastest as it passes through the central position, i.e. when $x = 0$.

$$v_{\text{max}} = -\omega\sqrt{r^2 - 0}$$

$$v_{\text{max}} = -\omega r \tag{2.60}$$

Examination of equation (2.57) shows that the projection P′ has maximum acceleration at the maximum value of x, i.e. at the amplitude of the motion, where $x = r$.

$$a_{\text{max}} = -\omega r^2 \tag{2.61}$$

Once you have calculated the acceleration of a body which moves with SHM, the force acting on the body at that instant in time can be found by applying Newton's second law of motion. That is,

$$F = ma \tag{2.62}$$

Key point

The circular frequency of a system is the angular velocity of the circular motion which would generate SHM with the same amplitude and periodic time.

Key point

The maximum velocity of a system which describes SHM occurs at the central position whilst the maximum acceleration occurs at the amplitude of the motion.

Example 2.16

A body of mass 2.5 kg describes SHM of amplitude 500 mm and periodic time of 2.5 s (Figure 2.42). Determine (a) its velocity and acceleration at a displacement of 250 mm from the centre, (b) its maximum velocity and acceleration, (c) the maximum force acting on the body.

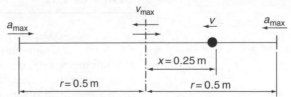

Figure 2.42 *Simple harmonic motion*

(a) Finding value of ω, the circular frequency of the motion:

$$T = \frac{2\pi}{\omega}$$

$$\omega = \frac{2\pi}{T} = \frac{2\pi}{2.5}$$

$$\boldsymbol{\omega = 2.51\ rad\ s^{-1}}$$

Finding velocity when $x = 0.25\ m$:

$$v = \omega\sqrt{r^2 - x^2} = 2.51 \times \sqrt{0.5^2 - 0.25^2}$$

$$\boldsymbol{v = 1.09\ m\ s^{-1}}$$

Finding acceleration when $x = 0.25\ m$:

$$a = \omega^2 x = 2.51^2 \times 0.25$$

$$\boldsymbol{a = 1.58\ m\ s^{-2}}$$

(b) Finding maximum velocity:

$$v_{max} = \omega r = 2.51 \times 0.5$$

$$\boldsymbol{v_{max} = 1.26\ m\ s^{-1}}$$

Finding maximum acceleration:

$$a_{max} = \omega^2 r = 2.51^2 \times 0.5$$

$$\boldsymbol{a_{max} = 3.15\ m\ s^{-2}}$$

(c) Finding maximum force acting on body:

$$F_{max} = m\, a_{max}$$

$$F_{max} = 2.5 \times 3.15$$

$$\boldsymbol{F_{max} = 7.88\ N}$$

Test your knowledge 2.9

1. What is the definition of SHM?
2. What is meant by the *amplitude* of vibration?
3. What is meant by the *circular frequency* of a vibrating system?
4. What is meant by the *periodic time* of a vibrating system?
5. Where in its path do maximum velocity and acceleration occur when a body moves with SHM?

Activity 2.9

A body of mass 5 kg moves with SHM whose amplitude is 100 mm and frequency of vibration is 5 Hz. Determine (a) the circular frequency of the motion, (b) the maximum values of velocity and acceleration, (c) the displacements at which the velocity and acceleration are half of the maximum values, (d) the maximum force which acts on the body.

Problems 2.8

1. A machine component of mass 2 kg moves with SHM of amplitude 250 mm and periodic time 3 s. Determine (a) the maximum velocity, (b) the maximum acceleration, (c) the acceleration and the force acting on the component when it is 50 mm from the central position.

$$[0.524\ m\ s^{-1},\ 1.1\ m\ s^{-2},\ 0.291\ m\ s^{-2},\ 2.2\ N]$$

2. A body moves with SHM in a straight line and has velocities of $24\,\mathrm{m\,s^{-1}}$ and $12\,\mathrm{m\,s^{-1}}$ when it is 250 mm and 400 mm respectively from its central position. Determine (a) the amplitude of the motion, (b) the maximum velocity and acceleration, (c) the periodic time and frequency of the motion.

$$[0.439\,\mathrm{m},\ 29.2\,\mathrm{m\,s^{-1}},\ 1941\,\mathrm{m\,s^{-2}},\ 0.0946\,\mathrm{s},\ 10.6\,\mathrm{Hz}]$$

3. The length of the path of a body which moves with SHM is 1 m and the frequency of the oscillations is 4 Hz. Determine (a) the periodic time of the motion, (b) the maximum velocity and acceleration, (c) the velocity and acceleration when the body is 300 mm from its amplitude.

$$[0.25\,\mathrm{s},\ 12.6\,\mathrm{m\,s^{-1}},\ 316\,\mathrm{m\,s^{-2}},\ 11.5\,\mathrm{m\,s^{-1}},\ 126.4\,\mathrm{m\,s^{-2}}]$$

4. A body of mass 50 kg describes SHM along a straight line. The time for one complete oscillation is 10 s and the amplitude of the motion is 1.3 m. Determine (a) the maximum force acting on the body, (b) the velocity of the body when it is 1 m from the amplitude position, (c) the time taken for the body to travel 0.3 m from the amplitude position.

$$[25.7\,\mathrm{N},\ 0.795\,\mathrm{m\,s^{-1}},\ 1.1\,\mathrm{s}]$$

5. A machine component of mass 25 kg describes SHM along a straight line path. The periodic time is 0.6 s and the maximum velocity is $5\,\mathrm{m\,s^{-1}}$. Calculate (a) the amplitude of the motion, (b) the distance from the central position when the force acting on the component is 750 N, (c) the time taken for the component to move from its amplitude to a point mid-way from the central position.

$$[0.477\,\mathrm{m},\ 0.274\,\mathrm{m},\ 0.1\,\mathrm{s}]$$

Mass-spring systems

A mass, suspended from the free end of a helical spring, will oscillate if displaced from its equilibrium position. To show that the oscillations are simple harmonic it is required to obtain an expression for the acceleration of the mass. If this is proportional to displacement from the equilibrium position, the definition is satisfied. Consider now a mass of m kg suspended from a spring of stiffness $S\,\mathrm{N\,m^{-1}}$.

When the mass, m is gently placed on the spring it produces a static deflection, d and rests in the equilibrium position as shown in Figure 2.43(b). The forces acting on the mass are its weight and

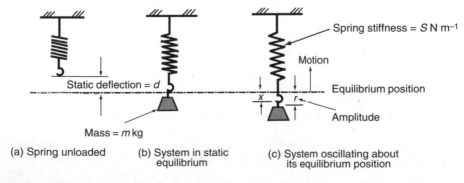

(a) Spring unloaded   (b) System in static equilibrium   (c) System oscillating about its equilibrium position

Figure 2.43 *Mass-spring system*

the spring tension which are equal and opposite as shown in Figure 2.44(a).

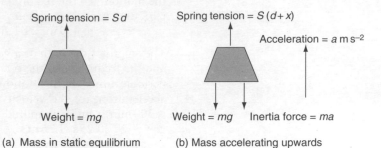

Spring tension = Sd

Spring tension = $S(d+x)$

Acceleration = a m s^{-2}

Weight = mg

Weight = mg Inertia force = ma

(a) Mass in static equilibrium   (b) Mass accelerating upwards

Figure 2.44 *Forces acting on mass*

When the system is in static equilibrium,

tension in spring = weight of mass

$$Sd = mg \qquad \text{(i)}$$

and

$$S = \frac{mg}{d}$$

That is,

$$\textbf{spring stiffness} = \frac{\textbf{static load}}{\textbf{static deflection}} \qquad (2.63)$$

If the mass is given a downwards displacement, r below its equilibrium position and then released, it will oscillate about the equilibrium position with amplitude, r as shown in Figure 2.43(c). At some instant when the mass is distance x below its equilibrium position and accelerating upwards, the forces acting on it are as shown in Figure 2.44(b). It is assumed that the mass oscillates with *free* or *natural vibrations*, i.e. that the air resistance is negligible and that there are no energy losses in the spring material. Equating forces gives,

tension in spring = weight + inertia force

$$S(d + x) = mg + ma$$
$$Sd + Sx = mg + ma$$

But from (i), $Sd = mg$, and so these terms cancel. This leaves

$$Sx = ma$$
$$a = \frac{Sx}{m} \qquad \text{(ii)}$$

The equation is of the form

acceleration = constant × displacement

This proves that the mass-spring system describes SHM. The acceleration is also given by the general formula

$$a = \omega^2 x$$

The natural circular frequency of the system, $\omega\,\mathrm{rad\,s^{-1}}$ is thus given by

$$\omega^2 = \frac{S}{m}$$

$$\boldsymbol{\omega = \sqrt{\frac{S}{m}}} \tag{2.64}$$

The periodic time of the natural vibrations is given by

$$T = \frac{2\pi}{\omega}$$

$$\boldsymbol{T = 2\pi\sqrt{\frac{m}{S}}} \tag{2.65}$$

The frequency of the vibrations known as the *natural frequency* of the system. It is the reciprocal of periodic time. That is,

$$f = \frac{\omega}{2\pi}$$

$$\boldsymbol{f = \frac{1}{2\pi}\sqrt{\frac{S}{m}}} \tag{2.66}$$

In practical mass-spring systems there is always some air resistance. There are also energy losses in the spring material as it is twisted and untwisted. They are known as *hysteresis losses*. The combined effect is known as *damping*, which eventually causes the vibrations to die away.

Example 2.17

A close-coiled helical spring undergoes a static deflection of 30 mm when a mass of 2.5 kg is placed on its lower end. The mass is then pulled downwards through a further distance of 20 mm and released so that it oscillates about the static equilibrium position. Neglecting air resistance, energy losses in the spring material and the mass of the spring determine (a) the periodic time and natural frequency of vibration, (b) the maximum velocity and acceleration of the mass.

(a) Finding stiffness of spring:

$$\text{spring stiffness} = \frac{\text{static load}}{\text{static deflection}}$$

$$S = \frac{mg}{d} = \frac{2.5 \times 9.81}{0.03}$$

$$\boldsymbol{S = 818 \ N\,m^{-1}}$$

Finding natural circular frequency of system:

$$\omega = \sqrt{\frac{S}{m}} = \sqrt{\frac{818}{2.5}}$$

$$\boldsymbol{\omega = 18.2 \ rad\,s^{-1}}$$

Finding periodic time:

$$T = \frac{2\pi}{\omega} = \frac{2\pi}{18.2}$$

$$\boldsymbol{T = 0.345 \ s}$$

Finding natural frequency of vibration:

$$f = \frac{1}{T} = \frac{1}{0.345}$$

$$f = 2.90\,\text{Hz}$$

(b) Finding maximum velocity of mass:

$$v_{max} = \omega r = 18.2 \times 0.02$$

$$v_{max} = 0.364\,\text{m s}^{-1}$$

Finding maximum acceleration of mass:

$$a_{max} = \omega^2 r = 18.2^2 \times 0.02$$

$$a_{max} = 6.62\,\text{m s}^{-2}$$

Simple pendulum

A simple pendulum consists of a concentrated mass, known as the pendulum 'bob', which is free to swing from side to side at the end of a light cord or a light rod. When it is given a small angular displacement and released, the pendulum bob appears to describe SHM. It doesn't travel along a perfectly straight line, but for small angular displacements it is approximately straight, with a central equilibrium position.

If the mass is given a small displacement, r from its equilibrium position and then released, it will oscillate about the equilibrium position with amplitude r, as shown in Figure 2.45(a). At some instant when the bob is at a distance x from its equilibrium position, and accelerating towards it, the forces acting on it will be as shown in Figure 2.45(b). It is assumed that the bob oscillates with *free* or *natural vibrations*, i.e. that the air resistance is negligible and that there are no energy losses at the point of suspension.

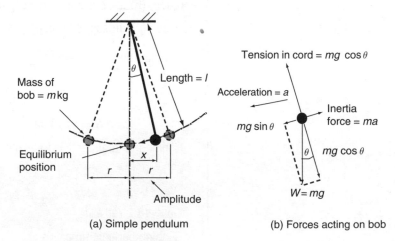

(a) Simple pendulum   (b) Forces acting on bob

Figure 2.45 *Simple pendulum*

Equating forces in the direction of motion at the instant shown gives inertia force = component of weight in direction of motion

$$ma = mg \sin \theta$$
$$a = g \sin \theta$$

But $\sin \theta = \dfrac{x}{l}$

$$a = g\frac{x}{l}$$

or

$$\boldsymbol{a = g\frac{x}{l}}$$

This is of the form

acceleration = constant × displacement

It shows that for small displacements, the simple pendulum describes SHM. You will recall that the acceleration is also given by the general formula,

$$a = \omega^2 x$$

The natural circular frequency of the system, $\omega\,\mathrm{rad\,s^{-1}}$ is thus given by

$$\omega^2 = \frac{g}{l}$$

$$\boldsymbol{\omega = \sqrt{\frac{g}{l}}} \qquad\qquad (2.67)$$

The periodic time of the natural vibrations is given by

$$T = \frac{2\pi}{\omega}$$

$$\boldsymbol{T = 2\pi\sqrt{\frac{l}{g}}} \qquad\qquad (2.68)$$

The natural frequency of the vibrations is the reciprocal of periodic time. That is,

$$f = \frac{\omega}{2\pi}$$

$$\boldsymbol{f = \frac{1}{2\pi}\sqrt{\frac{g}{l}}} \qquad\qquad (2.69)$$

As with the mass-spring systems, there is always some air resistance. There is also some energy loss at the point of suspension and the combined effect eventually causes the vibrations to die away.

Example 2.18

A simple pendulum describes 45 complete oscillations of amplitude 30 mm in a time of 1 minute. Assuming that the pendulum is swinging freely, calculate (a) the length of the supporting cord, (b) the maximum velocity and acceleration of the bob.

(a) Finding the periodic time:

$$T = \frac{\text{recorded time}}{\text{number of oscillations}} = \frac{60}{45}$$

$$T = 1.33\,\mathrm{s}$$

Finding length of pendulum:

$$T = 2\pi\sqrt{\frac{l}{g}}$$

$$T^2 = 4\pi^2 \frac{l}{g}$$

$$l = \frac{T^2 g}{4\pi^2} = \frac{1.33 \times 9.81}{4\pi^2}$$

$$l = 0.440\,\text{m}$$

(b) Finding the natural circular frequency of the system:

$$T = \frac{2\pi}{\omega}$$

$$\omega = \frac{2\pi}{T} = \frac{2\pi}{1.33}$$

$$\omega = 4.72\,\text{rad s}^{-1}$$

Finding maximum velocity of pendulum bob:

$$v = \omega r = 4.72 \times 0.03$$

$$v = 0.142\,\text{m s}^{-1}$$

Finding maximum acceleration of pendulum bob:

$$a = \omega^2 r = 4.72^2 \times 0.03$$

$$a = 0.668\,\text{m s}^{-2}$$

Experiment to verify that a mass-spring system describes simple harmonic motion

Apparatus: A light close-coiled helical spring, retort stand and clamps, hanger and slotted masses, metre rule, stop watch (Figure 2.46).

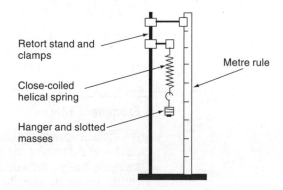

Figure 2.46 *Arrangement of apparatus*

Procedure:
1. Hang the spring on one of the retort stand clamps and position the metre rule alongside it, secured by the other clamp.

2. Place the hanger and a slotted mass on the lower end of the spring and note the static displacement.
3. Displace the mass from its equilibrium position and release it so that it oscillates freely.
4. Note the number of complete oscillations that the mass makes in a time of 1 min and calculate the periodic time of the motion.
5. Repeat the procedure with increasing values of the slotted mass until at least six sets of readings have been taken.
6. Calculate the square of the periodic time for each set of readings and tabulate the results.
7. Plot a graph of load against static deflection and from it determine the spring stiffness.
8. Plot a graph of mass against the square of periodic time, observe its shape and calculate its gradient.

Theory: Ploting static load, W against static deflection, d should give a straight line whose gradient is the stiffness, S of the spring (Figure 2.47).

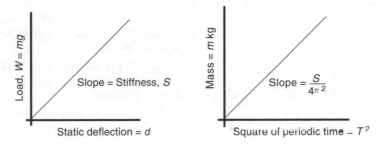

Figure 2.47 *Characteristic graphs for mass-spring system*

The periodic time of a mass-spring system which oscillates with SHM is given by the expression

$$T = 2\pi\sqrt{\frac{m}{S}}$$
$$T^2 = 4\pi^2\frac{m}{S}$$
or
$$m = \frac{S}{4\pi^2}T^2$$

If the mass-spring system is describing SHM, the graph of mass against the square of the periodic time will be a straight line whose gradient is $S/4\pi^2$.

Experiment to verify that a simple pendulum describes simple harmonic motion

Apparatus: Length of light cord, pendulum bob, retort stand and clamp, metre rule, stopwatch (Figure 2.48).

Procedure:
1. Attach one end of the cord to the bob and the other to the retort stand clamp.

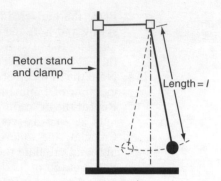

Figure 2.48 *Arrangement of apparatus*

2. Measure and record the length of the cord from the support to the centre of the bob.
3. Displace the bob from its equilibrium position and release it so that it oscillates freely.
4. Note the number of complete oscillations made by the bob in a time of 1 min and calculate the periodic time of the motion.
5. Repeat the procedure with different lengths of the cord until at least six sets of readings have been taken.
6. Calculate the square of the periodic time for each set of readings and tabulate the results.
7. Plot a graph of cord length against the square of the periodic time, observe its shape and calculate its gradient.

Theory: The periodic time of a simple pendulum which oscillates with SHM is given by the expression

$$T = 2\pi\sqrt{\frac{l}{g}}$$

$$T^2 = 4\pi^2\frac{l}{g}$$

or

$$l = \frac{g}{4\pi^2}T^2$$

If the pendulum is describing SHM, the graph of pendulum length against the square of the periodic time will be a straight line whose gradient is $g/4\pi^2$ (Figure 2.49).

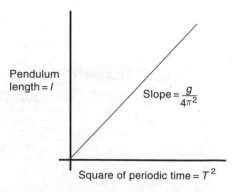

Figure 2.49 *Characteristic graph for simple pendulum*

Activity 2.10

The length of a close-coiled helical spring extends by a distance of 25 mm when a mass of 0.5 kg is placed on its lower end. An additional mass of 1.5 kg is then added and displaced so that the system oscillates with an amplitude of 50 mm. Determine (a) the periodic time and frequency of the oscillations, (b) the maximum velocity and acceleration of the mass, (c) the length of a simple pendulum which would have the same periodic time and frequency.

Problems 2.9

1. A mass of 25 kg is suspended from an elastic coiled spring which undergoes a 50 mm increase in length. The mass is then pulled downwards through a further distance of 25 mm and released. Determine (a) the periodic time of the resulting SHM, (b) the maximum velocity of the mass, (c) the maximum acceleration of the mass.

$$[0.45\,\text{s}, 0.35\,\text{m s}^{-1}, 4.9\,\text{m s}^{-2}]$$

2. A helical spring is seen to undergo a change in length of 10 mm when a mass of 1 kg is gently suspended from its lower end. A further 3 kg is then added, displaced from the equilibrium through a distance of 30 mm and released. Determine (a) the frequency of the oscillations, (b) the maximum velocity of the mass, (c) the maximum acceleration of the mass.

$$[2.49\,\text{Hz}, 0.469\,\text{m s}^{-1}, 7.34\,\text{m s}^{-1}]$$

3. A mass of 6 kg is suspended from a helical spring and produces a static deflection of 6 mm. The load on the spring is then increased to 18 kg and settles at a new equilibrium position. It is then displaced through a further 10 mm and released so that system oscillates freely. Determine (a) the frequency of the oscillations, (b) the maximum velocity and acceleration, (c) the maximum tension in the spring.

$$[3.7\,\text{Hz}, 0.233\,\text{m s}^{-1}, 5.42\,\text{m s}^{-2}, 247\,\text{N}]$$

4. An elastic spring of stiffness $0.4\,\text{kN m}^{-1}$ is suspended vertically with a load attached to its lower end. When displaced, the load is seen to oscillate with a periodic time of 1.27 s. Determine (a) the magnitude of the load, (b) the acceleration of the load when it is 25 mm away from the equilibrium position, (c) the tension in the spring when the load is 25 mm away from the equilibrium position.

$$[6.3\,\text{kg}, 1.58\,\text{m s}^{-2}, 71.8\,\text{N}]$$

5. A simple pendulum is made from a light cord 900 mm long and a concentrated mass of 2.25 kg. What will be the periodic time of the oscillations when the pendulum is given a small displacement and allowed to swing freely? What will be the stiffness of the spring which will have the same periodic time when carrying the same mass?

$$[1.9\,\text{s}, 24.5\,\text{N m}^{-1}]$$

Chapter 3 | Mechanical technology

Summary

Mechanical engineering embraces a wide field of activity. Its range includes power generation, land, sea and air transport, manufacturing plant and machinery and products used in the home and office such as the photocopier, computer printer and washing machine. The term *mechatronics* is often used to describe systems that incorporate mechanical devices, electrical and electronic circuits and elements of information technology. These are to be found in all of the above areas and it is the aim of this unit to investigate some of the more common mechanical systems and components.

Moving parts generally require lubrication and the first part of this chapter examines lubricant types and lubrication systems. Pressurised systems require seals and gaskets to prevent the escape of lubricants and other working fluids. Rotating parts require bearings and all mechanical systems incorporate fixing devices to hold the various components in position. These will also be examined in this chapter.

A prime purpose of mechanical systems is to transmit power and motion and the various ways in which this can be achieved is investigated. The chapter also provides an overview of hydraulic and pneumatic systems, steam plant, refrigeration and air-conditioning systems and mechanical handling equipment.

Lubricants and lubrication systems

In Chapter 2, we calculated the forces needed to overcome static and dynamic frictional resistance between surfaces in sliding contact. It is essential that this should be kept to a minimum in most mechanical systems. The exceptions are of course belt drives, clutches and braking systems that depend on frictional resistance for their operation. Generally however friction causes wear, generates heat and wastes power. Efficient lubrication will not eliminate these effects but it can minimise them.

Purpose and action of lubricants

A lubricant separates the sliding and rolling surfaces so that there is little or no metal to metal contact. In effect, it forms a cushion between the surfaces and if properly chosen it will keep the surfaces apart even when squeezed by high contact pressures. As the surfaces move over each other the lubricant is subjected to shearing and offers some resistance to the motion. This, however, is far less than that when the surfaces are in dry contact. In the case of liquid lubricants such as oils, the resistance depends on a property known as *dynamic viscosity*. This will be described in more detail in Chapter 5. It is a measure of the internal resistance of the liquid to being stirred, poured or, in the case of a lubricant, to being sheared. The SI units of dynamic viscosity are $N\,s\,m^{-2}$ and some typical values for different grades of lubricating oil are given in Table 3.1.

> **Key point**
>
> Dynamic viscosity is a measure of the resistance of a liquid to being stirred or poured.

Table 3.1 *Lubricating oils*

Group	Applications	Dynamic viscosity
Spindle oils	Lubrication of high speed bearings	Below $0.01\,N\,s\,m^{-2}$ at 60 °C
Light machine oils	Lubrication of machinery running at moderate speeds	0.010 to $0.02\,N\,s\,m^{-2}$ at 60 °C
Heavy machine oils	Lubrication of slow moving machinery	0.02 to $0.1\,N\,s\,m^{-2}$ at 60 °C
Cylinder oils	Lubrication of steam plant components	0.1 to $0.3\,N\,s\,m^{-2}$ at 60 °C

A lubricant fulfils purposes other than reducing frictional resistance and wear. A steady flow of lubricant can carry away heat energy and solid particles from the contact area. This is an important function of the cutting lubricants used in the machining of metal components. They reduce the friction between the cutting tool face and the metal being removed and they also carry away the heat energy and the metal swarf which is produced by the cutting tool.

Another function that a lubricant performs is to prevent corrosion which might occur from the presence of moisture or steam. It is important that a lubricant has good adhesion with the contact surfaces and does not drain away when they are at rest. It is then able to protect them from corrosion and maintain surface separation in readiness for start-up. A lubricant must not react with any other substances in the working environment. Its function is also to prevent contamination of the contact surfaces and prevent the ingress of solid particles that would damage them.

Lubricant types and applications

Lubricants may be broadly divided into mineral oils, synthetic oils, vegetable oils, greases, solid lubricants and compressed gas. Mineral oils and greases are perhaps the most widely used but each has its own particular characteristics and field of application. We will now examine some of these.

Mineral oils

Mineral oils are hydrocarbons. That is to say that they are principally made up of complex molecules composed of hydrogen and carbon. The chemists who specialise in this field refer to the hydrocarbon types as *paraffins, naphthenes* and *aromatics*. Mineral oils for different applications contain different proportions of these compounds. This affects their viscosity and the way that it changes under the effects of temperature and pressure.

Traditionally, lubricating oils have been grouped according to their different uses and viscosity values as shown in Table 3.1. Within the different types shown in Table 3.1, lubricating oils are graded according to the change in viscosity which takes place with a rise in temperature. This is indicated by the *viscosity index* number that has been allocated to the oil. It can range from zero to over one hundred as shown in Table 3.2. The higher the value, the less is the effect of temperature rise. A value of 100+ indicates that the viscosity of the oil is not much affected by temperature change.

Key point

The viscosity index of a lubricating oil gives an indication of how its viscosity changes with temperature rise.

Table 3.2 *Viscosity index groupings*

Viscosity index group	Viscosity index
Low viscosity index (LVI)	Below 35
Medium viscosity index (MVI)	35 to 80
High viscosity index (HVI)	80 to 110
Very high viscosity index (VHVI)	Above 110

Table 3.3 *Group analysis of mineral lubricating oils*

Group	Viscosity ($N\,s\,m^{-2}$)	Viscosity index (VI)	Paraffins (%)	Naphthenes (%)	Aromatics (%)
Spindle oils					
Low VI	2.7×10^{-3}	Below 35	46	32	22
Light machine oils					
Medium VI	3.9×10^{-3}	35 to 80	59	37	4
High VI	4.3×10^{-3}	80 to 110	68	26	6
Heavy machine oils					
Low VI	7.4×10^{-3}	Below 35	51	26	23
Medium VI	7.5×10^{-3}	35 to 80	54	37	8
High VI	9.1×10^{-3}	80 to 110	70	23	7
Cylinder oils					
High VI	26.8×10^{-3}	80 to 100	70	22	8

Table 3.3 shows the typical viscosity and viscosity index values for the different types of lubricating together with the likely percentage composition of paraffins, naphthenes and aromatic hydrocarbon molecules which might be expected. You might notice that within each group, the high viscosity index oils, which are least affected by temperature change, have the highest percentages of paraffin compounds. Also within each group, the oils with the lowest viscosity have the highest percentages of aromatic compounds. It might thus be stated as a general rule, that aromatic compounds lower the viscosity of an oil whilst paraffin compounds help it to retain its viscosity as the temperature rises.

There are other ways of measuring viscosity and if you go to buy oil for a motor car or motor cycle, the service manual might recommend SAE 10W-40 lubricating oil for the engine and gearbox. The letters stand for the Society of Automotive Engineers of America and the numbers indicate the viscosity and a measure of its variation with temperature. The system is widely used by the motor industry, but is not entirely suitable for other industrial applications where wider ranges of working temperatures, pressures, running speeds and service environments are to be found.

Additives and synthetic oils

Plain mineral oils are suitable for applications such as the lubrication of bearings, gears and slide-ways where the operating temperature is relatively low and the service environment does not contain substances that will readily contaminate or degrade the lubricant. Typical contaminants are air (which is unavoidable), ammonia, water, oil of another grade, soot, dust and wear particles from the lubricated components. Special chemicals are often added to plain mineral oils to improve their properties and prolong their life. Table 3.4 lists some of the more common additive types.

Table 3.4 *Additive types*

Additive types	Purpose
Acid neutralisers	To neutralise contaminating acids such as those formed by the combustion of sulphur in solid and liquid fuels
Anti-foam	To reduce the formation of surface foam where the lubricant is subjected to aggressive agitation
Anti-oxidants	To reduce the build-up of sludge and acidic products which result from oxygen from the air reacting with the oil
Anti-corrosion agents	To reduce the corrosion of ferrous metals, copper alloys and bearing metals
Anti-wear agents	To minimise surface contact and reduce wear under heavy loading conditions
Detergents	To clean away or reduce surface deposits such as those which occur in internal combustion engine cylinders

Polymers are sometimes added to mineral oils to enhance their properties. These are often referred to as synthetic oils although it is the mineral oil that forms the bulk of the lubricant. Various types of ester, silicone and other polymers are added, mainly to increase the viscosity index and wear resistance of the oil.

Some of the more expensive multi-grade motor oils contain polymers. They appear to be much thinner than the less expensive engine oils at normal temperature but undergo a much smaller change of viscosity as the temperature rises. This makes cold starting easier and gives improved circulation of the lubricant during the warming-up period. At running temperatures their viscosity is similar to that of the less expensive multi-grades but they are purported to have better wear resistant properties.

Key point

Modern multi-grade motor oils with synthetic additives give better cold starting and offer less resistance to the moving parts during engine warm-up.

Vegetable oils

Soluble vegetable oils are used as coolants and cutting tool lubricants in metal machining. The vegetable oils used have a high flash point. This is essential since high temperatures can be generated at the tip of a cutting tool. When mixed with water, the oil takes on a low viscosity milky appearance. It can be delivered under pressure to the cutting tool and assist in the clearance of metal particles. This is of particular importance in deep-hole drilling operations.

Castor oil has been a popular additive to the lubricating oil used for the highly tuned engines in motor cycle and formula car racing. Indeed, it gives its name to a popular brand of motor oil. When used, the exhaust gases have a distinctive smell that immediately identifies the additive.

Greases

A grease consists of a lubricating fluid which contains a thickening agent. It may be defined as a semi-solid lubricant. The lubricating fluid may be a mineral oil or a polymer liquid and the thickening agent may be a soap or a clay. Table 3.5 lists some of the more common thickeners. The main advantage of a grease is that it stays where it is applied and is less likely to be displaced by pressure or centrifugal action than oil. It can act as both a lubricant and a seal, preventing the entry of water, abrasive grit and other contaminants to the lubricated surfaces.

Table 3.5 *Greases*

Lubricating fluid	Soap or thickener	Temperature range	Uses
Mineral oil	Calcium (Lime)	−20°C to 80°C	General purpose, ball and roller bearings
Mineral oil	Sodium	0°C to 175°C	Glands, seals, low to medium speed ball and roller bearings
Mineral oil	Mixed sodium and calcium	−40°C to 150°C	High speed ball and roller bearings
Mineral oil	Lithium	−40°C to 150°C	General purpose, ball and roller bearings
Mineral oil	Clay	−30°C to 200°C	Slideways
Ester polymer	Lithium	−75°C to 120°C	General purpose, ball and roller bearings
Silicone polymer	Lithium	−55°C to 205°C	General purpose, ball and roller bearings
Silicone polymer	Silicone soap	−55°C to 260°C	Miniature bearings

Bearings such as the wheel bearings of motor vehicles can be pre-packed with grease and function for long periods of time without attention. The main disadvantages of grease are that it does not dissipate heat energy so well as a lubricating oil and because of its higher viscosity, it offers greater resistance to motion.

Key point

Greases do not dissipate heat energy like a lubricating oil but they have good adherence and are not easily displaced by gravity or centrifugal action.

Solid lubricants

There are many applications, such as at very low and very high working temperatures, where lubricating oils and greases are unsuitable. In such cases the use of a solid lubricant can reduce wear without adversely affecting the contact surfaces or the working environment. Three of the most common solid lubricants are graphite, molybdenum disulphide and polytetrafluoroethylene, which is better known as PTFE. You will be quite familiar with non-stick cooking utensils that are coated with PTFE. Lead, gold and silver are also used as dry lubricants for aerospace applications but their use is expensive.

Solid lubricants may be applied to the contact surfaces as a dry powder and you might recall that cast iron is self-lubricating because it already contains graphite flakes. They may be mixed with a resin and sprayed on the surfaces to form a bonded coating, or they may be used as an additive to oils and greases. Molybdenum disulphide, whose chemical formula is MoS_2, has been used in this way by car owners for a number of years. It can be purchased from motor accessory dealers in small tins as a suspension in mineral oil. Solid lubricants can also be added to molten metal in the forming process so that when solidified, the metal is impregnated with particles of the solid lubricant. Graphite is often added to phosphor-bronze in this way to improve its qualities as a bearing material.

Compressed gases

Compressed air and inert gases such as carbon dioxide have a very low viscosity compared to oils and greases. With compressed gas bearings, the contact surfaces are separated by a thin cushion of gas which offers very little resistance to motion. The main disadvantage is the cost. A gas delivery system is required and the stationary outer surface of a journal bearing or lower stationary slide-way, must be machined with evenly spaced exit orifices to provide a dry and clean gas cushion. The usual operating pressures are 2 to 5 bar with higher pressures up to 10 bar being used for heavy duty applications. The bearing surfaces may be coated with a solid lubricant to guard against dry running should the air supply fail.

Gas bearings are well-suited to high speed applications. They can operate at a speed up to 300 000 rpm and at temperatures well outside the range of oils and greases. A further advantage is that they do not allow dirt or moisture to enter the bearing and are surprisingly rigid. You might think that the air cushion could be easily be penetrated or displaced by high contact pressures and shock loads. This is not the case, however, as gas bearings have a self-correcting property. Shaft displacement causes the air gap to increase on one side and the pressure in that region to fall. At the same time, the air gap on the opposite side decreases and the pressure rises, forcing the shaft back to the central position.
See Questions 3.1 on p. 216.

Key point

Solid lubricants are able to function outside the working temperature range of oils and greases.

Key point

Air bearings are rigid and self-correcting under load, offer very little resistance to motion and can operate at high and low temperatures.

Test your knowledge 3.1

1. What are the purposes and functions of a lubricant?
2. What does the viscosity index of a lubricating oil indicate?
3. What are the main constituents of a grease?
4. Name two solid lubricants.
5. What are the advantages of air bearings?

Activity 3.1

Using only the raw materials available in your science laboratory and workshop, devise a method which can be used to compare the viscosity of different liquids and the effect of temperature rise on viscosity.

Lubrication systems

Oil lubricating systems can be divided into three categories – total loss, self contained and re-circulating. With total loss lubrication, the oil is applied to the moving parts by means of an oil can or an aerosol spray. Over a period of time the lubricant evaporates, drips away or takes in dust and dirt to become semi-solidified. The chain of a bicycle is lubricated in this way and periodically, after appropriate cleaning, the lubricating oil needs to be re-applied. An alternative method, which is used on larger items of equipment, is to supply lubricating oil or oil mist periodically from a central reservoir by means of a hand operated lever or an automated pump.

With self contained lubrication, the oil is contained in a reservoir. The gearbox of a car or a lathe is generally lubricated in this way. The gears are partly submerged in the oil as shown in Figure 3.1 and the process is known as *splash lubrication*. Oil is carried up to the parts that are not submerged, and an oil mist is created inside the gearbox. This too has a lubricating effect.

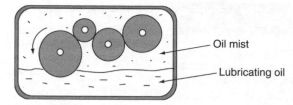

Figure 3.1 *Splash lubrication*

Ring oiling is another self-contained system in which oil from a reservoir is carried up to the rotating parts of a mechanism. This is shown in Figure 3.2 where a ring rotating with a shaft is partly submerged in the oil and carries it up to the shaft bearings.

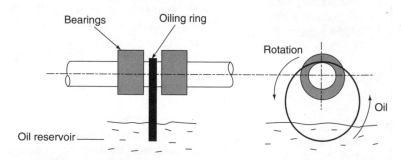

Figure 3.2 *Ring oiling*

Plane bearings and machine slides that are lightly loaded are sometimes lubricated by means of a pad or wick feed from a small reservoir as shown in Figure 3.3. This too is a self-contained method of lubrication. The oil travels down the wick by capillary action to the moving parts.

With re-circulating lubrication systems, oil from a reservoir is fed under pressure direct to the moving parts or delivered as a spray. The flow is continuous and after passing over the contact

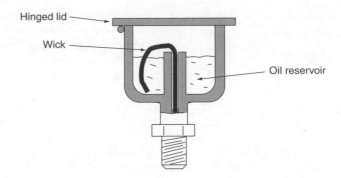

Figure 3.3 *Wick-feed lubrication*

surfaces, the oil runs back into the reservoir under the effects of gravity. A single pump may deliver lubricant to several locations as shown in Figure 3.4.

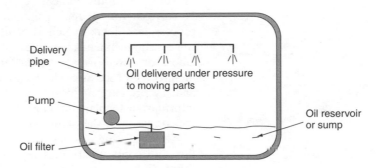

Figure 3.4 *Re-circulating oil system*

Grease lubrication systems fall into two categories, replenishable and non-replenishable. The bearings in some domestic appliances and power tools are examples of the non-replenishable type. They are packed with grease on assembly and this is judged to give sufficient lubrication for the life of the equipment. The wheel bearings on cars, caravans and trailers also fall into this category. They are packed with grease when new and only need to be repacked when stripped down after long periods of service.

Replenishable systems incorporate the grease nipples and screw-down grease cups shown in Figure 3.5. Lubrication is carried out at recommended periods by means of a grease gun or by screwing down the grease cup. As with lubricating oils, grease can

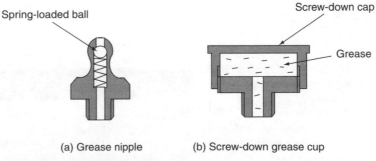

(a) Grease nipple   (b) Screw-down grease cup

Figure 3.5 *Grease lubricators*

be supplied under pressure from a central reservoir, by means of a hand-operated or motor-driven pump, to a number of points in a mechanism. Some trucks and buses are fitted with automated chassis lubrication which supplies grease to the suspension and steering linkages in this way. Unlike oil lubrication systems, all greasing systems are total loss.

As a general rule, oil lubrication systems should be checked weekly. This can be carried out by plant operators or vehicle drivers and needs very little skill. If necessary, the tank or reservoir should be topped up taking care not to let any dirt into the system during the process. Systems should not be over-filled as this might cause increased resistance to splash-lubricated parts. It might also lead to overheating of the oil due to excessive churning. Dip-sticks and sight gauges are usually provided to indicate the correct depth or level in reservoirs.

It should be noted that the oil level whilst a machine or engine is running might be different to when it is stationary. Servicing instructions should indicate clearly whether the equipment is to be stopped for oil replenishment. It is usually only continuous process machinery which is replenished whilst in operation. In some establishments, samples of lubricating oil are taken for laboratory analysis at regular intervals as part of a condition monitoring procedure. Here they are examined for contamination and the presence of solid wear particles. This can provide useful information to plant managers and maintenance engineers as to the condition of the plant and machinery.

All lubricating oils degrade over a period of time due to oxidation and contamination. They should be changed at the recommended intervals together with the filters in re-circulating systems. Lubricating oils contain chemicals that can cause skin irritation and in extreme cases, exposure can lead to skin and other forms of cancer. Safety equipment should be used, and protective clothing worn, at all times when handling lubricating oils. Oil spillage should be dealt with immediately. Supervision should be informed and the spillage treated with an absorbing material and removed. Water and a suitable detergent can then be used to remove all traces from the work area.

Waste oil should be disposed of in the approved manner. In some cases this might be controlled incineration at temperatures in excess of 1500 °C by specially licensed contractors. Some may be fit for re-cycling. After filtering, it may be used as a lower grade lubricant or in the production of greases. New supplies should be stored in a safe manner away from substances where cross-contamination might occur. They should be clearly marked as to their grade and type in accordance with COSHH regulations, i.e. The Care of Substances Hazardous to Health.
See Questions 3.2 on p. 217.

Key point

Safety equipment and protective clothing should always be used when replenishing or disposing of lubricating oils.

Test your knowledge 3.2

1. What is total loss lubrication?
2. How does a ring oiler function?
3. How does a wick-feed oiler operate?
4. What is provided to indicate the correct level of the oil in a supply tank or sump?
5. What do the initials COSHH stand for and how do they apply to lubricating oils?

Activity 3.2

Re-circulating lubrication systems require a pump to supply the oil under pressure to the different lubrication points. Describe the kind of pump that is most commonly used in car engines and the kind of filter that is used to remove solid particles from the oil.

Engineering components

There are certain components that are common to wide range of engineered products. Nuts, bolts, screws, rivets and other fixing devices are used to connect and hold the parts together. Seals and gaskets are used to prevent the escape of fluid and the ingress of dirt, and bearings are used to support rotating and reciprocating parts. We will now examine some of these components and their applications.

Seals and packing

The seals used on engineered products may be divided into stationary seals and dynamic seals. Dynamic seals may be further sub-divided into those used with reciprocating components and those used with rotating components. Other relevant terms which you might encounter are *gaskets*, which are a form of stationary seal, and *glands* which are assemblies containing seals or packing to maintain the pressure inside a system at the entry or exit point of a reciprocating rod or a rotating shaft.

Stationary seals are used between the flanged joints of pipes, on the cylinder heads of engines pumps and compressors, between crank cases and oil sumps and under inspection and access covers. Sometimes a liquid sealant is used and there are many proprietary brands that are resistant to oil and water. Some of them solidify after assembly whilst others are intended to remain tacky. Separation of the surfaces for maintenance can sometimes be a problem after which they have to be carefully cleaned before re-assembly.

Where there are possible irregularities in the contact surfaces, it is usual to employ a seal which can be compressed to accommodate them. A variety of materials are used. They include paper, cork, plastic, rubber, bonded fibre and copper. Laminated materials are also widely used. The cylinder head gaskets of a great many internal combustion engines are of this type. They consist a bonded fibre sandwiched between two thin sheets of copper. Rubber seals sometimes take the form of an O-ring which is seated in semi-circular recesses as shown in Figure 3.6.

Three of the most common dynamic seals are the rotational lip seal, the packed gland and the mechanical seal. They are used as

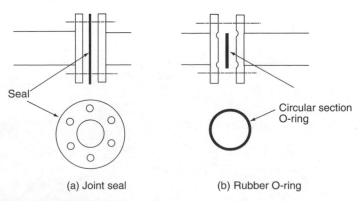

(a) Joint seal   (b) Rubber O-ring

Seal

Circular section O-ring

Figure 3.6 *Pipe joints*

a seal around the drive shafts to pumps and compressors, and around the input and output shafts of gearboxes.

Rotary lip seals are used to prevent the escape of fluids from systems where the pressure difference across the system boundary is relatively small. They are widely used on the input and output shafts to gearboxes to prevent the escape of lubricating oil and are suitable for moderate to high running speeds. A section through a lip seal is shown in Figure 3.7. A rubber garter surrounds the rotating shaft and is held in contact by a spring around its circumference.

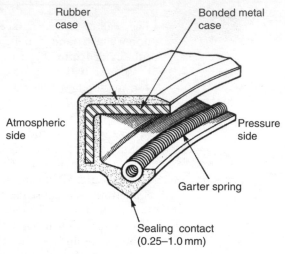

Figure 3.7 *Rotary lip seal*
Source: *From Drives and Seals by M.J. Neale. Reprinted by permission of Elsevier Ltd*

Additional contact force is provided by the internal pressure which pushes down on the garter. During assembly the seal is pressed into a machined recess in the system casing, usually alongside the shaft bearing, and covered by a retaining plate. Different grades of rubber and flexible plastic are used for different operating temperatures and contained fluids.

Packed glands, as shown in Figure 3.8, are able to withstand higher internal pressures than lip seals. A compression collar is

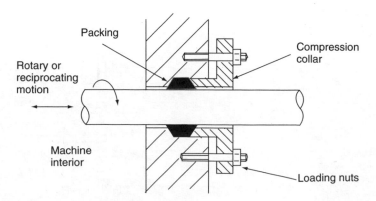

Figure 3.8 *Packed gland*

tightened so that packing material forms a tight seal around the shaft. A wide variety of packing materials is used. It includes compressed mineral and vegetable fibres such as asbestos, cotton, flax and jute. Compressed synthetic fibres such as nylon and PTFE are also used and sometimes they are impregnated with graphite to assist lubrication. Packed glands may be used as a seal for rotating and reciprocating shafts.

Packed glands need periodic adjustment and replacement of the packing material. They are still widely used with reciprocating shafts but to reduce the need for maintenance, the mechanical seal, such as that shown in Figure 3.9, has been developed for use with rotating shafts. There are a variety of designs but the operating principle is the same. The seal is formed between the stationary and rotating rings. The loading collar, spring, flexible sheath and rotating ring rotate with the shaft. The stationary ring is pressed into a recess in the machine casing. The loading collar is a tight fit on the shaft and the spring applies a contact force between the rings. Lubrication is from the interior of the machine.

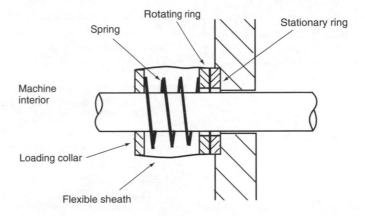

Figure 3.9 *Mechanical seal*

The pistons in engines and compressors are fitted with piston rings whose purpose is to prevent the leakage of high pressure gas around the piston and into the crankcase. Piston rings are made from a springy metal such as steel, cast iron and bronze. A typical arrangement is shown in Figure 3.10. When fitted into the

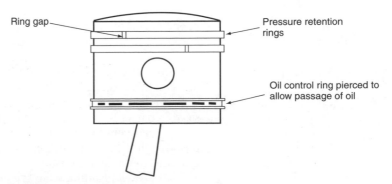

Figure 3.10 *Piston rings*

cylinder, the ends of the rings are sprung almost together to leave only a very small gap and provide firm contact with the cylinder walls. This ensures that the amount of compressed gas that is able to bypass the piston is negligible. The upper of the two rings provides the seal whilst the lower controls lubrication. As can be seen from Figure 3.10, the oil control ring has a channel section that is pierced with elongated slots. The groove in the piston also contains holes through which oil mist can pass from inside the piston. Its function is to control the lubrication of the cylinder walls and scrape away carbon deposits.

Care needs to be taken when fitting new lip seals, mechanical seals and piston rings. Lip seals and the fixed rings of mechanical seals need to be fitted using a press and special centralising tools to ensure that they are properly aligned with the machined recesses. Care also needs to be taken when fitting new piston rings and reassembling the pistons in the cylinders of an engine or compressor. Cast iron rings in particular are brittle and they should be opened only sufficiently for them to pass over the piston and be positioned in the turned grooves. When replacing the pistons in the cylinders, a compression sleeve should be placed around the rings. This can be tightened by means of a screw to compress the rings. The piston can then be pushed through the sleeve and into the cylinder without damage.

See Questions on 3.3 p. 217.

Test your knowledge 3.3

1. What do you understand by the terms *gasket* and *gland*?
2. What are the limitations of rotary lip seals?
3. How does a mechanical seal function?
4. What is the purpose of the different kinds of piston ring used in internal combustion engines?
5. What are the materials most commonly used for piston rings?

Activity 3.3

In addition to the oil seals which we have considered, there is a type known as a labyrinth seal. There is also a related type that incorporates a screw thread. It is known as a viscoseal, or sometimes a screw seal or wind-back seal. Describe with the aid of sketches how these oil seals function and applications to which they are suited.

Bearings

Rotating shafts need to be supported. Depending on the application, the support is provided by plain bearings, ball bearings or roller bearings. In addition to providing radial support, the bearings sometimes also need to accommodate axial forces or, *end thrust*, along a shaft. The plain bearings for steel shafts are very often made from phosphor bronze. This has good load carrying properties which can be enhanced by impregnating it with graphite, as will be described in Chapter 4. Nylon is sometimes used for light duty low speed applications. Figure 3.11 shows a plane bearing which is designed to give both radial and axial support.

Plain bearings are best suited to low rotational speeds and steady loads. It is essential that they are adequately lubricated with a relatively low viscosity oil. This is usually delivered under pressure in a re-circulating oil system. The oil also carries away heat and maintains the bearing at steady temperature. Very often a spiral groove, leading from the oil passage, is cut in the internal bearing surface to ensure even distribution of the oil.

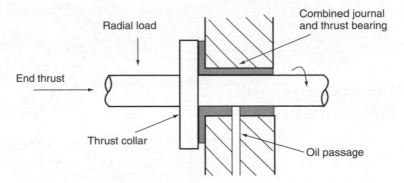

Figure 3.11 *Journal and thrust bearing*

Another type of plain bearing that is widely used in internal combustion engines and reciprocating compressors, is the split bearing with steel-backed white metal bearing shells. They are used for the crankshaft main bearings and the connecting rod big-end bearings. Figure 3.12 shows a connecting rod big-end bearing. The shells are made from mild steel coated with a soft tin-lead alloy which is able to absorb the shock loading of the power strokes.

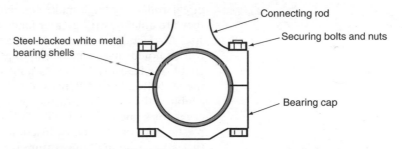

Figure 3.12 *A split big-end bearing*

As with other plain bearings, adequate lubrication is essential and oil is delivered under pressure through oil passages in the crankshaft. White bearing metal has a very low melting point and over-heating can be disastrous. Failure of the white metal layer is indicated by vibration and an unmistakable knocking sound. Great expense will then be incurred in acquiring and fitting a replacement engine.

Ball and roller bearings are classed as rolling element bearings. They offer less resistance to motion than plain bearings and are able to operate at higher speeds and carry greater loads. They are of course more expensive but when correctly fitted and lubricated, they give trouble-free service for long periods. Ball bearings may be contained in a cage which runs between an inner and outer ring or race, as shown in Figure 3.13. The balls and the rings are made from hardened and toughened steel which have been precision ground to size. The balls are not always contained in a cage however, and for some low-speed applications they are packed freely in grease between the rings. This is known as a *crowded assembly* and is to be found in the wheel and steering head bearings of some bicycles.

Because of their low-friction rolling action, grease lubrication is often sufficient for ball bearings and in recent years a range has been produced which are greased and sealed for life. As an

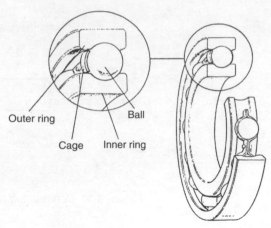

Figure 3.13 *Caged ball bearing*
Source: *From Mechanical Engineer's Reference Book by E.H. Smith. Reprinted by permission of Elsevier Ltd*

example of their speed capability, a ball race with 8 mm diameter balls is capable of speeds up to 32 000 rpm with grease lubrication, and this can be exceeded with a re-circulating oil system.

Ball bearings make point contact with the rings whereas cylindrical roller bearings make line contact. As might be expected, they are able to carry greater loads than ball bearings of the same size. A selection of caged roller bearing is shown in Figure 3.14. It will be noted that some of the rings are fitted with flanges. These enable the bearings to carry a limited amount of axial load. Where the inner ring has no flanges or one flange only, it can be removed enabling the two parts of the bearing to be fitted separately. The cages for some ball and roller bearings are made from reinforced plastic material to reduce friction. These perform well provided that a low running temperature is maintained. Pressed steel cages should be used for running temperatures above 100 °C.

For applications where high radial and axial loads are to be carried it is advisable to use tapered roller bearings. These are shown in Figure 3.15. Single row tapered roller bearings can carry axial loads in one direction only. If axial loads are present in both directions the bearings may be paired back to back as shown. The front wheel bearings in motor vehicles, which have to carry considerable radial loads and axial loads in both directions when cornering, contain tapered roller bearings that are paired in a similar way.

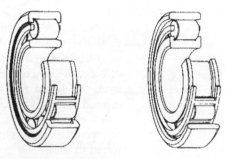

Figure 3.14 *Caged roller bearings*
Source: *From Mechanical Engineer's Reference Book by E.H. Smith. Reprinted by permission of Elsevier Ltd*

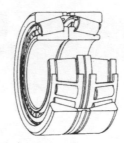

Figure 3.15 *Tapered roller bearings*
Source: *From Mechanical Engineer's Reference Book by E.H. Smith. Reprinted by permission of Elsevier Ltd*

Needle bearings are roller bearings with long, thin rollers. They are designed for applications where there is limited radial space. Very often the inner race is dispensed with so that the needle rollers run directly on the rotating shaft in a crowded assembly. They are intended to carry light loads at relatively low or intermittent rotational speeds.

Care needs to be taken when assembling bearing races in position. Very often the rings needs to be pressed onto a shaft or into a housing and as with oil seals, a special centralising tool should be used to ensure correct alignment and seating. A possible arrangement on an input or output shaft from a machine shown in Figure 3.16.

The slideways on machine beds and worktables may be classed as linear bearing surfaces. The hardened cast iron slideways on lathe beds are V-shaped as are the mating surfaces in the tailstock and saddle assemblies. Tool slides and worktable slides are generally of dovetail section. The contact materials are mostly steel and cast iron. Because the movement on these slides is relatively slow and intermittent, no special bearing materials are required and regular lubrication by hand is all that is required for smooth operation. Where the movement on a slide is rapid and continuous, as with the ram of a shaping machine, the slides are often faced with a bearing material such as phosphor–bronze. Continuous lubrication is also required either by a wick feed or by a forced feed from a pump. *See Questions 3.3 on p. 217.*

Test your knowledge 3.4

1. What is the usual material used for plain journal bearings?
2. Describe the bearing shells used in the main and big-end bearings of internal combustion engines?
3. What is meant by a crowded assembly of ball bearings?
4. What types of bearing are used to accommodate axial thrust?
5. What are needle bearings?

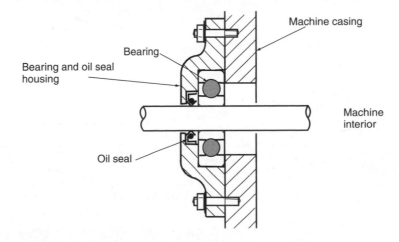

Figure 3.16 *Bearing and oil seal arrangement*

Activity 3.4

Use your resource centre or public library to find out about the components which make up the front wheel hub of a motor car. Make a labelled sketch which shows how the major components are arranged. What is the method of lubrication and the recommended lubricant in the particular model which you have researched?

Fastenings

Screw fastening and rivets are the most widely used semi-permanent fastenings in engineered products. Screwed fastenings on covers and internal components allow access and removal for maintenance and repair. Components joined by rivets are not intended to be separated, but if the need arises this can be achieved by grinding off the rivet heads or drilling out the rivets. Most countries throughout the world now use the International Organisation for Standardisation (ISO) metric screw thread. The USA is an exception where feet and inches are still in widespread use. Small sized screw fastenings, particularly those used in electrical and IT equipment, employ the British Association (BA) screw thread. Although this is British in origin, it is used internationally in these applications. The form of these screw threads is shown in Figure 3.17.

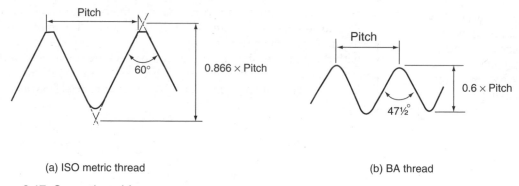

(a) ISO metric thread   (b) BA thread

Figure 3.17 *Screw thread forms*

The most commonly used screwed fastenings are nuts and bolts, setscrews, studs and self-tapping screws. Metric nuts and bolts are specified in a way that describes their shape of head and nut, nominal diameter, pitch and bolt length. Figure 3.18 shows a metric hexagonal head bolt and nut.

A hexagonal head nut and bolt might be specified as follows:

steel, hex hd bolt – M12 × 1.25 × 100

steel, hex hd nut – M12 × 1.25

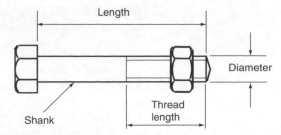

Figure 3.18 *Hexagonal head nut and bolt*

The M, specifies that it is metric, the 12 specifies the diameter in millimetres, the 1.25 specifies the pitch in millimetres and the 100 specifies the length in millimetres.

Bolts can be obtained with different lengths of thread for different applications. Bolts for general use are forged to shape and are generally known as *black bolts*. For applications where precision is required, fitted bolts can be obtained that have been accurately machined to size. Bolts made from high tensile steel are also supplied for heavy duty applications. Nuts and bolts are made from steel, brass, aluminium alloys and plastics. Steel nuts and bolts are sometimes cadmium plated to improve their resistance to corrosion.

Figure 3.19(a) shows two components that are joined by a nut and bolt. A bolt should always be selected such that the plain unthreaded part of its shank extends through the joint interface. This is to ensure that there will be no shearing force on the weaker threaded part. It will be seen that a washer has been placed under the nut. This is good practice. It prevents damage to the component

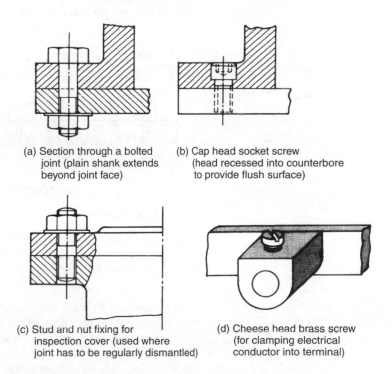

(a) Section through a bolted joint (plain shank extends beyond joint face)

(b) Cap head socket screw (head recessed into counterbore to provide flush surface)

(c) Stud and nut fixing for inspection cover (used where joint has to be regularly dismantled)

(d) Cheese head brass screw (for clamping electrical conductor into terminal)

Figure 3.19 *Typical screwed fastenings*
Source: *From Mechanical Engineering GNVQ Intermediate by Mike Tooley. Reprinted by permission of Elsevier Ltd*

face and has the effect of spreading the load. Figure 3.19(c) shows an alternative type of fastening using a stud and nut. A stud is a length of bar which has been threaded at each end as shown in Figure 3.20. The shorter threaded end is screwed into the major component, sometimes using a locking fluid to prevent it from easily becoming unscrewed. It is good practice to use studs where a component such as a machine inspection cover needs to be removed regularly for maintenance purposes. Because the studs are permanently in position in the machine casing, wear to the threaded holes is prevented. Any excessive wear will be at the nut end of the stud, which can be replaced.

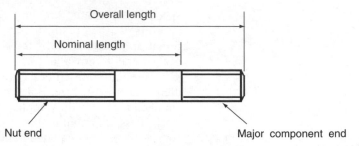

Figure 3.20 *Stud*

The unthreaded portion, in the middle of a stud should be a little shorter than the component through which it passes and the threaded nut end should protrude through by a distance greater than the thickness of the nut (Figure 3.21). As with bolted joints, it is good practice to place a washer under the nut.

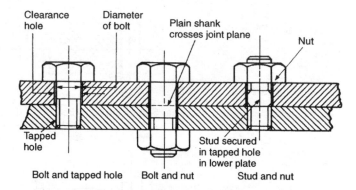

Figure 3.21 *Use of bolts and studs*
Source: *From Mechanical Engineering GNVQ Intermediate by Mike Tooley. Reprinted by permission of Elsevier Ltd*

Set screws, which are also called machine screws, do not use a nut. Figure 3.22 shows how a thin plate might be joined to a larger component using set screws. When joining sheet metal or thin plate components to a bulky object it is usual to use screws that are threaded over the whole of the shank.

Screwed fastenings that are subject to vibration require a locking device to prevent them working loose. There are two basic categories of locking device. Those which have a positive locking

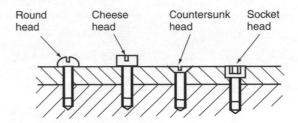

Figure 3.22 *Set screw types*

action and those that depend on friction. Figure 3.23 shows a selection of both types. The positive locking category includes castle nuts, secured with a split pin. The split pin passes through a hole which is drilled in the bolt and its ends are opened to ensure that it stays in position. Split pins should only be used once. After removal for maintenance they should be discarded and new ones used when the components are reassembled. Tab washers are also positive locking devices. After tightening the nut, one side is bent up against the nut face and the opposite side is bent over the component as shown in Figure 3.23. A positive locking method which is used with set screws is to drill a small hole across opposite faces of the head and secure two or more together with wire.

Among the locking devices that depend on friction are lock nuts, stiff nuts and spring washers. A lock nut is a secondary nut of smaller thickness that is tightened down on top of a plain nut. The combined frictional resistance is generally sufficient to withstand the effects of vibration. There are a variety of types of stiff

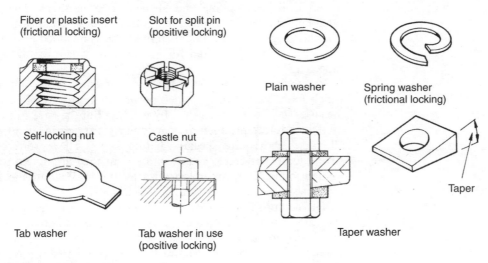

Figure 3.23 *Locking devices*
Source: *From Mechanical Engineering GNVQ Intermediate by Mike Tooley. Reprinted by permission of Elsevier Ltd*

nut. The type shown in Figure 3.23 can have a compressed fibre or a nylon insert. Those with a fibre insert are traditionally called Simmonds nuts and those with the nylon insert are often referred to as 'nyloc' nuts. Friction between these materials and the screw thread is sufficient to prevent them vibrating loose. As with split pins it is good practice to use new friction nuts when reassembling components after maintenance.

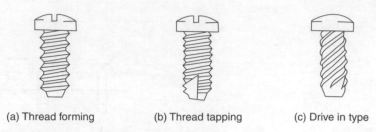

(a) Thread forming (b) Thread tapping (c) Drive in type

Figure 3.24 *Self-tapping screws*

Spring washers come in a variety of shapes. The one shown in Figure 3.23 is known as a split spring washer. When the nut is tightened down on it, the edges of the split resist anticlockwise turning. Other kinds of spring washer are stamped from spring steel to the shape of a many pointed star. The edges of the star are turned slightly so that like the split spring washer, they oppose anticlockwise turning of the nut. Figure 3.23 also shows how a wedge-shaped washer should be used when one of the joined components, such as a rolled steel section, has a tapering face.

Self-tapping screws are widely used with sheet metal and plastic components (Figure 3.24). The type used depends on the hardness of the receiving material. For each type, a pilot hole is drilled whose diameter is the same or slightly less than the root diameter of the thread. The thread forming type is used on soft materials such as thermoplastics. As it is screwed home it displaces the material to form its own mating thread. The thread cutting type is used on harder materials. It has a cutting edge at the start of its shank. This cuts a thread in the receiving material, rather like a screw cutting tap. The third type has a multi-start thread and is intended to be driven in by a hammer. It is used for applications where removal is not expected.

Rivets are often preferred to screw fastenings for components which are expected to be joined permanently. Many of their former uses have been taken over by welding but they are still widely used in aircraft production, particularly for joining light metal alloys which require special welding skills. They produce very strong joints and have the advantage that they are not so rigid as welded joints. They are able to flex a little under load and this is often desirable particularly with shock loads.

The composition of rivet material should always be as close as possible to that of the materials being joined. Otherwise, electrolytic corrosion of the type described in Chapter 4 might occur in the presence of moisture. Figure 3.25 shows a variety of rivet types for different applications. Rivets should not be loaded in tension. They should only be subjected to the single or double shearing loads described in Chapter 1. Some tensile stress will inevitably be present from the setting operation, particularly after hot riveting as the rivet cools. Round, conoidal and pan heads have the greatest strength whilst the countersunk and flat heads are used where a flush joint surface is required. Figure 3.26 shows the correct way in which rivets should be closed.

The pop rivets shown in Figure 3.27 are widely used with sheet metal fabrications where access is only possible from one side of the joint. After inserting the rivet in the joint, the head on the blind

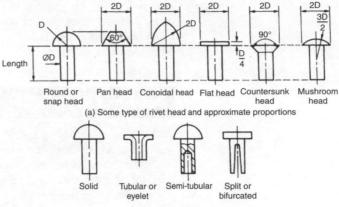

(a) Some type of rivet head and approximate proportions

(b) Types of rivets

Figure 3.25 *Rivets*
Source: *From Mechanical Engineering GNVQ Intermediate by Mike Tooley. Reprinted by permission of Elsevier Ltd*

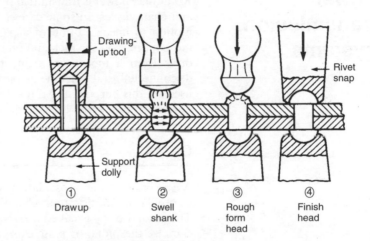

①	②	③	④
Draw up	Swell shank	Rough form head	Finish head

Figure 3.26 *Correct riveting procedure*
Source: *From Mechanical Engineering GNVQ Intermediate by Mike Tooley. Reprinted by permission of Elsevier Ltd*

Test your knowledge 3.5

1. What does the following specification indicate?

   Steel, hex hd M8 × 1 × 50

2. What are the two main categories of locking device used with screwed fastenings?

3. What kind of application are studs used for?

4. Under what circumstances might a riveted joint be preferred to welding?

5. For what application are pop rivets useful?

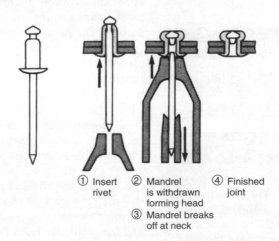

① Insert rivet

② Mandrel is withdrawn forming head

③ Mandrel breaks off at neck

④ Finished joint

Figure 3.27 *Pop riveting procedure*
Source: *From Mechanical Engineering GNVQ Intermediate by Mike Tooley. Reprinted by permission of Elsevier Ltd*

side is formed by pulling the central mandrel through the rivet using a special tool. As the rivet is formed, the mandrel breaks off at its head and is discarded.

See Questions 3.4 on p. 218.

Activity 3.5

(a) There are types of stiff nut other than the ones that we have considered. Describe two of these with the aid of sketches and the way that they function.

(b) What are shims, and how are they used in the assembly of engineered products?

Mechanical power transmission systems

The purpose of mechanical power transmission systems is to transmit force and motion from one point in a machine to another. Rotational power is transmitted by means of drive shafts, belt drives and chain drives. In some cases it is required to convert rotational motion to linear motion of a particular type. Cams, slider-crank mechanisms and the rack and pinion are used for this purpose. The drive from a motor or engine may need to be connected and disconnected to a machine as part of a working cycle. A clutch is used for this purpose and if it is required to bring the moving parts quickly to rest, some braking device is needed.

Cams

A *cam* is a rotating or oscillating body which imparts reciprocating or oscillatory motion to a second body which is in contact with it. The second body is called a *follower*. The most common types of cam are the radial or plate cam, the cylindrical cam and the face cam as shown in Figure 3.28.

Knife edge, flat-faced and roller followers are all used with plate and face cams. The knife edge type is only suitable for low speed applications where there is light contact force. The roller type is more wear resistant and is widely used in automated machinery. In automobile engines where the space is confined, the flat-faced follower is generally used, offset as shown in Figure 3.29(b). This causes it to rotate about its own axis so that wear is evenly distributed over the contact face. If a follower is stationary for a part of rotation cycle, it is said to *dwell* and the angle through which the cam turns during the dwell period is known as the *dwell angle*. Figure 3.29(b) shows a dwell angle which will occur when the follower is at its lowest position. Here the cam profile is circular with its centre on axis of rotation.

Cam profiles can be divided into two categories. Cams for which the cam profile is made up of circular arcs and tangents, as shown in Figure 3.29 and cams for which the required nature of the follower motion determines the cam profile, e.g. rise and fall with uniform velocity, rise and fall with SHM and rise and fall with uniform acceleration and retardation. The latter category are more expensive to produce and confined to special purpose applications. Performance graphs showing the rise and fall of a knife-edged

---

Key point

Dwell angle is the angle turned by a cam whilst the follower is stationary.

Key point

There are two categories of cam. Those where the chosen shape determines the motion of the follower and those where the specified follower motion determines the shape of the cam.

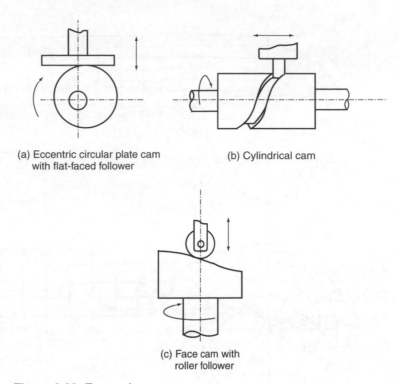

(a) Eccentric circular plate cam
with flat-faced follower

(b) Cylindrical cam

(c) Face cam with
roller follower

Figure 3.28 *Types of cam*

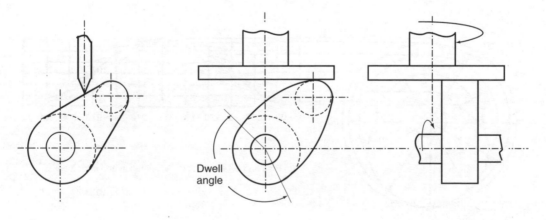

Dwell
angle

(a) Plate cam made up of circular arcs
and tangents with knife-edged follower
offset from axis of rotation

(b) Plate cam made up of circular arcs
with side offset flat-faced follower

Figure 3.29 *Cams with offset followers*

follower against angular position of the cam for these motions
are shown in Figure 3.30.

Dwell periods have been omitted from the cams in Figure 3.30
and it will be noted that the cams have a characteristic heart
shaped appearance. In certain applications, two or more of the
different motions may be required together with dwell periods
when the follower is at its extreme positions. Additionally, the
cam may have a roller or flat-faced follower with offset. The
projection construction process then becomes more complex and
is beyond the range of this unit.

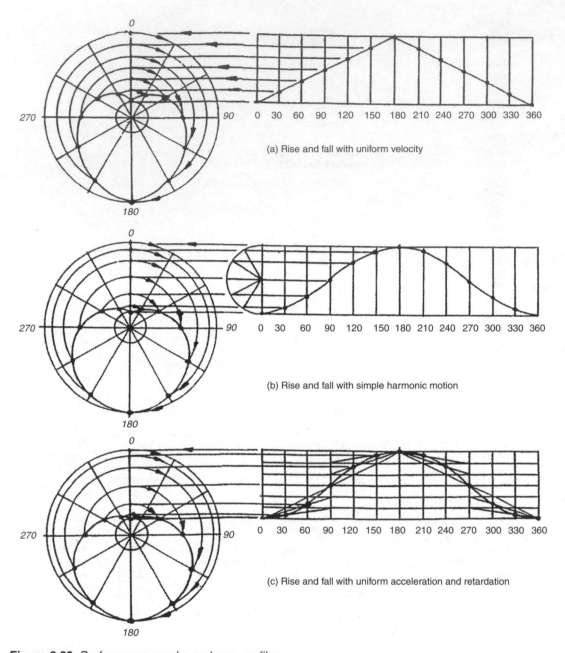

(a) Rise and fall with uniform velocity

(b) Rise and fall with simple harmonic motion

(c) Rise and fall with uniform acceleration and retardation

Figure 3.30 *Performance graphs and cam profiles*

Test your knowledge 3.6

1. What is a face cam?
2. How can flat-faced followers be made to rotate as they rise and fall?
3. What is a dwell angle?
4. What are the three main types of cam follower?

Activity 3.6

Plot the performance graph and cam profile for a plate cam with a knife-edged follower to the following specifications: (a) base circle diameter 25 mm, (b) rise 75 mm with SHM whilst rotating through an angle of 120°, (c) dwell for a further 60° of rotation, (d) fall 75 mm with uniform acceleration and retardation whilst rotating through an angle of 150°, (e) dwell for final 30° of rotation.

Linkage mechanisms

A mechanism may be defined as a *kinematic chain* in which one element is fixed for the purpose of transmitting or transforming motion. A kinematic chain is made up of a number of linked elements. Two elements which are linked together constitute a *kinematic pair*. If they are hinged, they are called a *turning pair*. If one element is constrained to slide through or around another, they are called a *sliding pair*. If one element rotates inside another in a screw thread, they are called a *screwed pair*.

Two of the most common mechanisms are the slider crank and the four-bar chain. Closely related to these are the slotted link and Whitworth quick return motion and Watt's parallel motion. When a mechanism is required to transmit power, the various links and joints have to be designed, with an appropriate factor of safety, to carry the forces to which they will be subjected. The mechanism, or a series of linked mechanisms, is then classed as a *machine*.

Slider-crank mechanisms

The slider-crank mechanism finds widespread use in reciprocating engines, pumps and compressors. It is made up of three links and a slider as shown in Figure 3.31. Link AB is the crank which in Figure 3.31(a) rotates at a steady speed. Link BC is the connecting rod or coupler, which causes the piston C to slide in the cylinder. The third link is the cylinder block itself, AD which is stationary and on which the crank rotates at A. As stated above, a mechanism is a chain of links, one of which is fixed. In this case it is the cylinder block AD.

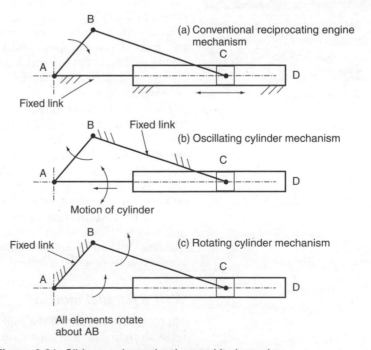

Figure 3.31 *Slider-crank mechanism and its inversions*

In Chapter 1, we used a graphical method to find the speed of the piston at a given instant in time, when the crank was at a particular angular position and rotating at a given speed. We now need to consider *inversions* of the mechanism. An inversion is obtained by holding another link in the fixed position. The inversion shown in Figure 3.31(b) is called an oscillating cylinder mechanism. It is obtained by holding the link BC in the fixed position so that the crank AB rotates about B. This imparts a combined rocking and oscillating motion to the cylinder which pivots about the stationary piston.

The inversion shown in Figure 3.31(c) is called a rotating cylinder mechanism. It is obtained by holding the link AB in the fixed position. Both the link BC and the cylinder AD rotate about the fixed link. The arrangement was used in the early days of aviation for rotating radial cylinder aero-engines. It also forms the basis for the Whitworth quick return motion which we shall shortly be considering.

Four-bar linkage mechanisms

The four-bar linkage is another mechanism which finds widespread use. It is to be found in applications such as windscreen wiper drives, vehicle suspension units and everyday uses such as the hinges on kitchen cupboard doors and squeeze-mop mechanisms. Two of the links rotate about fixed centres and are joined by a coupler link. The fourth link is formed by the frame or bed plate that contains the fixed centres of rotation.

It should be noted that the number of inversion of a mechanism is equal to the number of links, which in this case is four. In Figure 3.32, the links are all of different length and chosen so that the sum of the longest and shortest is less that the sum of the other two. That is,

$$AB + BC < AD + DC$$

When this is the case, three distinct types of mechanism can be obtained from the inversions. The inversions shown in Figures 3.32(a) and (c) are both crank-rocker mechanisms. It is the longest two links which are fixed whilst the input crank AB rotates at a steady speed, first about end A and then about end B. This imparts a rocking action to the output link CD.

With the inversion shown in Figure 3.32(b), the shortest link AB is fixed and this produces a double crank mechanism. Link BC rotates about B and link AD rotates about A. If BC is the input crank, rotating at a steady speed, the link AD will rotate at a varying speed and the links must be able to cross over certain times. The link CD will have in a complicated motion. It will rotate about the fixed link AB and also rotate about its own centre as the links BC and AD cross over each other.

With the inversion shown in Figure 3.32(d), the second shortest link CD is fixed and this produces a double rocker mechanism. The links BC and AD rotate about their fixed ends, C and D, but are unable to describe a full revolution.

Watt's parallel motion

This is an application of the four-bar linkage which gives approximate straight line motion to a particular point, P on the coupler link. The arrangement is shown in Figure 3.33.

Key point

An inversion of a mechanism is obtained by changing the link which is fixed.

Key point

The number of possible inversions of a mechanism is equal to the number of links.

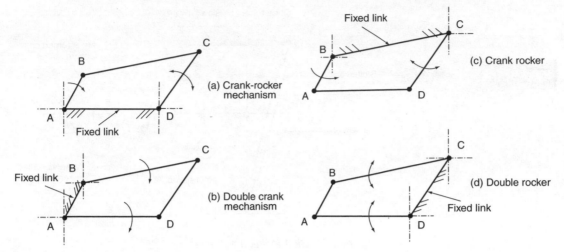

Figure 3.32 *Four-bar linkage mechanism inversions*

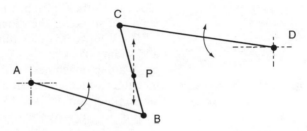

Figure 3.33 *Watt's parallel motion*

The point P travels in a line that is approximately straight between the positions where AB is horizontal and CD is horizontal. The position of the point P on the coupler link should be chosen so that

$$\frac{AB}{CD} = \frac{BP}{PC}$$

Quick return mechanisms

Slow forward and quick return mechanisms are to be found on some metal cutting machines, printers and scanners. The output gives a slow forward linear motion followed by a quick return to the start position along the same path. Two of the most common which are found in shaping, planing and slotting machines for metal cutting are the slotted link mechanism and the Whitworth quick return motion.

The slotted link mechanism is used on shaping machines where single point cutting tool is mounted on the front of the slider or ram, in a hinged tool post. The tool cuts on the slow forward stroke and lifts over the workpiece on the quick return stroke. The slotted link rocks from side to side, driven by the sliding block on the bull wheel.

The bull wheel rotates at a constant speed and as can be seen from Figure 3.34, the angle through which it rotates on the forward stroke is greater than the angle through which it rotates on the return stroke. This imparts the slow forward and quick return

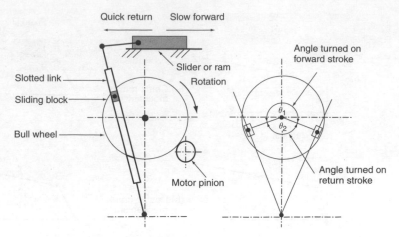

Figure 3.34 *Slotted link mechanism*

motion to the slotted link and slider. The distance of the sliding block from the centre of the bull wheel can be altered to vary the length of stroke of the slider.

The Whitworth quick return motion also employs a slotted link and sliding block as shown in Figure 3.35. The mechanism is used on planing machines, which are quite large, and on slotting machines which are small and compact. With slotting machines a single point tool is fixed to the front of the slider and is used for cutting fine grooves and key-ways. With planing machines the slider is the work table on which the workpiece is secured. This moves with slow forward and quick return motion beneath a stationary single point cutting tool.

The driving gear which contains the sliding block, rotates at a constant speed. The sliding block causes the slotted link to rotate but because it has a different centre of rotation, its speed is not constant. As can be seen from Figure 3.35, the angle through which the driving gear and slotted link rotate on the forward stroke is greater than the angle through which they rotate on the return stroke. This imparts the slow forward and quick return motion to the slider.

> **Key point**
>
> The slotted link and Whitworth quick return mechanisms are practical applications of the slider-crank inversion shown in Figure 3.31(c).

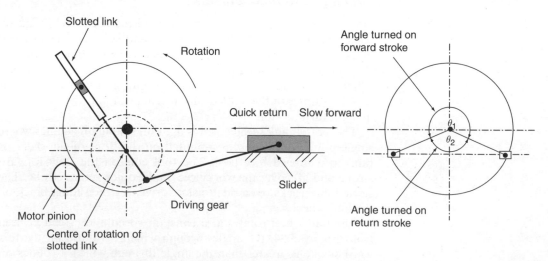

Figure 3.35 *Whitworth quick return motion*

See Questions 3.5 on p. 219.

Activity 3.7

When turning a corner in a car, the front wheel on the inside of the curve has to turn through a larger angle than the wheel on the outside of the curve. Find out the name and sketch the layout of the steering linkage used. Which of the mechanisms that we have described is it based upon?

Shafts, clutches and brakes

Transmission shafts are widely used to transmit power from prime movers such as electric motors and internal combustion engines. They are used to drive machinery, pumps, compressors and road vehicles. Transmission shafts are generally made from steel and may be of solid or tubular section.

Where the drive extends over long distances, it is often necessary to couple sections of shaft together and support them in bearings. Several different kinds of coupling are available and a clutch is often incorporated in the system to connect and disconnect the drive. Good alignment is desirable between the driving motor and driven machinery but where this is not possible, universal and constant velocity joints can be used to turn the line of drive into the required direction.

Joints and couplings

When two shafts are joined directly together, the joint is called a coupling. The coupling may be rigid or flexible. A rigid coupling is shown in Figure 3.36. It should only be used where the centres of rotation of the shafts are concentric. It consists of two coupling flanges keyed to the ends of the separate shafts and bolted together.

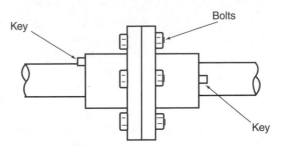

Figure 3.36 *Rigid coupling*

If the shafts are in line but there is likely to be shock loading or excessive vibration, a degree of flexibility is required. The same applies if one of the machines is likely to undergo a temperature change, causing a change in shaft height due to expansion. This is known as *thermal growth* and also requires a degree of flexibility. This can be achieved by the fitting of rubber or plastic bushes around the bolts or by separating the flange faces with flexible disc as shown in Figures 3.37(a) and (b). There are many variations of

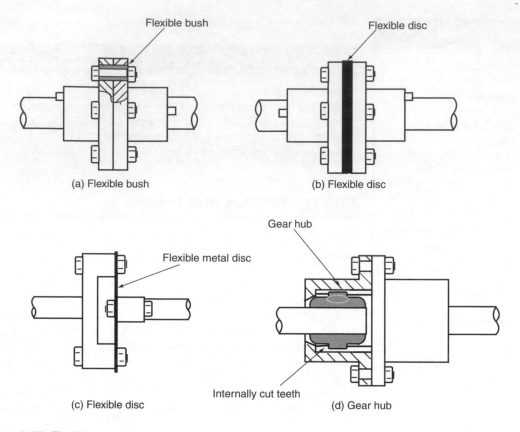

Figure 3.37 *Flexible couplings*

these designs where rubber bushes and pads are used to provide flexibility. They are generally known as *elastomeric* couplings.

The type shown in Figure 3.37(c) employs a metal disc to which the forked ends of the shafts are connected. In addition to allowing some flexibility, this can also function as a safety device which will fail in the event of an overload. The type shown in Figure 3.37(d) allows for a small degree of misalignment. Hubs with external gear teeth are keyed on the ends of the shafts. These locate in the flanged parts which have internally cut teeth. The length of the teeth on the hubs is relatively short which allows limited angular deflection of the shafts as they rotate.

The degree of accuracy required when aligning shafts that are to be coupled depends on the operating conditions and the type of coupling that is to be used. In the case where two shafts have only a small gap between them and a flexible coupling is to be used it may be sufficient to align them by means of a straight edge as shown in Figure 3.38. Assuming the machine carrying shaft A to be fixed, the position of shaft B is adjusted until the straight edge makes even contact with both shafts. The straight edge is then moved through 90° to the position shown by the dotted line and if required, further adjustments are made.

There are two kinds of misalignment. Parallel misalignment is when the shaft axes are parallel but not in line. Angular misalignment is when the axes intersect at some point. There may of course be a combination of the two requiring linear and angular adjustment of the movable shaft.

Key point

Misalignment is present if the shaft axes are parallel but not coincidental, if their axes intersect at some point and if there is a combination of these two faults.

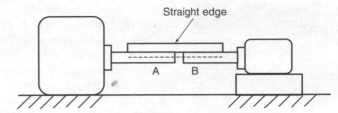

Figure 3.38 *Alignment using a straight edge*

If the shafts are to be joined by a rigid coupling, a more precise degree of alignment is required and if the gap between the shafts is small, a dial test indicator can be used as shown in Figure 3.39. The dial test indicator is mounted on a Vee-block and positioned as shown. The Vee-block is then moved along shaft A and deflection of the pointer denotes angular misalignment in that plane which can be corrected by turning the machine of B. When there is no movement of the pointer, the Vee-block is moved through 90° to the position shown by the dotted line. Once again it is moved along shaft B and further angular adjustments are made until there is no movement of the pointer.

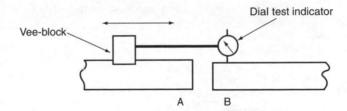

Figure 3.39 *Alignment using a single dial test indicator*

There should then be no angular misalignment but parallel misalignment may still be present. To check for this, the Vee-block is clamped to shaft A which is then rotated with shaft B stationary. Deflection of the pointer denotes parallel misalignment and the machine of shaft B is moved over or raised until there is no pointer movement. The Vee-block is then moved to shaft B with the dial test indicator in contact with shaft A, and the procedure is repeated.

An alternative method is to use two dial test indicators positioned on opposite sides of the two shafts as shown in Figure 3.40. The Vee-blocks are clamped to the shafts which are rotated together. The deflection of the two pointers at different angular positions enables judgements to be made as to the kind of misalignment present. Adjustments are then made until the shafts

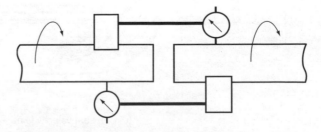

Figure 3.40 *Alignment using two dial test indicators*

190 *Mechanical technology*

Key point

The method which uses two dial-test indicators eliminates errors due to surface finish and ovality.

rotate without any movement of the pointers. The advantage of this method is that there is no sliding of the Vee-blocks or the dial test indicator plungers. Errors due to surface roughness and ovality of the shafts are thus eliminated.

In cases where alignment is required between machines that are separated by longer distances, it is modern practice to use laser equipment. The laser beam is used to ensure that the drive lines from the two machines lie in the same horizontal and vertical planes. One method is to mount the laser to one side of the drive line at the required shaft height, and set so that it can be swung from one machine to the other in a true horizontal plane. After the necessary adjustments, the process is repeated with the laser mounted above the machines and set so that it can be swung from one to the other in a true vertical plane. The laser can also be mounted to act along the drive line to enable the accurate positioning of bearings.

A *Hooke joint*, which is also called a *universal coupling*, can be used where it is required to turn the drive line of a shaft through an angle. Angles of up to 30° are possible with this kind of joint which is shown in Figure 3.41. It is widely used in cars and commercial vehicles at each end of the propeller shaft that connects the gearbox to the rear axle. The joint consists of two forked members that are attached to the ends of the shafts to be connected. The connection is made through a cross-shaped or cruciform member on

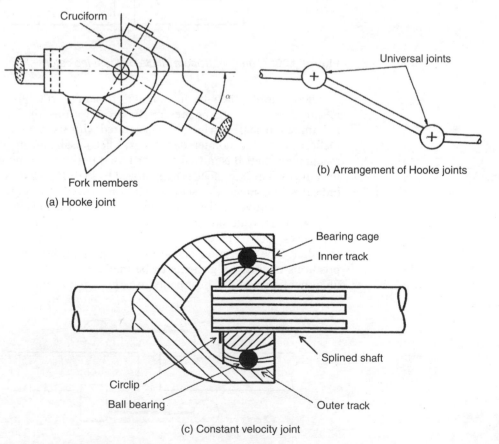

(a) Hooke joint

(b) Arrangement of Hooke joints

(c) Constant velocity joint

Figure 3.41 *Hooke coupling and constant velocity joint*

Key point

The condition for a universal coupling to transmit constant velocity is that the joint driving contacts must always be in a plane which bisects the angle between the connected shafts.

Test your knowledge 3.8

1. What is *thermal growth*?
2. What is an *elastomeric* coupling?
3. What is the advantage of the method of shaft alignment that uses two dial test indicators?
4. What is the main disadvantage of a Hooke joint?
5. What are the conditions required for a universal coupling to give a constant velocity ratio?

which the fork ends pivot. Needle roller bearings, packed with grease, are generally used on the pivot.

The main disadvantage of the joint is that the output speed fluctuates and the effect increases with the angle between the connected shafts. For this reason, it is usual to use two Hooke joints as shown in Figure 3.41(c). The fork ends at each end of the central shaft are positioned in the same plane and in this way, the speed fluctuations are cancelled out. The speed of the central connecting shaft will still fluctuate but the input and output shafts will rotate at a steady speed.

The condition for a universal coupling to transmit constant velocity is that the joint driving contacts must always be in a plane which bisects the angle between the connected shafts. In the case of the Hooke joint, the driving contacts are the arms of the cruciform member. As can be seen from Figure 3.41(a), the condition is not met. At the instant shown, the plane of the cruciform is at right angles to centre line of the shaft on the right. The *constant velocity joint* shown in Figure 3.41(b) overcomes the problem. The drive is transmitted through ball bearings which can move in tracks on the joint members. The geometry of the design is such that the plane in which the balls lie exactly bisects the angle between the shafts. The constant velocity joint is widely used in the drive between the gearbox to the wheels in front-wheel drive cars.

Clutches and brakes

Clutches are used to connect and disconnect the drive in mechanical power trains. The dog clutch is one of the most basic types which is sometimes regarded more as a coupling than a clutch. The two shafts are joined by engaging interlocking teeth on the input and output shafts. Three different kinds are shown in Figure 3.42. These are positive engagement clutches which should only be operated when the shafts are at rest.

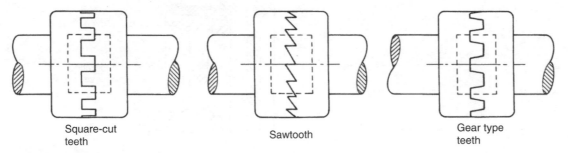

Square-cut teeth   Sawtooth   Gear type teeth

Figure 3.42 *Dog clutches*
Source: *From Mechanical Engineer's Reference Book by E.H. Smith, Reprinted by permission of Elsevier Ltd*

Key point

Dog clutches should only be engaged and disengaged when the shafts are stationary.

Flat plate clutches, similar to that shown in Figure 3.43, are widely used in cars and commercial vehicles. The single plate type is shown but multi-plate clutches are also widely used, particularly in motor cycles. The driven plate is lined with friction material and has a splined hole at its centre. This engages with the splines on the output shaft to which it transmits the drive. When the clutch is engaged, the driven plate in sandwiched between the driving member and the pressure plate. Coiled springs or a spring diaphragm

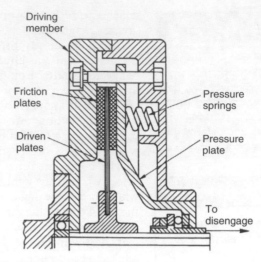

Figure 3.43 *Single plate friction clutch*
Source: *From Mechanical Engineer's Reference Book by E.H. Smith, Reprinted by permission of Elsevier Ltd*

provides the clamping pressure. To disengage the clutch, the pressure plate is pulled back against the springs by the release collar and thrust bearing which slides on the splines of the output shaft.

The conical clutch shown in Figure 3.44 is used in machine tools and also for some heavy duty applications in contractors plant. In a slightly different form it is used in synchromesh manual gearboxes and overdrive units. The clutch has a wedge action which reduces the spring force needed to hold the driving member and the driven cone in engagement.

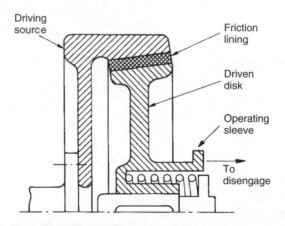

Figure 3.44 *Conical friction clutch*
Source: *From Mechanical Engineer's Reference Book by E.H. Smith, Reprinted by permission of Elsevier Ltd*

Key point

Centrifugal clutches engage and disengage automatically at a pre-determined speed.

The operation of centrifugal clutches was described in Chapter 1. Their advantage is that they are automatic in operation and enable motors with a low starting torque to commence engagement gradually without shock. The strength of the control springs determines the speed at which engagement begins (Figure 3.45).

Fluid couplings of the type shown in Figure 3.46 have many advantages. There is no mechanical connection between the input

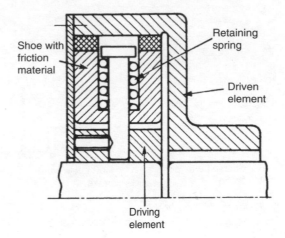

Figure 3.45 *Centrifugal clutch*
Source: *From Drives and Seals by M.J. Neale. Reprinted by permission of Elsevier Ltd*

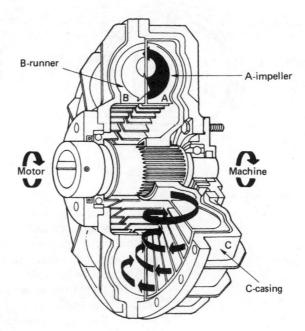

Figure 3.46 *Fluid coupling*
Source: *From Mechanical Engineer's Reference Book by E.H. Smith, Reprinted by permission of Elsevier Ltd*

and output shafts, the take up is smooth and gradual and the operation is fully automatic. There are two basic elements, the runner which is driven by the input shaft and the impeller which is connected to the output shaft. The assembly is filled with oil which is made to circulate as shown, by the vanes on the runner and impeller. As the runner speed increases, the oil transmits the drive to the impeller. The slippage between the runner and impeller speeds becomes less and less as the speed increases until eventually the oil transmits an almost solid drive to the output shaft. A similar form of fluid couplings is used in road vehicles with automatic gearboxes.

The purpose of friction brakes is the opposite to that of friction clutches. They are intended to slow down the rotating parts of a

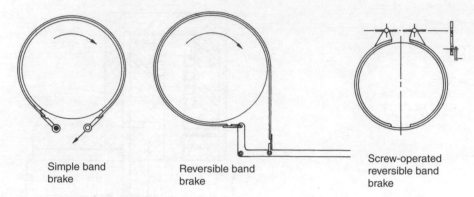

Simple band
brake

Reversible band
brake

Screw-operated
reversible band
brake

Figure 3.47 *External friction brakes*
Source: *From Mechanical Engineer's Reference Book by E.H. Smith. Reprinted by permission of Elsevier Ltd*

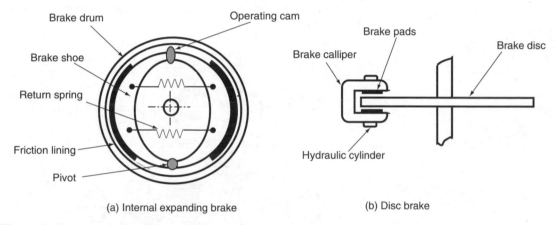

(a) Internal expanding brake

(b) Disc brake

Figure 3.48 *Internal expanding and disc brakes*

machine or a road vehicle. Figure 3.47 shows the operating principle of external band type brakes. Figure 3.48 shows the principle of internal expanding brakes and disc brakes of the type used on road vehicles. Brakes may be applied through a mechanical linkage but it is more usual for them to be activated by a pneumatic or hydraulic actuator.

A dynamometer is a particular type of brake that is used to measure the output torque from electric motors, internal combustion engines and gas turbines. The basic type of friction dynamometer is the rope brake shown in Figure 3.49. These are used only on low speed oil and gas engines, mainly for educational purposes. The brake drum is coupled to the engine output shaft and has an internal channel section. Whilst rotating, it is filled with water which is kept evenly distributed around the internal circumference by centrifugal force. This absorbs the heat generated by friction between the rope and brake drum. A dead weight is hung on the lower end of the rope and the upper end is attached to a spring balance. A safety chain to the dead weight hanger prevents the weight being flung off should the engine stall or back-fire.

The tangential force, F, acting on the brake drum is the difference between the dead weight and the spring balance reading. That is,

$$F = W - S \, \mathrm{N}$$

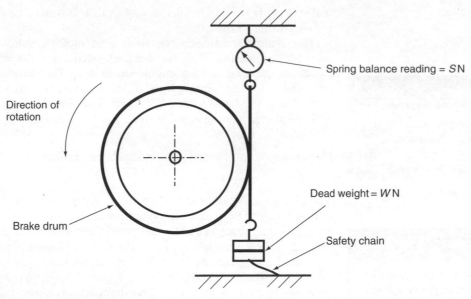

Figure 3.49 *Rope brake*

The braking torque, T, is the product of the tangential force and the effective radius, r, of the brake. That is,

$$T = Fr$$
$$T = (W - S)r$$

The hydraulic and electrical dynamometers shown in Figure 3.50 are widely used by test and research departments in industry. The hydraulic dynamometer, through which there is a steady flow of water, functions in the same way as a fluid coupling. Its casing is mounted on trunnions but is prevented from rotating by the dead weight and spring balance attached to the torque arm. The rotor, which is driven by the engine or motor under test, contains vanes that rotate alongside similar stationary vanes in the casing. The

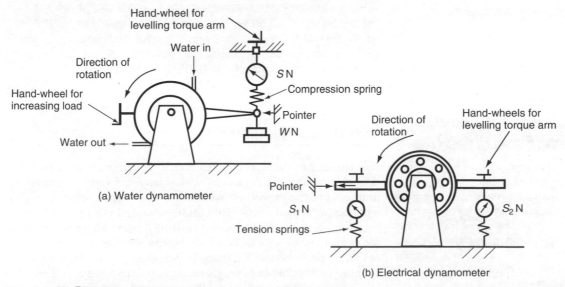

Figure 3.50 *Dynamometers*

water is made to swirl in such a way as to transmit force from the moving vanes to the casing.

The clearance between the fixed and moving vanes can be adjusted by means of a hand-wheel. Reducing the clearance increases the load on the engine or motor. The torque arm is adjusted to the horizontal position, indicated by the fixed pointer, using the upper hand-wheel. If W and S are the dead weight and spring balance reading and r, is the torque arm radius, the braking torque, T is given by

torque = force on torque arm × torque arm radius

$$T = (W + S)\,r$$

The electrical dynamometer is in fact a generator. It functions in a similar way to the hydraulic dynamometer except that the reaction torque is magnetic. The electrical power generated is usually dissipated as heat through banks of electrical resistors. The load is decreased and increased by switching additional resistors in and out of the circuit. The type shown in Figure 3.50(b) has two spring balances reading S_1 and S_2 N. The hand-wheels above the spring balances are adjusted so that for a particular load, the torque arm is horizontal. If r, is the torque arm radius to the spring balances, the braking torque is given by

torque = force on torque arm × torque arm radius

$$T = (S_2 - S_1)\,r$$

Routine maintenance for couplings, clutches, brakes and dynamometers chiefly involves checking coupling bolts for tightness, cleaning, lubricating and checking brake bands and friction linings and pads for wear. Fluid couplings should be checked for signs of leakage and the hydraulic fluid replaced at the recommended intervals.

See Questions 3.6 on p. 219.

Key point

The torque arm should always be adjusted to the horizontal position before any readings are taken.

Test your knowledge 3.9

1. What is a *dog clutch*?
2. Which of the clutches described are automatic in operation?
3. Why does a conical clutch require less spring force than the equivalent flat-plate clutch?
4. What is the purpose of a dynamometer?
5. To what position should the torque arm be adjusted when using fluid and electrical dynamometers?

Activity 3.8

Two single plate clutches as shown in Figure 3.43, have the same pressure springs and the same outer diameter. The inner diameters of the friction liner are, however, different. The clutch with the smaller contact surface is found to transmit more power before slipping. Explain why should this be so?

Key point

The maximum power that a flat or V-section belt can transmit before slipping depends on the initial tension setting. This should not be so high that maximum allowable stress in the belt material is exceeded before slipping occurs.

Belt and chain drives

Belt drives are used to connect shafts which may be some considerable distance apart. The belts may be flat, V-section or toothed and the pulleys on which they run can be selected to give a particular velocity ratio. Flat and V-section belts rely on friction between the belt and pulley to transmit power. There is a limit to the power that they can transmit before slipping occurs and this has been described in Chapter 1. Toothed or *synchronous* belts, running on toothed pulleys, give a more positive drive. If the belt is in good condition and correctly tensioned, slipping should not

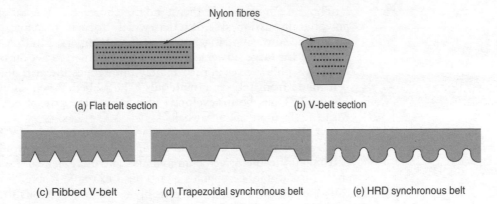

Nylon fibres

(a) Flat belt section   (b) V-belt section

(c) Ribbed V-belt   (d) Trapezoidal synchronous belt   (e) HRD synchronous belt

Figure 3.51 *Belt sections and profiles*

occur. The limiting factor on the power that can be transmitted is the allowable stress in the belt material. Figure 3.51 shows some typical belt sections and profiles.

Belts are generally made from synthetic neoprene rubber with nylon or sometimes metal, reinforcing fibres. Flat belts are used mainly for transmitting light loads. Because they are flexible, they are suitable for applications where there is some misalignment between shafts and as can be seen in Figure 3.52, they may be crossed to give opposite directions of rotation to the pulleys. They can also be twisted to connect shafts which are not in the same plane.

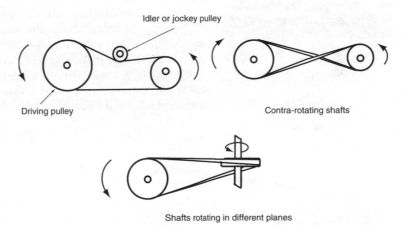

Idler or jockey pulley

Driving pulley

Contra-rotating shafts

Shafts rotating in different planes

Figure 3.52 *Belt drive configurations*

The pulleys should be slightly wider than the belt with a convex crown on which the belt will ride. High speeds are possible, but centrifugal effects as the belt passes round the pulleys can result in loss of grip at belt speeds in excess of around 18–$20\,\mathrm{m\,s^{-1}}$. In applications where the shaft centre distance is fixed, idler pulleys positioned on the slack side, are the usual means of tensioning flat belts.

The V-section belt is the standard choice for a great many power transmission systems. It is widely used in machine tools, automobiles, washing machines and tumble dryers. Multiple belts are used for large power transmissions. The wedge action of the belt in its pulley gives approximately three times the grip of a flat belt made from the same materials. The ribbed V-belt shown in Figure 3.51 has been developed to combine the grip of a V-belt with the flexibility of a flat belt.

Synchronous belts have a similar power capacity to V-belts. They give a positive drive and are used where slipping would be detrimental to the driven components. The drive belt to the camshaft on automobile engines is a typical example where slippage can alter the valve timing and possibly result in the pistons striking the valves. The HRD belt, which stands for *high torque drive*, has been developed from the trapezoidal synchronous belt and runs on toothed pulleys in the same way. It is said to be smoother and quieter and can transmit greater loads. The recommended speed limit for synchronous and HRD belts is around $60\,\text{m}\,\text{s}^{-1}$.

Routine maintenance on belt drives involves checking for wear, checking the tension setting and checking tensioning devices. The failure of synchronous belts can have serious consequences, particularly in automobile engines, and they should be changed at the recommended service intervals. Like toothed belts, chain drives provide a positive means of transmitting power between parallel shafts. The standard type is the bushed roller chain of the kind used on bicycles (Figure 3.53). Other types, intended to provide a better rolling action, have been designed but the roller chain is by far the most widely used. Belts can tolerate some misalignment between the shafts and pulleys, but a higher degree of alignment is required for chain drives. Misalignment can produce sideways bending which will strain the links and cause rapid wear. Chain drives can transmit larger loads than belts without the possibility of slipping. Multiple chains are used for very large load requirements. Until the introduction of toothed belts, roller chains were universally used for the camshaft drive on automobile engines.

Chain drives need to be adequately lubricated as shown in Figure 3.54, to slow down the rate of wear. The method of lubrication depends on the chain speed and the power which it is required

> **Key point**
>
> Good alignment of the shafts and sprockets is essential with chain drives to prevent excessive strain on the links.

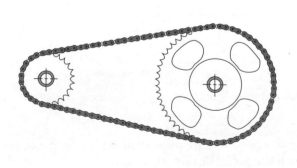

(a) Roller chain and sprockets

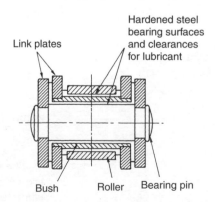

(b) Section through a roller chain

Figure 3.53 *Roller chain*
Source: *From Drives and Seals by M.J. Neale and Mechanical Engineer's Reference Book by E.H. Smith. Reprinted by permission of Elsevier Ltd*

(a) Manual lubrication for low power and speed

(b) Drip lubrication for low power and medium speed

(c) Oil bath for medium power and speed

(d) Continuous circulation from pump for high power and speed

Figure 3.54 *Chain drive lubrication*
Source: *From Drives and Seals by M.J. Neale. Reprinted by permission of Elsevier Ltd*

Test your knowledge 3.10

1. Why can a V-belt transmit more power than a flat belt of the same material and tension setting?
2. What is the advantage of a ribbed V-section belt?
3. What are synchronous belts used for?
4. Why is accurate alignment more important for chain drives than belt drives?
5. How should a chain drive that transmits high power at high speed be lubricated?

---

to transmit. Tensioning devices are often incorporated, consisting of an idler sprocket or a spring loaded friction pad acting on the slack side of the chain. Routine maintenance involves cleaning exposed chains, lubricating or replenishing lubricants, checking the operation of tensioning devices and checking the teeth on the sprockets for wear. The chain length should also be checked periodically. A 2% increase in length due to wear in the rollers indicates that replacement is due.

Activity 3.9

Two parallel shafts have a centre distance of 400 mm and are to be joined by a roller chain. The sprockets on the shafts have effective diameters of 200 mm and 100 mm and the chain has links of pitch 20 mm. What will be the length of chain required?

Note: A roller chain can only be lengthened or shortened by adding or taking off two links.

Gear trains

Gear trains form an essential part of a great many power transmission systems. By far the most common type used in engineering are gears with an *involute* tooth profile. An involute curve is generated by the end of a cord as it is unwound from around the surface of a cylinder as shown in Figure 3.55.

The reason for adopting the involute tooth form is that the teeth are strong, the velocity ratio between mating gears is constant and the teeth can be accurately machined with modern

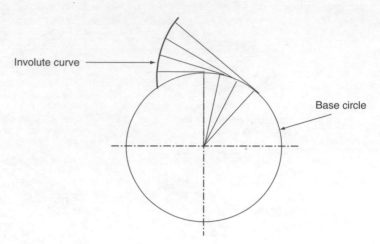

Figure 3.55 *Involute to a circle*

gear cutting machinery. A pure rolling action between meshing teeth would be the ideal situation but no such tooth form has yet been discovered. All gear teeth mesh with a rolling and sliding action, which is why good lubrication is essential.

An alternative tooth form based on the Russian Novokov gear has aroused some interest in recent years. It has been developed by Westland helicopters under the name of the *conformal* gear profile and is claimed to be superior for some applications. However, it has still to find widespread approval and we will concentrate only on the involute form.

A term commonly used with gears is the *pitch circle diameter*. This is the effective diameter of a gear. The pitch circle diameters of two mating gears are the diameters of the discs which would transmit the same velocity ratio by frictional contact alone. They are shown in Figure 3.56.

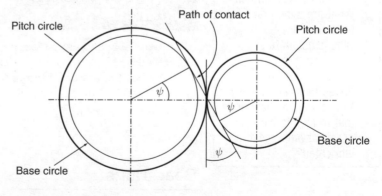

Figure 3.56 *Pitch and base circles*

The *base circles*, from which the involute teeth are generated, are smaller than the pitch circles. Their common tangent passes through the point where the pitch circles touch and makes an angle ψ (Psi) with the common centre line. This is known as the *pressure angle* of the gears. It is along part of this line that the point of contact between the gear teeth passes. In modern gears the pressure angle has been standardised at $\psi = 20°$. For a gear of pitch circle diameter D, the base circle diameter is given by

$$\text{base circle diameter} = D \cos \psi \qquad \text{(i)}$$

The *circumferencial* or *circular pitch* of a gear is the length of the arc of the pitch circle between the same point on successive teeth. If t, is the number of teeth, the circular pitch, p is given by the formula

$$p = \frac{\pi D}{t} \qquad \text{(ii)}$$

A parameter which is of great importance is the *module* of a gear. The module, m is obtained by dividing the pitch circle diameter by the number of teeth. It is this quantity which determines the size of the teeth. It is essential for mating gears to have the same module, otherwise they will not mesh.

$$m = \frac{D}{t} \qquad \text{(iii)}$$

The *addendum* of a gear tooth is its height above the pitch circle and this is equal to the module. The *dedendum* is the depth of the tooth below the pitch circle to the root circle. This is generally 1.25 times the module to give sufficient clearance between the root and the tip of mating teeth (Figure 3.57).

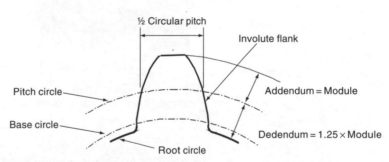

Figure 3.57 *Gear tooth geometry*

Some of the different types of gear are shown in Figure 3.58. External spur gears have straight teeth which are cut or moulded parallel to the gear axis. Amongst other materials, they are made from hardened steel, cast iron, phosphor–bronze and nylon. The meshing gears rotate in opposite directions. Spur gears give good results at moderate speeds but tend to be noisy at high speeds. Spur gears with internally cut teeth are known as annulus gears. They are used in epicyclic gear trains which we will shortly be describing. When meshing with an external spur gear as shown in Figure 3.58, both gears rotate in the same direction. Automatic gear boxes in cars contain epicyclic gear trains.

Helical gears are used to connect parallel shafts in the same way as spur gears, but have a superior load carrying capacity. With helical teeth, the point of contact moves across the tooth and it is possible for more than one pair of teeth to be simultaneously in

(a) External spur gears

(b) Internal spur gears

(c) Helical gears

(d) Straight bevel gears

(e) Spiral bevel gears

(f) Hypoid bevel gears

Figure 3.58 *Gear types*
Source: *From Drives and Seals by M.J. Neale. Reprinted by permission of Elsevier Ltd*

mesh. This makes for smoother and quieter running, particularly at high speeds. The one disadvantage with helical teeth is that they produce end thrust. To eliminate this, double helical or herring-bone teeth are sometimes used but these are expensive to manufacture.

Straight bevel gears are used to connect shafts whose axes intersect at some angle. The gear teeth radiate outwards from the point of intersection of the axes. Considerable end thrust is developed by bevel gears under load and this tends to push the teeth apart. Spiral bevel gears fulfil the same purpose but have a better load carrying capacity. As with helical gears, the teeth mesh gradually give smoother and quieter running. Hypoid bevel gears are closely related spiral bevels but are used where the axes of the two shafts do not intersect, as shown in Figure 3.58. Cars with rear wheel drive generally use hypoid bevel gears to transmit power from the propeller shaft to the road wheels. The velocity ratio of all of these gear combinations is given by the formula:

$$\text{velocity ratio} = \frac{\text{number of teeth on output gear}}{\text{number of teeth on input pinion}}$$

Key point

Helical gears are quieter and smoother running than spur gears but produce end thrust along the axis of the shafts.

Figure 3.59 *Worm gears*
Source: *From Drives and Seals by M.J. Neale. Reprinted by permission of Elsevier Ltd*

Worm gears consist of a worm shaft and worm wheel (Figure 3.59). Their axes do not intersect and are usually at right angles. The worm shaft may have one single helical tooth, rather like a screw thread, or multi-start teeth. The worm wheel teeth are specially machined to mesh with the worm. Very high velocity ratios are possible with single start worms since one complete revolution will only move the worm wheel through an angle subtended by the circumferential pitch, i.e. it will only move it forward by one tooth. Its velocity ratio is given by the formula:

$$\text{velocity ratio} = \frac{\textbf{number of teeth on worm wheel}}{\textbf{number of starts on worm}}$$

It is usual to make the two parts from different materials. The worm shaft is often made from hardened steel whilst the outer ring of the worm wheel is made from phosphor–bronze. Worm gears are sometimes used in preference to hypoid bevel gears in the driving axles of heavy slow moving vehicles.

In Chapter 1, we discussed the use of simple and compound gear trains in gear winches. We will now consider two different kinds of gearbox which might be used in a power transmission system. They might be used to reduce or increase the input speed depending on the numbers of teeth on the gears. The conventional gearbox shown in Figure 3.60 might contain spur or helical gears. The cluster ACE can be moved to the left and right along splines so that three velocity ratios can be obtained, i.e. A driving B, C

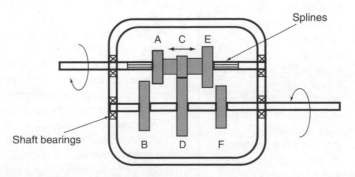

Figure 3.60 *Conventional gearbox*

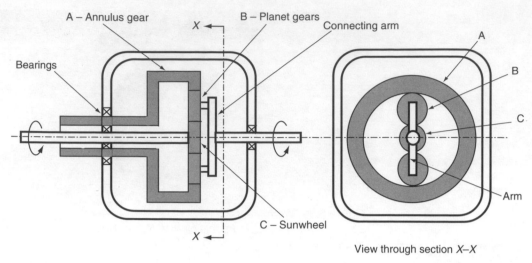

A – Annulus gear

B – Planet gears

Connecting arm

Bearings

A

B

C

Arm

C – Sunwheel

View through section X–X

Figure 3.61 *Epicyclic gearbox*

driving D and E driving F. Consider now the epicyclic gearbox shown in Figure 3.61.

This type of gear train and gearbox gets its name from the curve traced out by a point on a circle as it rolls inside a circle of larger diameter. It is called an *epicycloid*. The gear A in Figure 3.61 has internally cut teeth and is the annulus gear. The gear C is known as the sunwheel and the gears B, which can rotate about the sunwheel, are called planet gears. The two planet gears are connected to the arm which can rotate around the sunwheel with them. The gears are in mesh all of the time and three different velocity ratios can be obtained by holding different gears stationary as follows:

1. With the annulus fixed, input to the sunwheel C, and output from the shaft connected to the planet arm.
2. With the sunwheel C fixed, input to the annulus A, and output from the shaft connected to the planet arm.
3. With the planet arm fixed, input to the sunwheel C, and output from the annulus A.

The three velocity ratios will of course depend on the number of teeth on each gear. The values are not so easy to calculate as with conventional gear trains and the method is best left for study at a higher level. Epicyclic gear trains are to be found in automatic gearboxes, hub-reduction gears on heavy commercial vehicles, three-speed hubs on bicycles and speed reduction gearing for turbines. Epicyclic gear trains have the advantage of being compact and there is no radial force on the bearings other than that exerted by the weight of the gears. Their main disadvantage is cost. A greater degree of precision is required in producing the gears and the gear change mechanism, which is required to clamp the different gears in the stationary position, is more complex than that on a conventional gearbox.

The routine maintenance of geared systems generally involves the checking and replenishment of lubrication levels and the cleaning or replacement of oil filters. Excessive noise, vibration and overheating should be reported for expert attention.

See Questions 3.7 on p. 220.

Cut two circles of diameters 100 mm and 50 mm from a piece of stiff cardboard. Fix the larger circle to a flat surface using blue-tack and place the smaller one touching it. Mark the point of contact on both circles. Now, roll the smaller circle around the circumference of the larger stationary circle in the same way that a planet gear rotates around a stationary sunwheel. Count the number of turns that the planet circle rotates on its own axis whilst making one orbit of the sunwheel circle. Discuss, and give a reason for your findings.

Plant equipment and systems

A great many engineering and manufacturing processes make use of pneumatic and hydraulic devices for mechanical handling and positioning, material forming and process control. Steam is widely used for processes that require sustained high temperatures. It is also used for space heating and the bulk of our electricity is generated using steam. Refrigeration systems are used where sustained low temperatures are required for storage and processing. Refrigeration is also used in air-conditioning systems to provide comfortable working conditions. We will now describe how some of these systems operate, their major components, safety aspects and the routine maintenance duties required.

Hydraulic and pneumatic systems

Hydraulics and pneumatics are widely used in engineering processes and servo-control to convey energy from one location to another. The enclosed fluids most commonly used are specially formulated hydraulic oils and compressed air. Both kinds of system are used to transmit force and produce linear and rotational motion.

Air has a low density and is compressible whilst hydraulic oil has a much higher density and is almost incompressible. As a result, hydraulic systems are able to operate at much higher pressure and deliver the very large positive forces which are required in applications such as hydraulic presses and lifts. The major components of a hydraulic actuation system are shown in Figure 3.62.

The heart of the system is the motor-driven pump which draws filtered oil from the reservoir and delivers it via a pressure regulator to the points where it is required. The pump runs continuously and the excess oil which is not required for operations is diverted back to the reservoir by the pressure regulator. It should be noted that the system usually serves a relatively small work area in the locality of the pump and reservoir. It is not practicable to supply oil under pressure over large distances because of pressure drop and the need for a return pipe. A manual or automatic control valve supplies oil to the actuation cylinder and directs return oil to the reservoir.

Pneumatic systems have a softer action and are not able to deliver such large forces. They do however have certain advantages.

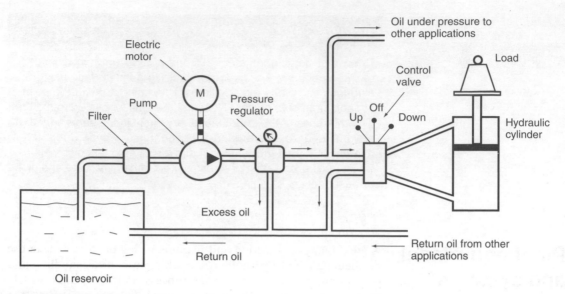

Figure 3.62 *Hydraulic system*

Compressed air is readily available in many industrial installations, being supplied as a service to the operational areas. Furthermore, it can be supplied over greater distances and is vented to the atmosphere after use. The major components of a typical pneumatic system are shown in Figure 3.63.

The compressor takes in filtered air and delivers it via an after-cooler to the compressed air receiving vessel. Compressors come in a variety of sizes. They may be single stage as shown, or two-stage. With two-stage compressors the air undergoes initial compression in the larger diameter low pressure cylinder. It then passes through an inter-cooler to the smaller diameter high pressure cylinder for further compression before delivery through the after-cooler. For some applications where the air must be perfectly dry, the system also contains a moisture separator.

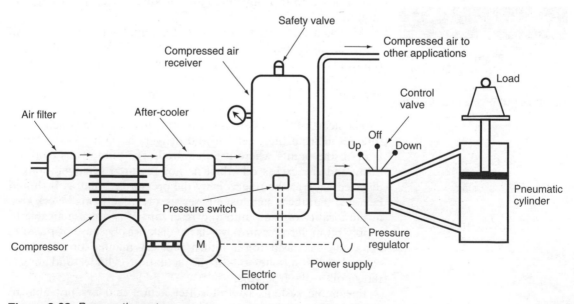

Figure 3.63 *Pneumatic system*

The receiving vessel is equipped with a pressure switch which cuts out the compressor motor when the supply pressure is reached and restarts it when the pressure falls. In addition, it is fitted with a safety valve which will open should the pressure switch fail to operate. A pressure regulator adjusts the supply pressure to that required at the point of application. As with hydraulic systems, a manual or automatic control valve directs compressed air to the actuation cylinder from which it is exhausted to the atmosphere.

Leaks from hydraulic systems can be both messy and dangerous whilst those from compressed air systems pose less of a problem for maintenance engineers. Routine maintenance involves checking the systems for leaks, the replenishment or replacement of hydraulic fluid and the cleaning or replacement of filters. In compressed air systems the lubricant in the compressor should be checked periodically and replenished. Also, the safety valve in the air receiver should be checked periodically for correct operation by over-riding the pressure switch until the blow-off pressure is reached.

Test your knowledge 3.12

1. What are the advantages and disadvantages of hydraulic systems?
2. Which system is able to exert the greater controlled force?
3. Why is it not practical to supply hydraulic pressure over long distances?
4. What is the purpose of the pressure switch in a compressed air receiver?
5. What kind routine maintenance activities should the operators of pneumatic and hydraulic equipment carry out?

Activity 3.11

The brakes on all makes of cars are hydraulically operated whilst those on heavy commercial vehicles are operated by compressed air. What is the reason for this?

Steam plant for power generation

By far the greater part of our electricity is produced by power stations in which the generators are powered by steam turbines. An approximate breakdown of the generating capacity in the United Kingdom is 37% coal fired, 31% gas fired, 25% nuclear, 2% oil fired and 5% from renewable sources such as hydro-electric and wind power. Whatever the heat source, steam generating plants have the same major components. A typical arrangement is shown in Figure 3.64.

Boilers and superheaters

In coal, gas and oil fired systems, the fuel and air enter the boiler where the hot gases from combustion heat the feed water to produce wet steam. There are two basic kinds of boiler, the *fire tube* type and the *water tube* type. In fire tube boilers, the hot gases from combustion pass through a system of tubes around which water is circulating. These are usually to be found in small installations where hot water or low pressure steam is required for industrial processes and space heating. In water tube boilers, the hot gases from combustion circulate around a system of tubes containing water. This is the type used in power stations for producing large quantities of high pressure steam.

The wet steam passes through a system of tubes in the *superheater* where additional heat energy is supplied from the combustion gases. This dries it out and raises its temperature to produce superheated steam. Every possible unit of heat energy is extracted

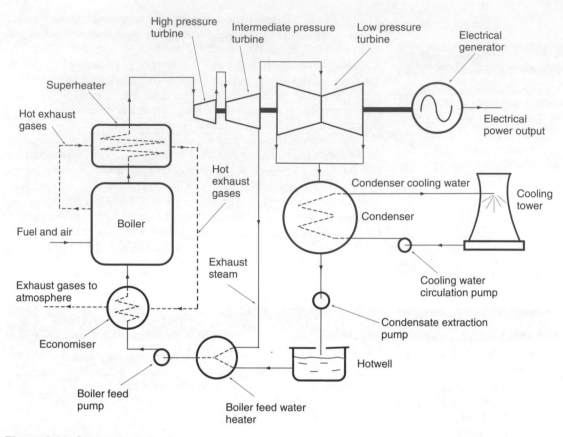

Figure 3.64 *Steam plant circuit*

from the exhaust gases and before escaping to the atmosphere, they are used to heat up the boiler feed water in the *economiser*. They are also used to pre-heat the incoming air, but this is not shown in Figure 3.64.

In nuclear installations, the heat source is enriched uranium. This is bombarded with neutrons in the reactor vessel causing some of the uranium atoms to split and release heat energy. The process is called *nuclear fission*. There are many different designs of reactor in operation throughout the world. In Britain, pressurised carbon dioxide is used to transfer heat energy from the reactor to the boilers, superheaters and economisers, whilst in American designs, pressurised water is preferred.

Turbines

The superheated steam passes to the *high pressure turbine* where it expands and does work on the rotor blades. It then passes to the *intermediate pressure turbine* where the blades have a larger diameter. Here it continues to expand and do work. You will note that some of the exhaust steam from the intermediate pressure turbine is fed to the boiler feed water heater where it is injected into the feed water from the hotwell. The remainder of the steam from the intermediate pressure turbine passes to the *low pressure turbine*. Here the blades are of a still larger diameter and arranged so that the steam enters centrally, as shown in Figure 3.64, and expands outwards through the two sets of low pressure blades. All three turbines are connected by a common shaft which drives the electrical generator.

Condensers and feed water heaters

The exhaust steam from the low pressure turbine passes to the condenser as low pressure wet steam. There are two basic types of condenser. In the *spray type*, cooling water is injected into the steam causing it to condense. In the *surface type*, the steam condenses on the surface of a system of pipes through which there is a flow of cooling water. The surface type is used in all large power stations. Sea water and river water are used for cooling wherever practical and a great many of our nuclear and gas fired power stations are sited on the coast. Coal fired power stations tend to be sited inland near the remaining coal fields, to reduce transportation costs. Here the cooling water for the condensers is generally re-circulated through cooling towers as shown in Figure 3.64.

When the steam condenses, it occupies a much smaller volume and the pressure in the condenser falls to below atmospheric pressure. This is beneficial because it creates as large a pressure drop as possible across the low pressure turbine, allowing steam to expand freely and do the maximum possible amount of work. The condensed steam must, however, be extracted from the condenser by the *condensate extraction pump*.

The condensate passes to a reservoir called the *hotwell* where make-up water is added for evaporation losses. The feed water from the hotwell is heated first in the *feed water heater* by exhaust steam, and then in the economiser by the exhaust gases from the boiler. The *boiler feed pump* delivers the feed water through the economiser to the boiler. The objective of the feed water heaters is to raise the temperature of the water to as close to its boiling point as possible before it enters the boiler.

There are a great many operational and maintenance duties on a steam plant. Routine maintenance includes the lubrication of moving parts, checking for leaks, checking that valves and steam traps are working properly and checking that pressure gauges and temperature measuring instruments are giving accurate readings. Steam traps are automatic devices used to drain off the water which sometimes collects in steam pipes, without allowing the steam itself to escape. The presence of high temperatures and pressures can make this work hazardous and permits to work are generally required in some parts of a plant. Maintenance should only be carried out by qualified personnel working to set procedures. Some parts of a steam plant such as the boilers and turbines, require specialised maintenance and this is usually sub-contracted to outside firms. *See Questions 3.8 on p. 221.*

Key point

The pressure inside a condenser is below atmospheric pressure and so a pump is required to extract the condensate.

Key point

A feed heater takes heat from exhaust steam whilst an economiser takes heat from the boiler exhaust gases.

Test your knowledge 3.13

1. What are the two basic types of boiler used in steam plant?
2. What is the purpose of a superheater?
3. What are the two basic types of condenser used in steam plant?
4. What is an economiser and how does it function?
5. What is the purpose of a cooling tower?

Activity 3.12

The steam generated in a boiler always contains water droplets. It is known as wet steam and when it is used for industrial processes, water tends to collect in the steam pipes. Steam traps are used to drain off water without allowing the steam to escape. The three main types are thermostatic traps, float traps and impulse traps. Select any particular one and with the aid of a diagram, describe how it functions.

Refrigeration systems

There are two basic types of refrigeration system. They are the vapour-compression system and the vapour-absorption system. Both types are used in commercial applications and domestic refrigerators and both work on the principle that when a liquid evaporates, it takes in latent heat from its surroundings. The liquids used in refrigerators and freezers are called *refrigerants*. They are made to evaporate at a temperature below 0 °C and in doing so, they take in latent heat and maintain the cold space at a sub-zero temperature.

A refrigerant must have a low freezing point so that it does not solidify or form slush in the low temperature part of the refrigeration cycle. Also it should have a high value for its latent heat of vaporisation to maximise the transfer of heat energy during the cycle. Until recently, refrigerants were only judged on how well they performed when taking in and giving out heat. The range of CFC (chlorofluorocarbon) refrigerants developed during the last century were thought to be quite suitable and safety was only considered in relation to the danger of poisonous leaks, fires and explosions. All this changed however when it was discovered that CFC's were damaging the earth's ozone layer and their use was banned under the Montreal Protocol of 1987.

One of the oldest, and still widely used refrigerants is ammonia, NH_3. Luckily, it is ozone friendly and is not affected by the protocol. Its main disadvantage is its toxicity but the installation of leak detection equipment makes it reasonably safe to use. Ammonia is used mostly in industrial refrigeration systems and in the small absorption refrigerators used in caravans and boats. Refrigerant R12, diclorodifluoromethane, CCl_2F_2, better known as Freon, was one of the most widely used CFC's in domestic and commercial refrigerators. This has now been banned and Refrigerant R134a, tetrafluoroethane, CH_2FcF_3, is used in its place. The basic circuit for a vapour-compression refrigerator is shown in Figure 3.65.

The main components of a vapour-compression system are the compressor, the throttling valve, the condenser and the evaporator. You will find the condenser grid at the rear of your domestic refrigerator. When you touch it, it feels quite warm. The evaporator grid is inside, or surrounds the cold chamber where you can store ice cubes. The high pressure liquid refrigerant passes through the throttling valve where it undergoes a rapid fall in pressure and temperature. Temperatures in the range −10 °C to −15 °C are quite common. This is well below that of the refrigerator contents and as it passes through the evaporator grid, it takes in latent heat and evaporates.

The refrigerant vapour passes from the evaporator to the compressor. During the compression process, its pressure increases and its temperature rises to above that of the surroundings. Temperatures of 30 °C to 35 °C are common. This is well above the normal temperature in countries with a temperate climate, such as the United Kingdom. As it passes through the condenser grid the vapour condenses, giving up its latent heat to the atmosphere. Large industrial installations sometimes have two-stage compressors with an intercooler between the stages. They might also be equipped with

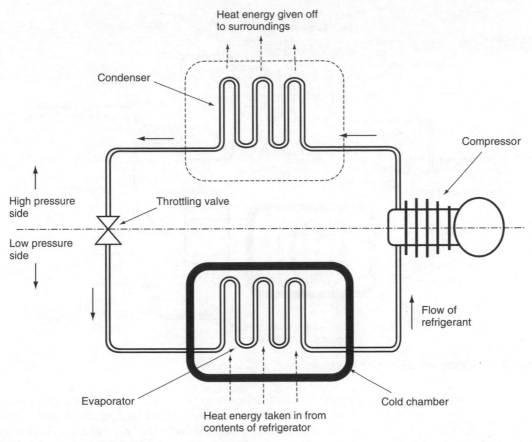

Heat energy given off
to surroundings

Condenser

Compressor

High pressure
side

Throttling valve

Low pressure
side

Flow of
refrigerant

Evaporator

Cold chamber

Heat energy taken in from
contents of refrigerator

Figure 3.65 *Vapour-compression refrigeration circuit*

an oil separator to filter out any lubricating oil which has become mixed with the refrigerant during the compression process.

Vapour-absorption refrigerators have been in use for some time in large industrial installations and in the small refrigerators used in caravans and boats. The refrigerants that they use are ozone friendly and they can be operated using waste heat from hot water or exhaust steam from manufacturing processes. The basic circuit for a vapour-absorption refrigerator is shown in Figure 3.66.

A vapour-absorption refrigerator functions with a condenser, throttling valve and evaporator in the same way as a vapour-compression refrigerator. The difference between the two systems is that the compressor is replaced by a vapour absorber, pump and vapour generator. These contain a secondary liquid which readily absorbs the refrigerant at low temperature and pressure and releases it at a later stage when the temperature and pressure have risen. The most common combination of liquids for industrial installations is water and lithium bromide. Water is the refrigerant and lithium bromide the absorber. The combination used for the small caravan refrigerators is ammonia and water, where ammonia is the refrigerant and water the absorber.

After taking in latent heat in the evaporator the refrigerant vapour enters the absorber at low temperature and pressure where it is readily absorbed by the secondary liquid. The refrigerant-absorber solution is then pumped up to the vapour generator which is at a higher pressure. Here it receives heat energy from an

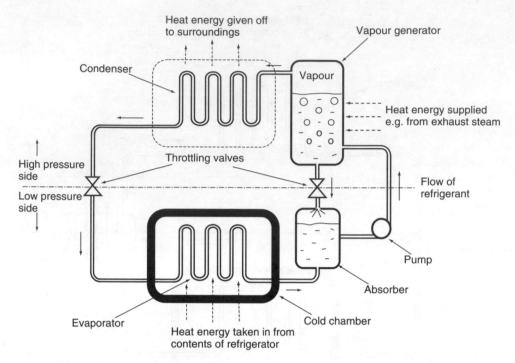

Figure 3.66 *Vapour-absorption refrigeration circuit*

external source, its temperature rises and the refrigerant vaporises out of solution. The absorber liquid passes back to the absorber vessel through the throttling valve between the two vessels, and the refrigerant vapour passes through the condenser. Here it gives up its latent heat of vaporisation to the atmosphere as it condenses before passing through the throttling valve to begin the cycle again. The pump is dispensed with in caravan and boat refrigerators. Here, gravity alone provides the circulation in the form of a convection current.

The routine maintenance of refrigeration systems involves inspection for leaks and signs of corrosion in the pipework, replenishment of the lubricant in the compressor and the periodic replacement of valves and seals.

Air-conditioning

Air-conditioning is the full mechanical control of the indoor environment to maintain comfortable and healthy conditions. Its objective is to provide clean, fresh air at a temperature and humidity level that is comfortable to the occupants. The general arrangement of an air-conditioning system are shown in Figure 3.67.

The major components of a modern air-conditioning system are an air filter, cooling coil, heater, humidifier, circulation fans, flow control dampers, room monitoring equipment and a processing unit. The processing unit controls the mixing of fresh air and return air. It is programmed to maintain the required standard of freshness, humidity, temperature and flow rate whilst keeping the energy input to a minimum. After mixing, the air is filtered and then heated or cooled to the required temperature. It is heated if the fresh air temperature is lower than that required, and the

Key point

A full air-conditioning system controls the temperature, humidity cleanliness and freshness of the air.

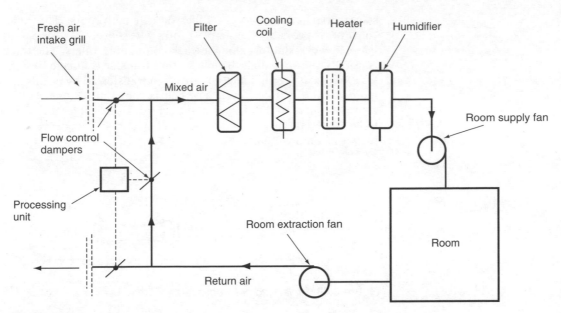

Figure 3.67 *Air-conditioning system*

system is then said to be working on a winter cycle. If the fresh air temperature is higher than that required, the mixture is cooled and the system is then said to be working on a summer cycle. In both cases the humidity, or moisture content of the air, may need to be adjusted to the required comfort level. If the air is too moist, the occupants of the room will tend to perspire, especially if they are doing physical work. If the air is too dry, the skin will feel dry and they will tend to feel thirsty.

The temperature of the air may be raised in the heater by the flow of hot water or steam or it may be heated by an electrical element. The temperature of the cooling coil is maintained by the flow of refrigerated water. Routine maintenance involves the cleaning or replacement of filters, lubrication of the moving parts and inspection for leaks.

Mechanical handling and positioning equipment

The range of mechanical handling, lifting and positioning equipment used in the engineering and manufacturing industries is very wide and in some cases it is highly specialised. The transfer of materials, components and assemblies through the production stages often takes place on roller or belt conveyors. The roller conveyer is probably the simplest form where products are passed between work stations along a track containing rollers. The track is usually set on a slight incline so that transfer is effected by the force of gravity. Bulk materials are often transferred by means of a motor driven belt conveyer. The belt is sometimes supported on concave rollers so that it sags in the centre. This enables it to carry loose materials in granular form. Roller and belt conveyers are shown in Figure 3.68.

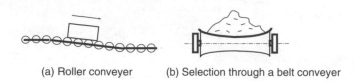

(a) Roller conveyer   (b) Selection through a belt conveyer

Figure 3.68 *Roller and belt conveyers*

The assembly of cars and other mass production items is generally carried out on slow moving track conveyers. At some stages of assembly the track may be set at ground level whilst at others it is raised so that parts can be fitted from below. Raised tracks, on which the components can be hung, are used for the transfer of components to the assembly stages and also for the transfer of components through washing, de-greasing and paint spraying booths.

A wide range of lifts, hoists and cranes are used for the vertical movement of products and materials. Passenger lifts and the lifts used to transport materials between the floors of industrial installations are generally raised and lowered by means of a cable system driven by an electric motor. A counter balance weight is sometimes included to reduce the power requirement. Fail-safe devices are always included in the design in case of a cable or power failure. Hydraulic lifts are used where heavy loads need to be raised through a comparatively small distance. The vehicle lifts in motor repair workshops are often of this type in which a pump supplies oil under pressure to the ram in a hydraulic cylinder. The principle of operation is similar to that of the hydraulic jack described in Chapter 5.

Heavy loads need to be lifted by a crane. The types used in engineering and manufacturing workshops are mainly jib and gantry cranes. Gantry cranes bridge the workshop or workshop bay. They run on overhead rails at either side of the service area as shown in Figure 3.69. The carriage which carries the driving motor and lifting drum runs on cross rails and is able to pick up and re-position loads from any part of the service area. In some installations the gantry rails extend outside the building so that the crane can be used for loading and unloading operations. Gantry cranes with a large lifting capacity are usually controlled by an operator seated on the crane. Lighter versions may be controlled from ground level by means of a handset.

Mobile jib cranes are often used in outdoor storage areas for loading, unloading and repositioning materials. Static jib cranes

Key point

A gantry crane can lift and re-position loads from any part of its workshop or workshop bay.

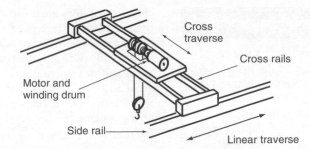

Figure 3.69 *General arrangement of a gantry crane*

such as that shown in Figure 3.70 are sometimes installed in indoor work areas for lifting heavy components at a work station. All types of crane and lifting device should be clearly marked with the safe working load, (SWL) e.g. 500 kg. This must never be exceeded. Routine maintenance on cranes is generally limited to cleaning and lubricating the moving parts and inspecting the wire ropes and electrical cables for wear or damage.

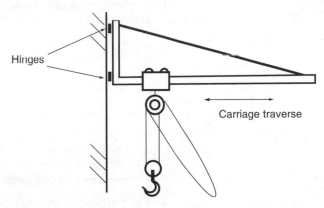

Figure 3.70 *Work station jib crane*

Fork-lift trucks find widespread use for transporting products and materials loaded on pallets. They should be clearly marked with the loading capacity when the forks are extended at maximum lift. Cranes and fork-lift trucks should be used only by operators who have received the appropriate training. Large storage warehouses with high rise storage racks often have automated positioning and retrieval lifts which work on the fork-lift principle. They run on tracks between the storage racks and can be directed to any position within the matrix from a remote control centre.

The automated manipulation, positioning and securing of components has seen many advances in recent years. CNC machining and automated assembly and packaging operations make widespread use of pneumatic and hydraulic systems and stepper motors. Sometimes these are incorporated in robot arms that enable components to be securely gripped, rotated, lifted and placed with great accuracy during manufacturing processes.

Stepper motors are a form of electric motor that rotates through a precise angle on receipt of a digital signal. They are widely used in CNC machining processes and robotic systems. Automated systems are generally well guarded with safety interlocks that ensure the system will not operate unless the guards are in place. Guard rails should also be provided around the working envelope of robot arms. Routine maintenance usually involves no more than cleaning and the replenishing of lubricants and coolants.
See Questions 3.9 on p. 221.

Activity 3.14

Automated handling and positioning equipment often contains proximity sensors to detect whether or not a component is present on the work station. Describe two such sensors which confirm the presence of a component by sending an electrical signal to a control unit so that the processing cycle can begin.

Questions 3.1

1. A measure of the resistance of a liquid to shearing forces is given by its

   (a) density
   (b) solubility
   (c) viscosity
   (d) resistivity.

2. The viscosity index of a lubricating oil is an indication of its

   (a) change of viscosity with pressure
   (b) resistance to shearing forces
   (c) change of viscosity with temperature
   (d) resistance to contamination.

3. The purpose of an anti-oxidant additive is to

   (a) reduce the formation of surface foam
   (b) give increased corrosion resistance
   (c) neutralise the effects of contaminating acids
   (d) reduce the effects of reaction with atmospheric oxygen.

4. Polytetrafluoroethylene is widely used as

   (a) a solid lubricant
   (b) an acid neutraliser
   (c) a detergent additive
   (d) an anti-foam agent.

5. Lithium is used as

   (a) an anti-oxidant additive to lubricating oil
   (b) a thickener in greases
   (c) an additive to increase the viscosity index of lubricating oil
   (d) a non-stick surface coating.

Answers: (c), (c), (d), (a), (b)

Questions 3.2

1. A lubrication system in which the lubricating oil is free to drain away is known as a

   (a) self-contained system
   (b) non-replenishable system
   (c) re-circulating system
   (d) total loss system.

2. COSHH regulations apply to the

   (a) specifications of lubricating oils
   (b) handling and storage of lubricating oils
   (c) periodic replenishment of lubricating oils
   (d) recommended uses of lubricating oils.

3. Motor vehicle gearboxes generally incorporate

   (a) splash lubrication
   (b) total loss lubrication
   (c) screw-down grease cup lubrication
   (d) wick-feed lubrication.

4. A re-circulating oil lubrication system is incorporated in

   (a) washing machines
   (b) diesel engines
   (c) electric motors
   (d) power tools.

5. Wick-feed lubrication utilises

   (a) a circulation pump
   (b) the force of gravity
   (c) capillary action
   (d) grease nipples.

Answers: (d), (b), (a), (b), (c)

Questions 3.3

1. The cylinder head seal that is widely used on automobile engines is a

   (a) composite metal gasket
   (b) mechanical seal
   (c) rotary lip seal
   (d) packed gland.

2. Rotary lip seals are used

   (a) between pipe flanges
   (b) around pistons
   (c) around rotating shafts
   (d) around reciprocating shafts.

3. The type of bearing best able to carry axial thrust is the

   (a) plane journal bearing
   (b) needle bearing

(c) caged ball bearing
(d) tapered roller bearing.

4. The crankshaft and big end bearings of a multi-cylinder internal combustion engine are generally

(a) phosphor–bronze
(b) steel-backed white metal
(c) hardened steel
(d) cast iron.

5. Needle roller bearings that run directly on a rotating shaft are mainly used for applications where there is

(a) high speed and high axial thrust
(b) low speed and light radial loading
(c) high speed and high radial loading
(d) low speed and high axial thrust.

Questions 3.4

1. The British Association screw thread is widely used for

(a) self-tapping screws
(b) electrical and IT applications
(c) accurately machined fitted bolts
(d) heavy engineering applications.

2. The screwed fastening made from a short length of bar and threaded at both ends is called a

(a) setscrew
(b) black bolt
(c) stud
(d) machine screw.

3. Which of the following is classed as a positive locking device?

(a) a tab washer
(b) a lock nut
(c) a spring washer
(d) a friction nut.

4. For applications where rivet heads must be flush with the joint surface, it is usual to use

(a) conoidal headed rivets
(b) round headed rivets
(c) pan headed rivets
(d) countersunk headed rivets.

5. For applications where access is only possible from one side of a riveted joint, it is usual to use

(a) flat headed rivets
(b) pan headed rivets

(c) pop rivets
(d) round headed rivets.

Answers: (b), (c), (a), (d), (c)

Questions 3.5

1. The angle turned through by a cam whilst its follower is stationary is called the

   (a) pressure angle
   (b) angle of lap
   (c) dwell angle
   (d) angle of friction.

2. An inversion of a plane mechanism can be obtained by

   (a) changing the input speed
   (b) changing the lengths of the links
   (c) reversing the input motion
   (d) changing the link that is fixed.

3. Two links in a mechanism that are hinged or pin-jointed together constitute a

   (a) turning pair
   (b) sliding pair
   (c) screwed pair
   (d) fixed pair.

4. A practical application of a four-bar linkage is the

   (a) Whitworth quick return motion
   (b) oscillating cylinder mechanism
   (c) slotted link quick return motion
   (d) Watt's parallel motion.

5. A flat follower is sometimes offset to the side of the cam in order to

   (a) rotate the follower and equalise wear
   (b) increase the dwell angle
   (c) reduce the travel of the follower
   (d) increase the speed of the follower.

Answers: (c), (d), (a), (d), (a)

Questions 3.6

1. Alignment for a connecting shaft between machines that are some distance apart is best carried out using

   (a) a straight edge
   (b) a single dial test indicator
   (c) optical equipment
   (d) two dial test indicators.

2. That the driving contacts of a universal coupling must be in a plane that bisects the angle between the coupled shafts is a condition for

(a) maximum power transmission
(b) a constant velocity ratio
(c) accurate alignment
(d) a secure joint.

3. Automatic engagement at a pre-determined speed is a characteristic of a

(a) dog clutch
(b) flat plate clutch
(c) conical clutch
(d) centrifugal clutch.

4. A dynamometer is a form of brake that is used

(a) to measure output torque
(b) on motor vehicles
(c) to measure angular velocity
(d) on power transmission shafts.

5. Fluid clutches are widely used in motor vehicles in conjunction with

(a) manually operated gearboxes
(b) power steering systems
(c) automatic gearboxes
(d) anti-lock braking systems.

Answers: (c), (b), (d), (a), (c)

Questions 3.7

1. The effective diameter of a gear is called the

(a) base circle diameter
(b) root circle diameter
(c) pitch circle diameter
(d) outer diameter.

2. The pitch circle diameter of a gear divided by the number of teeth is its

(a) module
(b) circumferential pitch
(c) pressure angle
(d) dedendum.

3. The maximum power that a belt drive can transmit before slipping depends primarily on

(a) the pulley diameters
(b) the centre distance between the pulleys
(c) the initial tension setting
(d) the mass of the belt.

4. Hypoid bevel gears are used where the axes of the connected shafts

   (a) lie in the same plane
   (b) do not intersect
   (c) intersect at right angles
   (d) are parallel.

5. A synchronous belt is used in applications where

   (a) a positive drive is required
   (b) the pulleys must rotate at the same speed
   (c) the belt is required to slip at a pre-determined speed
   (d) the shaft axes intersect at right angles.

Answers: (c), (a), (c), (b), (a)

Questions 3.8

1. An advantage of a hydraulic actuation system is that it

   (a) can operate over long distances
   (b) vents to the atmosphere
   (c) can transmit large positive forces
   (d) does not require a return pipe.

2. The purpose of a pressure switch in a pneumatic system is to

   (a) regulate the supply pressure
   (b) control the pneumatic actuator
   (c) stop and re-start the compressor
   (d) display the air pressure.

3. The purpose of a superheater in a steam plant is to

   (a) heat up the boiler feed water
   (b) dry out the steam and raise its temperature
   (c) pre-heat the air supply to the boiler
   (d) control the fuel supply to the boiler.

4. Cooling towers are installations that are used to

   (a) cool the exhaust gases from the boiler
   (b) condense the low pressure steam
   (c) release exhaust steam to the atmosphere
   (d) cool the water circulating through the condenser.

5. An economiser heats up the boiler feed water using

   (a) exhaust gases from the boiler and superheater
   (b) low pressure exhaust steam
   (c) electrical energy that is generated
   (d) low cost fuel.

Answers: (c), (c), (b), (d), (a)

Questions 3.9

1. The purpose of the compressor in a vapour-compression refrigerator is to

   (a) take in latent heat
   (b) release heat energy to the atmosphere

(c) raise the temperature and pressure of the refrigerant
(d) allow the refrigerant to condense.

2. In a refrigeration cycle, the high pressure refrigerant expands to a lower pressure and temperature on passing through the

(a) throttling valve
(b) evaporator matrix
(c) compressor
(d) condenser matrix.

3. The mixing of return air and fresh air in a modern air-conditioning system is controlled by the

(a) humidifier
(b) air filter
(c) processing unit
(d) cooling coil.

4. Loads can be repositioned in any part of a workshop bay if it is served by a

(a) roller conveyer
(b) work station jib crane
(c) belt conveyer
(d) gantry crane.

5. Stepper motors are able to rotate through a precise angle on receipt of a

(a) analogue signal
(b) pneumatic signal
(c) digital signal
(d) hydraulic signal.

Answers: (c), (a), (c), (d), (c)

Chapter 4 Engineering materials

Summary

Design and manufacturing engineers need to be aware of the range of materials available for use in engineered products. They need to know about the properties of the different materials, their cost, their availability and how they can be processed. This enables them to select materials which are fit for their purpose and which can be processed at a reasonable cost.

Maintenance engineers also need to have a knowledge of materials. They need to know how they are affected by service conditions and how they might be protected to prolong their service life. Replacement components must be made of the same material, or from a suitable alternative which has similar properties. When a component fails in service, engineers must be able to identify the cause. A design change or a change to a more suitable material can then be implemented to improve the quality of the product.

The aim of this unit is to provide you with a basic knowledge of the most common engineering materials. This will include their structure, properties, the effects of processing, their relative costs and forms of supply. It will also direct you to sources of information which will enable you to find out more about the range of available materials.

Structure and classification of materials

There are 92 naturally occurring chemical elements. A chemical element cannot be split into other substances. It is made up of atoms that are all the same. Sometimes, groups of two or more atoms combine together to form molecules. These too are identical in a pure element. Atoms are made up of particles. The heaviest of these are protons and neutrons which are roughly the same size and form the central core, or nucleus, of an atom.

A cloud of orbiting electrons surrounds the nucleus. These have only about two-thousandths of the mass of protons and neutrons and it is at the nucleus that the mass of an atom is concentrated. Protons carry a small positive electrical charge. The orbiting electrons, which are equal in number to the protons, carry a negative electrical charge. The combined effect is to make an atom electrically neutral.

The atoms of the different chemical elements have different numbers of protons, neutrons and electrons. Hydrogen has the smallest and lightest atoms. They have just one proton in the nucleus and one orbiting electron. Uranium has the largest and heaviest naturally occurring atoms with 92 protons and electrons and an even larger number of neutrons. The electrons in the different atoms are to be found orbiting in distinct shells. There are seven naturally occurring shells which are given the letters K, L, M, N, O, P and Q. The maximum number of electrons which each shell can hold is given by the general formula $N = 2n^2$, where n is the number of the shell. The first four shells can thus hold 2, 8, 18, and 32 electrons respectively.

The lighter elements only have electrons in the inner shells. The heavier the element, the more shells are occupied but the outer shells are never fully occupied in naturally occurring elements. Hydrogen atoms, with one proton in the nucleus, have just one electron orbiting in the K shell. Iron atoms have 26 protons and 20 neutrons in the nucleus and 26 orbiting electrons. They are distributed 2, 8, 14 and 2 in the K, L, M and N shells. Atoms with eight electrons in their outer shell are very stable. Although the L, M and N shells can hold more than eight electrons, atoms with their outer electrons in these shells will readily shed or share electrons in order to empty the shell or to achieve the stable number of eight.

The forces that bind the atoms of a material together arise from the electrical charges carried by their protons and electrons. Atoms of an element with a small number of electrons in the outer shell will readily donate them to the atoms of another element whose outer shell needs them to give a stable number (see Figure 4.1).

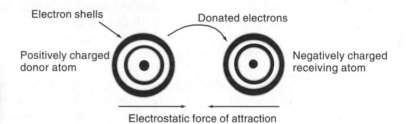

Figure 4.1 *Ionic or electrovalent bonding*

The donor atom then becomes positively charged and the receiving atom becomes negatively charged. The result is that the atoms become joined together by electrostatic forces to form a compound molecule. The process is called ionic or electrovalent bonding and is one of the ways in which chemical compounds are formed. Sodium and chlorine bond together in this way. Sodium has only one electron in its outer shell whilst chlorine has seven. Their atoms combine to form molecules of sodium-chloride, NaCl, which is better known as common salt.

Another way in which atoms can combine together to form molecules is by sharing their outer electrons to give a stable number in a common outer shell (see Figure 4.2). These are called valence electrons. Sharing them produces strong bonds which hold the atoms together. The process is called covalent bonding. The atoms of hydrogen combine in this way to form H_2 molecules. Similarly, atoms of nitrogen form covalent bonds with three atoms of hydrogen to form molecules of ammonia, NH_3.

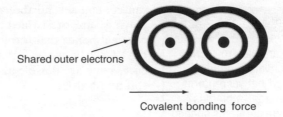

Shared outer electrons

Covalent bonding force

Figure 4.2 *Covalent bonding*

Metals have a small number of electrons in the outer shell which easily become detached and are shared between all the atoms in the material. They are known as *free electrons* which circulate between and around the atoms in a random fashion. The effect is to bond the atoms together with a form of covalent bond. It is sometimes referred to as *metallic bonding* which is quite strong (Figure 4.3).

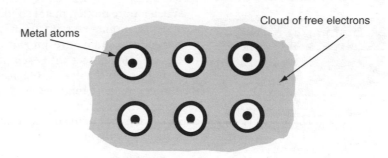

Cloud of free electrons

Metal atoms

Figure 4.3 *Metallic bonding*

All of the materials used to make engineered products are thus made up of atoms and molecules which are held together by strong bonds. A closer examination reveals that in different materials, the atoms and molecules are arranged in different ways. The structure of a solid material may be *amorphous*, *crystalline* or *polymer*. In amorphous materials the atoms or molecules are not arranged in any particular pattern. They are disordered rather like those in a liquid. Glass is one of the few truly amorphous materials used in engineering.

All metals, and a great many ceramics, are crystalline. That is to say that when they solidify from being molten, crystals start to form in different parts of the melt. These multiply and grow until the material is completely solidified. Within the crystals or *grains* the atoms pack themselves into a regular geometric pattern. The size of the grains depends on the rate of cooling. Slow cooling usually gives large grains. Grain size affects the properties of a material. Large grains tend to increase the brittleness and reduce the strength of a metal.

Sometimes you can see the grains in a metal with the naked eye, such as the grains of zinc on galvanised steel. In other cases you will need to polish the surface of the metal and then etch it with chemicals to show up the grain boundaries. The grains can then be seen using a magnifying glass or a microscope.

Polymer materials include plastics and rubbers. Here the atoms are arranged in long molecular chains, made up mainly of carbon and hydrogen atoms, which are known as *polymers*. Each polymer

Key point

An alloy is a mixture of metals or a mixture of a metal and other substances which results in a material which displays metallic properties.

can contain several thousand atoms. In plastics and rubbers the polymers are intertwined, rather like spaghetti. The strength of the covalent bonds that form between the polymers affect the properties of plastic materials and rubbers. As you know from experience, some are flexible and easy to stretch whilst others are quite stiff and brittle.

Metals

The metals used in engineering may be sub-divided into two main groups: (1) Ferrous metals, which contain iron as a major constituent, and (2) non-ferrous metals which do not contain iron, or in which iron is only present in small amounts. Some metals such as copper and lead are used in an almost pure form. Others are mixed to form *alloys*. An alloy can be a mixture of metals. It can also be a mixture of a metal and other substances so long as the resulting material displays metallic properties.

When a pure molten metal solidifies, latent heat is given off and crystals, or grains, start to form at different points in the fluid. These embryo grains are known as *dendrites*, which grow until they come into contact with each other. Solidification is then complete and within the grains, the atoms are found to have arranged themselves in a regular geometric pattern. The orientation of the grains is different and they come into contact at irregular angles. This is why some of them seem darker than others when you view them (see Figure 4.4).

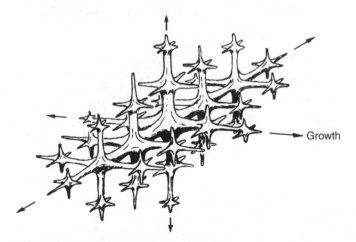

Figure 4.4 *Growth of a dendrite*
Source: *From Properties of Engineering Materials by R.A. Higgins. Reprinted by permission of Elsevier Ltd*

The pattern in which the atoms arrange themselves is known as the *crystal lattice structure*. It is held together by the metallic bond produced by the cloud of free electrons. The atoms take up positions of minimum potential energy and in most engineering metals they arrange themselves in one of the three possible lattice structures.

The most open packed is the *body-centred cubic* (BCC) formation (see Figure 4.5). The pattern is continuous throughout the

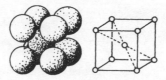

Figure 4.5 *Body-centred cubic structure*
Source: *From Properties of Engineering Materials by R.A. Higgins. Reprinted by permission of Elsevier Ltd*

crystal but it is convenient to consider a unit cell where the atoms are arranged at the eight corners of a cube, surrounding another atom at the centre of the cube.

This structure is taken up by iron at normal temperatures and also by chromium, tungsten, niobium, molybdenum and vanadium. The open packed planes of atoms which make up this structure do not easily move over each other when external forces are applied. This is the main reason why the metals listed above are hard to deform in comparison to some others.

Some of the softest and more easily deformed metals take up a *face-centred cubic* (FCC) formation (see Figure 4.6). This is made

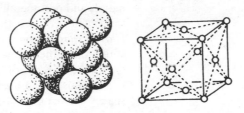

Figure 4.6 *Face-centred cubic structure*
Source: *From Properties of Engineering Materials by R.A. Higgins. Reprinted by permission of Elsevier Ltd*

up of planes of atoms that are more closely packed. The unit cell consists of a cube with atoms at the eight corners and atoms at the centre of each face.

This structure is taken up by aluminium, copper, silver, gold, platinum, lead and also iron when heated to a temperature of about 800 °C. These are some of the softest and easily deformed metals. The closely packed planes of atoms move more readily over each other when external forces are applied. They do not actually slide over each other. It is thought to be more of a rippling action which occurs due to imperfections in the lattice, but the effect is the same.

Iron has a BCC structure at normal temperatures but at temperatures approaching 800 °C, the atoms rearrange themselves in the solid material to take up a FCC structure. The iron then becomes much softer and easier to shape by forging and hot pressing. Materials such as iron, which can exist with different crystal lattice structures, are said to be *allotropic*.

The third type of crystal lattice formation that is to be found in engineering metals is the *close-packed hexagonal* (CPH) structure. This is shown in Figure 4.7. It is made up of layers of atoms which are just as tightly packed as those in a FCC structure. A good example of this packing is the way in which the red snooker balls

Key point

Metals with a BCC crystal lattice structure tend to be hard and difficult to deform whereas metals with a FCC crystal lattice structure tend to be softer and easier to deform.

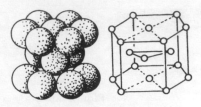

Figure 4.7 *Close-packed hexagonal structure*
Source: *From Properties of Engineering Materials by R.A. Higgins. Reprinted by permission of Elsevier Ltd*

are packed in the triangular frame at the start of a game. The unit cell of the CPH consists of a hexagonal prism running through three layers of atoms. It has atoms at the twelve corners of the prism and an atom at the centre of each hexagonal face. Between these are sandwiched another three atoms from the middle layer.

The difference between CPH and FCC structures is that every third layer is displaced by half an atomic distance. This has the effect of making the metal a little more difficult to deform. Zinc, magnesium, beryllium and cadmium are some of the metals which take on this formation.

Alloys are formed when different metals, and sometimes also metals and non-metallic substances, are mixed together in the molten state. After cooling, the resulting *solid solution* may be an *interstitial alloy*, a *substitutional alloy* or an *intermetallic compound*. In an interstitial alloy, the atoms of one of the constituents are relatively small compared to those of the other. As the molten mixture solidifies and the crystals start to form, the smaller atoms are able to occupy the spaces between the larger atoms (see Figure 4.8). The material with the larger atoms is called the *solvent* and that with the smaller atoms is called the *solute*. This is what occurs with some of the carbon atoms when carbon is mixed with iron to form steel. The effect is generally to enhance the strength and toughness of the parent metal.

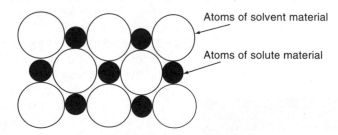

Atoms of solvent material

Atoms of solute material

Figure 4.8 *Interstitial alloy*

In a substitutional alloy, two or more materials whose atoms are of roughly the same size are mixed together in the molten state (see Figure 4.9). During the cooling process, atoms of the solute material replace atoms of the solvent in the crystal lattice structure. Brasses, bronzes and cupro–nickel are substitutional alloys.

The difference in size of the atoms tends to distort the crystal lattice structure. This makes it more difficult for the planes of atoms to slip over each other when external forces are applied. As a result the alloy is generally stronger and tougher than its main constituent, the solvent metal.

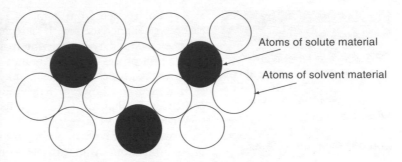

Atoms of solute material

Atoms of solvent material

Figure 4.9 *Substitutional alloy*

Sometimes a chemical reaction takes place between the solvent and solute materials. This results in the formation of an intermetallic compound. An example of this is the reaction which takes place between some of the carbon and iron atoms in steel. The iron can only take in a relatively small number of carbon atoms in the interstitial spaces. The remainder combine with some of the iron atoms to form iron carbide, Fe_3C.

Intermetallic compounds are usually quite different in their appearance and properties to the two parent materials. Iron carbide is hard and brittle but its presence in some of the grains of steel contributes to its hardness and toughness. Another example of an intermetallic compound is the combination of antimony and tin in some bearing metals to form tin-antimonide, $SbSn$. Its presence is found to improve the load-carrying properties of the alloy when it is used as a bearing material.

Ferrous metals

As has been stated, these are metals in which iron is the main constituent. Pure iron is a relatively soft metal. It is not easy to machine it to a good surface finish because it tends to tear when being cut. When molten, it does not have good fluidity and as a result, it is difficult to cast. When small amounts of carbon are added to it however, this greatly improves its strength and machinability. The resulting alloys are called plain carbon steels. The different grades, over which the carbon content ranges from 0.1% to 1.4%, are shown in Table 4.1.

Table 4.1 *Plain carbon steels*

Classification	Carbon (%)	Tensile strength	Applications and uses
Dead mild steel	0.1–0.15	400 MPa	Steel wire, nails, rivets, tube, rolled sheet for the production of pressings
Mild steel	0.15–0.3	500 MPa	Bar for machining, plate for pressure vessels, nuts, bolts, washers, girders and stanchions for building and construction purposes
Medium carbon steel	0.3–0.8	750 MPa	Crankshafts, axles, chains, gears, cold chisels, punches, hammer heads
High carbon steel	0.8–1.4	900 MPa	Springs, screwcutting taps and dies, woodworking tools, craft knives

As can be seen from Table 4.1, the tensile strength of plain carbon steel increases with carbon content. Small amounts of other elements may be present from the iron smelting and steel production process and others may be added in small amounts to improve the properties of the metal. The main constituents of plain carbon steel are however iron and carbon and it is the carbon content which has most effect on its properties.

When mild steel is viewed under a microscope it can be seen to have two distinct types of grain. One type appears white and is made up of iron atoms with carbon atoms absorbed in the interstitial spaces. These are known as *ferrite* grains. The other type has a layered mother-of-pearl appearance and is made up of alternate layers of ferrite and the intermetallic compound iron carbide, Fe_3C which is known as *cementite*. These are known as *pearlite* grains. It is in the pearlite grains that most of the carbon content resides. The higher the carbon content of steel, the more pearlite grains are present, and as a result the steel is harder and tougher (see Figure 4.10).

Grains of ferrite

Grains of pearlite

Figure 4.10 *Microstructure of mild steel*

Dead mild and mild steel are general purpose materials which are used for a host of components in the engineering and the building industries. The surface hardness of mild steel can be increased by heat treatment processes known as case carburising and case hardening. These give improved wear resistance and will be discussed later.

The increased carbon content of medium carbon steel gives it increased toughness and impact resistance, making it suitable for power transmission components and tools such as hammers, cold chisels and rivet snaps. The hardness and toughness of both medium and high carbon steel can be enhanced by heat treatment processes known as hardening and tempering and these will also be discussed later.

The carbon content in high carbon steel, together with appropriate heat treatment, results in a material which is very hard and wear resistant. This makes it most suitable for cutting tools such as files, tin-snips, woodwork tools and craft knives where the cutting process takes place at normal temperatures and at a relatively slow speed. Alternative heat treatment can produce a material which is less hard but very tough and resilient. It is widely used for coil and leaf springs.

Steels in which elements other than carbon are present in relatively large amounts are referred to as *alloy steels*. The elements added are selected to give the steel special properties. Three common alloy steels that are produced for their tensile strength, corrosion resistance and for high speed cutting operations are listed in Table 4.2.

Table 4.2 *Alloy steels*

Classification	Carbon and other elements (%)	Tensile strength	Applications and uses
Nickel–chrome–molybdenum steel	0.4% Carbon 1.5% Nickel 1.2% Chromium 0.55% Manganese 0.3% Molybdenum	1080 MPa to 2010 MPa depending on degree of heat treatment	Highly stressed machine components where a high resistance to fatigue and shock loading is required
Stainless steel	0.04% Carbon 14.0% Chromium 0.45% Manganese	600 MPa	Food processing, kitchen and surgical equipment, pressings, decorative trim
High-speed steel	0.7% Carbon 18.0% Tungsten 4.0% Chromium 1.0% Vanadium	Not relevant	Twist drills, milling cutters, turning tools, hacksaw blades

Key point

Grey cast iron has good fluidity when molten, it is self lubricating and easy to machine, it is strong in compression but weak in tension.

Test your knowledge 4.3

1. What is a plain carbon steel?
2. What are high carbon steels used for?
3. How does the carbon content affect the properties of plain carbon steel?
4. Give an example of an alloy steel.
5. What are the properties of grey cast iron?
6. How is the carbon content distributed in grey cast iron?

When carbon is added to molten iron in quantities greater than 1.7%, it cannot all be absorbed interstitially or as iron carbide. After the mixture has cooled down, the excess is seen to be present as flakes of graphite between the grains. The material is then known as *grey cast iron*. A carbon content of 3.2% to 3.5% is usual which gives the molten metal good fluidity. This enables it to be cast into intricate shapes. It is easy to machine without the use of a cutting fluid since the graphite flakes have a self-lubricating effect. Grey cast iron is strong in compression but tends to be weak and brittle in tension. The graphite flakes also act as vibration absorbers which make it an ideal material for machine beds.

The properties of cast iron can be improved by heat treatment processes to produce what are known as *malleable cast irons*. The effect is to modify the graphite content so that the material is less brittle and better able to withstand shock and impact. The properties can also be improved by adding certain alloying elements, particularly small amounts of magnesium. This has the effect of producing small spheres of graphite instead of graphite flakes. The material is then known as *spheroidal graphite iron*. Spheroidal graphite or SG iron, has the fluidity of the grey cast iron but is much tougher and stronger when solidified.

Cast irons are used for a variety of engineering components. They are used for internal combustion engine cylinder blocks and cylinder heads, brake drums, pump and turbine casings, water pipes, manhole covers and gratings. In recent years, the popularity of cast iron as a decorative material has returned. In particular, there has been an increase in its use for lamp standards, garden furniture and decorative structural castings in buildings and public places.

Activity 4.1

Investigate modern methods of iron and steel production and prepare a report on your findings. Make particular reference to: (a) the different kinds of iron ore and their suitability for processing, (b) the way in which iron ore is processed to extract the iron, (c) modern methods of steel production.

Table 4.3 *Non-ferrous metals*

Metal	Tensile strength	Density	Melting point	Applications and uses
Copper	220 MPa	8900 kg m^{-3}	1083 °C	Water pipes, heat exchangers, vehicle radiators, electrical wire and cable, major constituent of brasses, bronzes and cupro–nickel alloys
Zinc	110 MPa	7100 kg m^{-3}	420 °C	Major constituent of brasses and some die-casting alloys. Used for galvanising, i.e. as a protective coating for steel products
Tin	15 MPa	7300 kg m^{-3}	232 °C	Major constituent of tin–bronze alloys, soft solders and white bearing metals. Used as a protective coating for mild steel
Lead	Very low	11300 kg m^{-3}	327 °C	Major constituent of soft solders and white bearing metals. Used for protective cladding on buildings and protective lining in tanks
Aluminium	60 MPa	2700 kg m^{-3}	660 °C	Major constituent of aluminium alloys and aluminium bronzes. Used for electrical cables and conductors and domestic utensils
Titanium	216 MPa	4500 kg m^{-3}	1725 °C	Major constituent of titanium alloys used in aircraft production and for gas turbine compressors blades

Non-ferrous metals

Non-ferrous metals do not contain iron except in relatively small amounts. In particular, it is included in some cupro–nickel alloys to improve their properties. The major non-ferrous base metals used in engineering are copper, zinc, tin, lead, aluminium and titanium. Their properties and uses are listed in Table 4.3.

Copper is an excellent conductor of heat and electricity and like most non-ferrous metals, it is corrosion resistant. Although it does not have a high tensile strength, it is malleable and ductile and easily drawn out into wire and tube. Zinc is used to give a protective coating to mild steel products by a process known as galvanising. Other than this, it is not widely used alone because it is rather brittle and has a relatively low tensile strength.

Tin has a very low tensile strength and melting point. It is very soft and malleable and highly corrosion resistant. It is used as a protection for mild steel sheet which is then known as tinplate. Its other main use is as a constituent of soft solder.

Lead is very soft and malleable. It has a low tensile strength and being relatively heavy, it can creep and fracture under its own weight. Properly supported, it is widely used for weather protection on buildings and also for lining tanks containing chemicals that are corrosive to iron and steel. Until fairly recently, lead was used for domestic water pipes and these are still to be found in some older properties. This use has now been discontinued because of the danger from lead poisoning. Most of the lead piping has now been replaced by copper and plastic materials.

Like copper, aluminium is an excellent conductor of heat and electricity. It is relatively light in weight and has good corrosion resistance. In its pure form it has a low tensile strength but is very malleable and ductile. It is widely used for domestic utensils and as the core of electrical power transmission cables. Titanium is also light in weight, ductile, highly corrosion resistant and has a high melting point. It is however expensive and this has limited its use as an engineering material. It is most widely used in the aerospace industries.

The main non-ferrous alloys used for engineering components are brasses, tin–bronzes, cupro–nickels, aluminium–bronzes, aluminium alloys and white bearing metals. There are of course many other special purpose alloys but these are the ones that you will encounter most frequently.

Brasses are alloys in which the main constituents are copper and zinc. As a general rule, brasses with the higher copper content are the more ductile and suitable for cold forming operations such as pressing and drawing. Those with the higher zinc content are less ductile and more suitable for hot forming operations such as casting, forging, extrusion and hot stamping. Some of the more common brasses alloys are listed in Table 4.4.

> **Key point**
>
> Brass is an alloy of copper and zinc. Brasses with a high copper content are the more ductile and suitable for cold forming. Brasses with a high zinc content are less ductile and more suitable for hot forming operations.

Table 4.4 *Brasses*

Type of brass	Composition	Properties	Applications and uses
Cartridge brass	70% Copper 30% Zinc	Very ductile, suitable for cold forming operations such as deep drawing	Cartridge cases, condenser tubes
Standard brass	65% Copper 35% Zinc	Ductile, suitable for cold working and limited deep drawing	Pressings and general purposes
Naval brass	62% Copper 37% Zinc 1% Tin	Strong and tough, malleable and ductile when heated, corrosion resistant	Hot formed components for marine and other structural purposes
Muntz metal	60% Copper 40% Zinc	Strong and tough, malleable and ductile when heated, good fluidity when molten	Hot rolled plate, castings extruded tubes and other sections

> **Key point**
>
> The main constituents of tin–bronzes are copper and tin. The addition of zinc improves fluidity when molten and the addition of phosphorous gives good load bearing properties.

Tin–bronzes are alloys in which the main constituents are copper and tin. Phosphorous is sometimes added in small amounts to prevent the tin from oxidising when molten. This gives rise to the name phosphor–bronze which is very malleable and ductile and widely used for plane bearings. Tin–bronzes which contain zinc are known as gunmetal. They have good fluidity when molten and can be cast into intricate shapes. When cooled they are strong, tough and corrosion resistant. Small amounts of lead are also sometimes added to tin–bronzes to improve their machinability. Four of the more common tin–bronzes are listed in Table 4.5.

Cupro–nickel alloys contain copper and nickel as the main constituents with small quantities of manganese and sometimes iron, added to improve their properties. Cupro–nickels are strong, tough and corrosion resistant. They find use in chemical plant and marine installations and are widely used throughout the world as 'silver' coinage. Two of the more common cupro–nickel alloys are listed in Table 4.6.

Aluminium is also alloyed with copper to produce a range of metals known as aluminium–bronzes. They also contain nickel

Table 4.5 *Tin–bronzes*

Type of tin–bronze	Composition	Properties	Applications and uses
Low tin–bronze	96% Copper 3.9% Tin 0.1% Phosphorous	Malleable and ductile. Good elasticity after cold forming	Electrical contacts, instrument parts, springs
Cast phosphor–bronze	90% Copper 9.5% Tin 0.5% Phosphorous	Tough, good fluidity when molten, good anti-friction properties	Plane bush and thrust bearings, gears
Admiralty gunmetal	88% Copper 10% Tin 2% Zinc	Tough, good fluidity when molten, good corrosion resistance	Pump and valve components, marine components, miscellaneous castings
Bell metal	78% Copper 22% Tin	Tough, good fluidity when molten, sonorous when struck	Church and ships bells, other miscellaneous castings

Table 4.6 *Cupro–nickel alloys*

Type of cupro–nickel	Composition	Properties	Applications and uses
Coinage metal	74.75% Copper 25% Nickel 0.25% Manganese	Strong, tough, corrosion and wear resistant	'Silver' coins
Monel metal	29.5% Copper 68% Nickel 1.25% Iron 1.25% Manganese	Strong, tough and highly corrosion resistant	Chemical, marine and engineering plant components

and manganese and are highly corrosion resistant. Two of the more common aluminium–bronzes are listed in Table 4.7.

There is a wide range of alloys in which aluminium is the major constituent. Aluminium in its pure form is soft, malleable and ductile with a low tensile strength. The addition of copper, silicon, nickel, silver, manganese, and magnesium in various small quantities greatly enhances its properties. Some of the resulting aluminium alloys are produced for cold working whilst others are produced for casting. They all retain the low weight characteristic of aluminium and some may be hardened by a heat treatment process known as precipitation hardening. Some of the most common types are listed in Table 4.8.

Table 4.7 *Aluminium bronzes*

Type of aluminium–bronze	Composition	Properties	Applications and uses
Wrought aluminium–bronze	91% Copper 5% Aluminium 2% Nickel 2% Manganese	Ductile and malleable for cold working excellent corrosion resistance at high temperatures	Boiler and condenser tubes, chemical plant components
Cast aluminium–bronze	86% Copper 9.5% Aluminium 1% Nickel 1% Manganese	Strong, good fluidity when molten, corrosion resistant	Valve and pump parts, boat propellers, gears, miscellaneous sand and die cast components

Table 4.8 *Common aluminium alloys*

Type of aluminium alloy	Composition	Properties	Applications and uses
Casting alloy BS 1490/LM4M	92% Aluminium 5% Silicon 3% Copper	Good fluidity when molten, moderately strong	Miscellaneous sand and die cast components for light duty applications
Casting alloy BS 1490/LM6M	88% Aluminium 12% Silicon	Good fluidity and strength	Miscellaneous sand and die cast components for motor vehicle and marine applications
'Y' Alloy	92% Aluminium 2% Nickel 1.5% Manganese	Good fluidity when molten and able to be hardened by heat treatment	Internal combustion engine pistons and cylinder heads
Duralumin	94% Aluminium 4% Copper 0.8% Magnesium 0.7% Manganese 0.5% Silicon	Ductile and malleable in the soft condition. Tough and strong when heat treated	Motor vehicle and aircraft structural components
Wrought alloy BS 1470/5:H30	97.3% Aluminium 1% Magnesium 1% Silicon 0.7% Manganese	Ductile in the soft condition, good strength when heat treated	Ladders, scaffold tubes, overhead electric power lines

Key point

Pure aluminium is soft, malleable and ductile. It has a low tensile strength but is a good conductor of both heat and electricity. Some aluminium alloys are produced for cold working whilst others are produced for casting.

Test your knowledge 4.4

1. What are the main uses and applications of zinc?
2. What are the main constituents of brass?
3. What is cast phosphor–bronze widely used for?
4. What is the non-ferrous alloy from which 'silver' coins are made?
5. What are the main constituents of aluminium–bronze?
6. Which aluminium alloy is widely used for aircraft and motor vehicle structural components?

Activity 4.2

In the early years of the last century, £1 gold coins, known as sovereigns, were in general circulation and the accompanying silver coins were really made from silver. Nowadays, gold sovereigns are only minted for special occasions and have a value which is much greater than £1. Take a modern £1 coin, a 10p piece and a 1p piece and examine them.

What materials do you think the modern coins are made from? Why do you think these materials have been selected? Which crystal lattice structure would you expect gold and silver to have? Give your reasons.

A little research in the chemistry or metallurgy section of your library or on the Internet, will tell you if you have made the correct choices.

See Questions 4.1 on p. 272.

Polymers

Polymer materials include plastics and rubbers whose atoms are arranged in long molecular chains known as *polymers*. The use of natural rubber, obtained from the rubber tree, began in the nineteenth century. Unfortunately natural rubber degrades or 'perishes' over a relatively short period of time and is readily attacked by solvents. It is not used much in engineering today except when mixed with synthetic rubbers. Developments over the last hundred years have seen the introduction of many new synthetic rubbers and plastics with many different properties. The raw materials that are used to make them come from oil distillation, coal and from animal and vegetable substances.

The name 'plastic' is a little misleading. Plasticine, chewing gum and linseed oil putty are examples of materials which are truly plastic. They are easily deformed by tensile, compressive and shearing forces and retain whatever shape they are moulded into. Polymer materials are in fact quite elastic at normal temperatures and the name 'plastics' refers to their condition when they are being formed into shape. At this stage the raw material is often a resin which can be poured or injected into a mould. Alternatively the raw material may be a powder or granules which are heated into a liquid prior to moulding.

Polymers are made up mainly of carbon and hydrogen atoms but may also have oxygen, chlorine, and fluorine atoms attached to them depending on the type of plastic. Each polymer can contain several thousand atoms. If fully extended, they can be anything up to a millimetre long and possibly more. The polymers are intertwined and held together by covalent bonds. The strength of the bonds affects the properties of plastic materials and rubbers. Some are very flexible and easy to stretch whilst others are hard and brittle. Some will melt and burn at relatively low temperatures whilst others are heat resistant. The polymers of two common plastics, polyethylene (polythene) and polychloroethene (PVC), are shown in Figures 4.11 and 4.12.

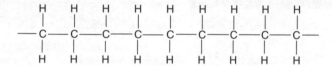

Figure 4.11 *The polythene polymer*

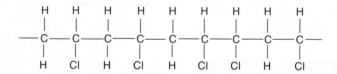

Figure 4.12 *The PVC polymer*

Key point

Plastic and rubbers are made up of intertwined chains of molecules called polymers. Polymers are mainly composed of carbon and hydrogen atoms.

Plastics can be part amorphous, where the polymers are randomly entangled, and part crystalline where the polymers become arranged in a geometric pattern. These regions are called *crystallites* (Figure 4.13). The relative proportions of amorphous regions and crystallites varies with different plastics and the way in which they have been processed. Forces of attraction are set up between the polymers as they form. They are known Van der Waal forces and it is these that hold plastics such as polythene and PVC together.

Plastics can be sub-divided into *thermoplastics*, *thermosetting plastics* and *rubbers*. Thermoplastics can be softened and remoulded by heating whereas thermosetting plastics cannot. Rubbers are polymer materials known as *elastomers* which have the property of being able to return to their original shape after large amounts of deformation.

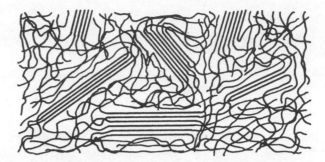

Figure 4.13 *Crystalline and amorphous regions*
Source: *From properties of Engineering Materials by R.A. Higgins.
Reprinted by permission of Elsevier Ltd*

Thermoplastics

Some thermoplastic materials are soft and flexible at normal temperatures whilst others are hard and brittle. When heated the hard thermoplastics will eventually soften and become flexible. The temperature at which this occurs is known as the *glass transition temperature* denoted by the symbol T_g. Further heating eventually causes the material to reach its melting temperature which has the symbol T_m. Figure 4.14 shows how the properties of a thermoplastic change with temperature. Some common thermoplastic materials and their uses, are listed in Table 4.9.

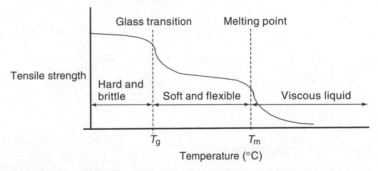

Figure 4.14 *Change of thermoplastic properties with temperature*

Thermosetting plastics

Thermosetting plastics, which are also called *thermosets*, undergo a chemical change during the moulding process. Strong covalent bonds are formed between the polymers that are known as *cross-links*. The cross-links may consist of a single atom or a small chain of atoms that form in the mould during what is known as the *curing process*. As a result, the bonding forces are much stronger than the Van der Waal forces and once formed, the cross-links cannot be broken. Thermosetting plastics are generally harder and more rigid than thermoplastics and cannot be softened or melted by heating.

Filler materials are very often mixed in with the raw materials for thermosetting plastics prior to moulding. These improve the mechanical properties and include wood flour, shredded textiles, paper and various other fibres and powdered materials. Some common thermosetting plastics and their uses, are listed in Table 4.10.

Table 4.9 *Common thermoplastic materials*

Polymer	Common name	Properties	Applications and uses
Polyethene	Low-density polythene	Tough, flexible, solvent resistant, degrades if exposed to light or ultra-violet radiation	Squeeze containers, packaging, piping, cable and wire insulation
Polyethene	High-density polythene	Harder and stiffer than low-density polythene with higher tensile strength	Food containers, pipes, mouldings, tubs, crates, kitchen utensils, medical equipment
Polypropene	Polypropene	High strength, hard, high melting point, can be produced as a fibre	Tubes, pipes, fibres, ropes, electronic components, kitchen utensils, medical equipment
Polychloroethene	Polyvinyl chloride or PVC	Can be made tough and hard or soft and flexible, solvent resistant, soft form tends to harden with time	When hard, window frames, piping and guttering. When soft, cable and wire insulation, upholstery
Polyphenylethene	Polystyrene	Tough, hard, rigid but somewhat brittle, can be made into a light cellular foam, liable to be attacked by petrol based solvents	Foam mouldings used for packaging and disposable drinks cups. Solid mouldings used for refrigerators and other appliances
Methyl-2-methylpropenoate	Perspex	Strong, rigid, transparent but easily scratched, easily softened and moulded, can be attacked by petrol based solvents	Lenses, corrugated sheets for roof lights, protective shields, aircraft windows, light fittings
Polytetra fluoroethylene	PTFE	Tough, flexible, heat-resistant, highly solvent resistant, has a waxy low-friction surface	Bearings, seals, gaskets, non-stick coatings, tape
Polyamide	Nylon	Tough, flexible and very strong, good solvent resistance but does absorb water and deteriorates with outdoor exposure	Bearings, gears, cams, bristles for brushes, textiles
Thermoplastic polyester	Terylene	Strong, flexible and solvent resistant, can be made as a fibre, tape or sheet	Textile fibres, recording tape, electrical insulation tape

Table 4.10 *Common thermosetting plastics*

Polymer	Common name	Properties	Applications and uses
Phenolic resins	Bakelite	Hard, resistant to heat and solvents, good electrical insulator and machinable, colours limited to brown and black	Electrical components, vehicle distributor caps saucepan handles, glues, laminates
Urea–methanal resins	Formica	As above, but naturally transparent and can be produced in a variety of colours	Electrical fittings, toilet seats, kitchen ware, trays, laminates
Methanal–melamine resins	Melamine	As above, but harder and with better resistance to heat. Very smooth surface finish	Electrical equipment, table ware, control knobs, handles, laminates
Epoxy resins	Epoxy resins	Strong, tough, good chemical and thermal stability, good electrical insulator, good adhesive	Container linings, panels, flooring material, laminates, adhesives
Polyester resins	Polyester resins	Strong, tough, good wear resistance, and resistance to heat and water	Boat hulls, motor panels, aircraft parts, fishing rods, skis, laminates

Rubbers

The polymers in rubbers are known as *elastomers*. They are longer and tend to be more complex than other polymers (Table 4.11). Elastomers can be made up of more than 40 000 atoms and when the material is unloaded, they are intertwined and folded over each other in a random fashion.

Table 4.11 *Common types of rubber*

Type of rubber	Properties/origins	Applications and uses
Natural rubber	Obtained as latex from the rubber tree. Perishes with time and is readily attacked by solvents	Not used for engineering purposes except when mixed with synthetic rubbers
Styrene rubber	Synthetic rubber developed in USA during Second World War as a substitute for natural rubber. Also known as GR-S rubber. Resistant to oils and petrol	Blended with natural rubber and used for vehicle tyres and footwear
Neoprene	Synthetic rubber with close resemblance to natural rubber. Resistant to mineral and vegetable oils and can withstand moderately high temperatures	Used in engineering for oil seals, gaskets and hoses
Butyl rubber	Synthetic rubber with good resistance to heat and temperature. Impermeable to gases	Used in engineering for tank linings, moulded diaphragms, inner tubes and air bags
Silicone rubber	Synthetic rubber with good solvent resistance and which retains its properties over a wider temperature range than other types. Remains flexible from $-80\,°C$ to $300\,°C$	Widely use in chemical and process plant for seals and gaskets where a wide variation in temperature is likely. Also in aircraft where low temperatures are experienced at high altitude

When loads are applied to rubber, the elastomers behave rather like coiled springs which become aligned as the material is stretched. This gives the material its elastic properties. If the loading is excessive, the forces between the polymers may be insufficient to prevent them from sliding over each other. This will ultimately lead to failure or produce permanent deformation. In this condition the rubber is said to be both elastomeric and thermoplastic (Figure 4.15).

When sulphur or certain metal oxides are added to the rubber during manufacture, cross-links are formed between the polymers. These are similar to those formed in thermosetting plastics and the process is known as *vulcanising*. The cross-links enable the rubber to retain its elasticity and return to its original shape when loaded

Unloaded elastomers

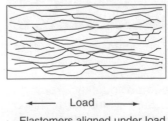

⟵ Load ⟶

Elastomers aligned under load

Figure 4.15 *Behaviour of elastomers*

almost to the point of failure. The addition of larger quantities of sulphur causes the rubber to loose some of its elasticity. If sufficient is added, it becomes hard and brittle and its properties are then very similar to the hard thermosetting plastics.
See Questions 4.2 on p. 272.

Ceramics

The term 'ceramic' comes from the Greek word for potters clay. In engineering, it covers a wide range of materials such as bricks, tiles, cement, furnace linings, glass and porcelain. It also includes the abrasive grits used for grinding wheels and some cutting tools. The main ingredients of ceramics are sand and clays.

Sand contains silica which is a compound of silicon and oxygen. It also contains feldspar which is a compound of aluminium, silicon and oxygen. Clays also contain compounds silicon and aluminium. Other elements such as potassium, calcium and magnesium might be present in combination with silicon. These compounds are called *silicates*.

Ceramics may be sub-divided into amorphous ceramics, crystalline ceramics, bonded ceramics and cements. Their main properties are that they are hard wearing, good electrical insulators and strong in compression when used as structural materials. In addition, many of them can withstand very high temperatures which makes them suitable for lining furnaces and kilns. Ceramics do however tend to be weak in tension, brittle and unable to withstand impact loading.

Amorphous ceramics

In amorphous ceramics, the molecules are not arranged in any geometrical pattern. They include the various types of glass whose main ingredient is silica, or silicon oxide, whose chemical symbol is SiO_2. This is contained in sand which is heated until molten and then rapidly cooled. Other ingredients can include sodium carbonate, sodium borate and oxides of magnesium, calcium and boron. If the mixture is allowed to cool slowly it crystallises and the resulting solid is opaque. When it is cooled rapidly however, there is no

Table 4.12 *Common types of glass*

Type of glass	Ingredients/properties	Applications and uses
Soda lime glass	Silica (72%) plus oxides of sodium, calcium and magnesium. Can be toughened by heat treatment	Accounts for about 95% of all glass production. Used for windows, windscreens, bottles and jars
Lead glass	Silica (60%), lead oxide (25%) plus oxides of sodium, potassium and aluminium. High electrical resistance	Used for lamps and cathode ray tubes
	Silica (40%), lead oxide (47%) plus oxides of sodium, potassium and aluminium. High refractive index	Used for lenses, and crystal glass tableware
Borosilicate glass	Silica (70%) boron oxide (20%) plus oxides of sodium, potassium and aluminium. Low expansivity, good resistance to chemicals	Trade name 'Pyrex'. Used for laboratory equipment, kitchen glassware and electrical insulators

crystallisation and the resulting solid is amorphous and transparent. Some of the more common types of glass are listed in Table 4.12.

Crystalline ceramics

This group includes the hard, abrasive grits used for surface preparation and material removal. The most widely used crystals are aluminium oxide (emery), beryllium oxide, silicon carbide and boron carbide. Singly or in combinations, they are used to make emery cloth and paper. They are mixed with oil to form grinding paste and they are bonded together to make grinding wheels.

Also included in this group are tungsten carbide, zirconium carbide, titanium carbide and silicon nitride. These crystals are bonded together in a metal matrix to produce very hard cutting tool tips.

Bonded clay ceramics

The ingredients for this group are natural clays and mixtures of clays to which crystalline ceramics are sometimes added. After mixing and moulding, the products are fired in kilns and ovens. During the firing, a process of *vitrification* occurs in which the crystals become bonded together. Bonded clay ceramics may be sub-divided into whiteware, structural clay products and refractory ceramics.

Whiteware includes china pottery, porcelain, earthenware, stoneware, decorative tiles and sanitary products. Firing takes place at temperatures between 1200 °C and 1500 °C after which some of the products are glazed with a thin layer of molten glass. Selected and refined clays are used for whiteware to which crystalline ceramics such as sodium borate (borax) and sodium aluminium fluoride (cryolite) have been added. These assist the bonding process by acting as a flux and they also lower the vitrification temperature.

Structural clay products are made from common clays and are fired at a higher temperature. They include bricks, drain and sewer pipes, terracotta products and flooring and roofing tiles. Natural impurities in the clay act as a flux during vitrification. Structural

Key point

Aluminium oxide (emery) is one of the most widely used crystalline ceramic grits in the manufacture of grinding paste, abrasive wheels, abrasive cloth and paper.

Key point

Vitrification is the bonding together of the crystals in bonded clay ceramics. The vitrification temperature is the firing temperature at which this occurs.

Test your knowledge 4.7

1. What is the main ingredient of all types of glass?
2. What is the commonly used name for borosilicate glass?
3. What is 'emery'?
4. What occurs in ceramic materials during the vitrification process?
5. What are refractory ceramics used for?

clay products have relatively large interstitial spaces in their crystal lattice structure. As a result they tend to absorb water and to prevent this, it is usual for drain and sewer pipes to be given a glazed finish.

Refractory ceramics are made from fireclays which have a high content of silicon oxide (silica) or aluminium oxide (emery). They are used to make firebricks and to line the inside of furnaces and kilns. The high silica or aluminium oxide content makes them able to withstand sustained high temperatures and chemical attack from combustion gasses. They are also used to line the buckets and ladles that are used for carrying molten metal.

Cements

Cements contain metal oxides such as those of calcium, silicon, and magnesium which react chemically with water. The mixture solidifies into a hard crystalline structure with good compressive strength and bonding properties. Portland cement mixed with sand is similar in colour to Portland stone and is widely used in the building industry. Stone aggregates are added to it to make concrete which has much greater compressive strength and toughness. Other types of cement for special applications are silicate cements, which resist chemical attack, and high alumina cements, which resist attack from sea water.

Activity 4.4

Silica, whose chemical name is silicon (IV) oxide, or SiO_2, is an ingredient in many ceramics. Find out how the atoms are arranged in its crystal lattice structure and the effect this has on its properties.

The hob plates of many electric cooker are now made of a ceramic material. Find out what this is and why its structure makes it suitable for this purpose.

Composites

Composites are made up of two or more materials which are bonded together. They may be sub-divided into laminated, particulate and fibrous composites.

Laminates

Laminates consist of two or more materials which are bonded together. Perhaps, the oldest example is plywood which is made up of three or more bonded layers of wood with their grain directions at right angles to each other. Other wood laminates include blockboard and laminated chipboard. Blockboard is made up of strips of wood bonded together and sandwiched between two thin outer layers of wood. Laminated chipboard is made up of compressed wood shavings and sawdust which are bonded together and sandwiched between two thin outer layers. The bonding agent is usually a thermosetting resin. The resulting composites are resistant to warping and do not have any weakness due to grain direction.

Laminates made from a thermosetting polymer resin and a filler material are widely used for working surfaces in the home, in restaurants and in the workplace. Formica and melamine are probably the best known trade names that come from the thermosetting resins used to make them. The filler material may be paper or cloth that is impregnated with resin and partly cured. That is to say that the cross-linking process has started, but is not complete. Alternate layers are then stacked together and placed in a hot press. The heating allows the cross-linking to proceed within and between the layers and the pressure ensures that there are no cavities present. The laminates produced are hard wearing, tough, heat resistant and resistant to staining.

Combinations of polymer materials and metals are increasingly being used to produce composites which are light in weight but extremely rigid. They have a low density core of plastic foam or a metal honeycomb made from aluminium or titanium. This is sandwiched between two high strength laminate skins. Composites of this kind are widely used in aircraft where low weight and high strength are essential requirements.

Particulate composites

In particulate composites, one material acts as a matrix that surrounds the particles of another. Concrete is perhaps the oldest example of a particulate composite where stone aggregate and sand are bonded in a matrix of cement. Steel reinforcing rods may also be included to give the concrete added strength. Particle reinforced polymers in which thermoplastics and thermosetting plastics are strengthened and toughened with silica, glass beads and rubber particles, are also particulate composites. High impact polystyrene, which contains rubber particles for added toughness, is typical of this type.

Sometimes the particles are bonded in a matrix of metal. Aluminium oxide particles contained in aluminium, increase the strength of the material, particularly its tensile strength at high working temperatures. Graphite and PTFE particles contained in phosphor–bronze, improve its properties as a bearing metal by reducing frictional resistance.

In recent years a range of composite materials known as *cermets* has been developed. These are formed by mixing together powdered ceramic and metal particles which are compressed at high temperatures. The process is known as *sintering* and results in a hard and wear resistant composite in which the ceramic particles become surrounded by a metal matrix. The cemented carbides used for cutting tool tips are produced in this way. The metal is usually cobalt and the ceramic powders are tungsten, titanium, silicon and molybdenum carbides. Tungsten carbide is perhaps the oldest and best-known cermet and has been used on tipped lathe tools and masonry drills for many years.

An alternative type of particulate composite has oxidised metal particles contained in a matrix of ceramic material. These materials have excellent thermal insulation properties and are used in aerospace applications for heat shields.

Key point

Sintering is a process used to make cermets such as tungsten carbide. The constituents are subjected to high temperatures and pressures at which they become bonded together.

Test your knowledge 4.8

1. Describe the composition of laminated chipboard.
2. How are formica laminates produced?
3. What particulate materials are added to phosphor–bronze to improve its properties as a bearing material?
4. What is a cermet?
5. Which thermosetting resins are used to make GRP products?

Material properties and effects of processing

Fibrous composites

The most common fibrous composites are those in which glass, carbon, kevlar, silicon carbide and aluminium oxide fibres are contained in a matrix of thermosetting plastic. Epoxy and polyester resins are the usual matrix materials. The fibres increase the strength and the stiffness of the plastic materials. Glass fibre composites or GRP, which stands for 'glass reinforced plastic', have been in use for many years and are widely used for boat hulls and motor panels.

Carbon fibre composites are not quite as strong as those reinforced with glass fibres but they are very stiff and light in weight. They are widely used in aircraft and also for fishing rods. Kevlar, silicon carbide and aluminium oxide fibres are used in composites which are required to have high impact resistance. Kevlar is an extremely strong polymer. It is used in the frames of tennis racquets, in bullet-proof body armour and to reinforce rubber in conveyer belts and tyres.
See Questions 4.3 on p. 273.

When we describe the physical properties of a material, we are describing how it is likely to behave whilst it is being processed and when it is in service. We have already mentioned properties such as strength, hardness, toughness and corrosion resistance. These now need to be carefully defined so that wherever possible, they can be measured. Physical properties can be sub-divided into mechanical properties, thermal properties, electrical and magnetic properties and durability.

Mechanical properties

These include density, tensile strength, hardness, toughness, ductility, malleability, elasticity and brittleness. They are defined as follows.

Density

Density is the mass in kilograms, contained in a cubic metre of a material or substance. When used in calculations, it is usually denoted by the Greek letter ρ (rho) and its units are kilograms per cubic metre ($kg\,m^{-3}$). The density of steel is around $7800\,kg\,m^{-3}$ and that of cast iron is around $7300\,kg\,m^{-3}$. The densities of the major non-ferrous metals used in engineering are listed in Table 4.3. When designing aircraft, engineers need to be able to estimate the weight of the different components and then the total weight. Knowing the densities of the different materials helps them do this.

Tensile strength

The tensile strength of a material is a measure of its ability to withstand tensile forces. The Ultimate Tensile Strength (UTS) is the tensile stress that causes a material to fracture. If you have completed the unit Science for Technicians, you will know that tensile stress is the load carried per square metre of cross-sectional area. Its units are thus Newtons per square metre or Pascals (Pa).

The UTS of mild steel is around 500 MPa. The values of UTS for other ferrous and non-ferrous metals are given in Tables 4.1,

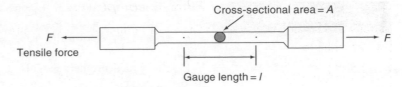

Figure 4.16 *Tensile test specimen*

4.2, and 4.3. Sometimes material suppliers quote the UTS of a material in Newtons per square millimetre ($\mathrm{N\,mm^{-2}}$). It is useful to know that the value is numerically the same in both megapascals and Newtons per square millimeter, i.e. the UTS of mild steel can be written as 500 MPa or $500\,\mathrm{N\,mm^{-2}}$.

The Ultimate Tensile Strength of a material can be determined by carrying out a destructive tensile test on a prepared specimen (Figure 4.16). The specimen is gripped in the chucks of a tensile testing machine and increasing values of tensile load are applied up to the point of fracture. Depending on the type of machine, the load is applied by means of hydraulic rams, a lever system or a lead-screw. Its value is read off from an analogue or a digital display.

The extension of the specimen can be measured by means of an extensometer that is attached to the specimen. This can be of the Lindley type, which records the extension on a dial test indicator, or the Monsanto type, which operates on the micrometer principle. There is also an electrical type, which is a linear variable differential transformer. This can be interfaced with an *x–y* plotter or a computer to produce a graph of load v. extension (Figure 4.17).

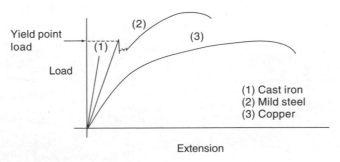

Figure 4.17 *Load v. extension graphs*

The extension of a specimen is usually measured over a gauge length of 50 mm. The initial diameter is checked using a micrometer and the initial cross-sectional area A, is calculated. The UTS of the material can then be calculated using the formula

$$\mathrm{UTS} = \frac{\text{maximum load}}{\text{initial cross-sectional area}} \tag{4.1}$$

Ductility

The ductility of a material is a measure of the amount by which it can be drawn out in tension before it fractures. As can be seen from the typical load v. extension graphs in Figure 4.17, copper has a high degree of ductility and this enables it to be drawn out into long lengths of wire and tube. There are a number of different

Key point

The Ultimate tensile Strength (UTS) of a material is the stress at which it will break. It has the same numerical value when measured in MPa and $\mathrm{N\,mm^{-2}}$.

ways in which ductility can be measured. One method is to carry out a bend test in which a sample of material is bent through an angle and observed for the appearance of cracks. The angle at which cracking or breaking occurs gives a measure of ductility.

Ductility measurement can also be incorporated in a tensile test. Before a tensile test, the gauge length over which extension is measured is centre punched. After the test, the two pieces of the fractured specimen are placed in contact and the elongation of the gauge length is measured. The ductility can then be calculated as the percentage increase in the gauge length.

$$\% \text{ increase in length} = \frac{\text{original gauge length}}{\text{extended gauge length}} \times 100 \qquad (4.2)$$

Alternatively, the diameter at the point of fracture can be measured and the cross-sectional area calculated. The ductility can then be calculated as the percentage reduction in area.

$$\% \text{ reduction in area} = \frac{\text{cross-sectional area at fracture point}}{\text{original cross-sectional area}} \times 100$$

$$(4.3)$$

When a ductile material fails, the appearance of the fracture is as shown in Figure 4.18.

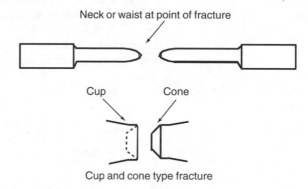

Figure 4.18 *Ductile fracture characteristics*

As the maximum load is approached, a neck or waist is seen to form at the point where fracture will occur. When the fracture surfaces are examined it is found that they have a characteristic cup and cone appearance which is typical of ductile materials.

When ductile materials are loaded, a point is reached known as the *elastic limit*. If this is exceeded, elastic failure is said to have occurred. This is followed by plastic deformation up to the point of fracture. If a material is unloaded within the plastic extension range, it does not return to its original shape and there is permanent deformation (Figure 4.19).

Mild steel has the characteristic load v. extension graph shown in Figure 4.17. Its elastic limit is followed closely by the yield point where the material suddenly gives way under the load. Elastic failure is then said to have occurred. If the material is loaded beyond this point there will be some permanent deformation when the load is removed.

Mild steel behaves in this way because of the interstitial carbon atoms. They are said to 'pin' the planes of iron atoms within the

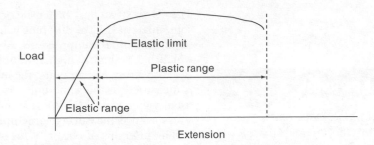

Figure 4.19 *Elastic and plastic extension*

grains, which stops them from slipping under load. Eventually they can no longer hold the iron atoms and yielding takes place. For a short time, extension occurs at a reduced load but the material soon becomes work hardened and the load v. extension graph then rises in a curve until the maximum load is reached. It is important to know the stress at which yielding takes place in steel components so that they can be designed to operate below the yield point load.

Brittleness

This is the opposite of ductility. When metals are loaded in tension they first undergo elastic deformation which is proportional to the load applied. When unloaded from within the elastic range, the material returns to its original shape. Brittle materials tend to fail within the elastic range, or very early in the plastic range, before very much deformation has taken place. Cast iron is brittle, as can be seen from its load v. extension graph in Figure 4.17. Brittle materials very often fracture across a plane at right angles to the direction of loading. Sometimes however, the fracture plane is at 45° to the direction of loading as shown in Figure 4.20. This indicates that the material has a low shear strength.

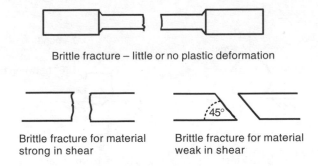

Figure 4.20 *Brittle fracture characteristics*

Elasticity

The elasticity of a material is a measure of its ability to withstand elastic deformation. If you have completed the unit Science for Technicians you will know that the modulus of elasticity of a material is given by the formula

$$\text{modulus of elasticity} = \frac{\text{direct stress}}{\text{direct strain}} \qquad (4.4)$$

The direct stress may be tensile or compressive and it is found that some materials have different values of the modulus when loaded in tension and compression. Aluminium is one such example, which has a slightly higher value in compression than in tension. Because strain has no units, the modulus of elasticity of a material is measured in the same units as stress, i.e. Pascals (Pa) or more usually gigapascals (GPa). It should not however be confused with stress, as it is measuring something entirely different.

The modulus of elasticity for mild steel is around 200 GPa and that of some aluminium alloys is around 100 GPa. This tells you that if identical specimens are subjected to the same tensile load, the elastic extension of the aluminium will be double that of the steel, i.e. the higher the modulus of elasticity of a material, the harder it will be to stretch.

Key point

In a perfectly elastic material, the deformation produced is proportional to the load applied and when the load is removed, the material immediately returns to its original shape.

Malleability

Whereas ductility is the ability of a material to be drawn out in tension, malleability is the ability of a material to be deformed or spread in different directions. This is usually caused by compressive forces during rolling, pressing and hammering operations. Copper is both ductile and malleable but the two properties do not necessarily go together. Lead is extremely malleable but not very ductile, and soon fractures when loaded in tension.

Hardness

Hardness is the ability of a material to withstand wear and abrasion. The surface hardness of a material is usually measured by carrying out a non-destructive indentation test. Depending on the material, a steel ball or a pointed diamond indentor is pressed into the surface of the material under a controlled load. There are three such methods of inspection in common use for engineering metals. They are the Brinell Hardness test, the Vickers Pyramid Hardness test and the Rockwell Hardness test (Figure 4.21).

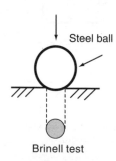
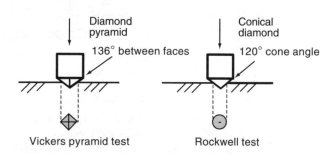

Figure 4.21 *Indentation tests*

The Brinell hardness tester uses a hardened steel ball indentor and is used for relatively soft materials. The Vickers hardness tester uses either a hardened steel ball or a pyramid shaped diamond indentor and can be used for both soft and hard materials. The Rockwell hardness tester, which is American in origin, uses a hardened steel ball or a conical diamond indentor and can also be used for both soft and hard materials.

Each test enables a hardness number to be calculated. For the Brinell and Vickers tests, it is given by dividing the applied load by the surface area of the indentation. This requires careful measurement of the indentation, using a calibrated microscope, after which the Brinell Hardness Number (BHN) or the Vickers Hardness Number (VPN) can be calculated. More usually however, it is read off by referring the measurements to a chart supplied with the testing machine.

For the Rockwell test, the hardness number is a function of the depth of penetration. The machine senses this automatically and the hardness number is read off directly from a display. It is important to note that these tests will give different hardness numbers for the same material. The hardness number required, and the test which should be used, is specified on the engineering drawings for components where hardness is critical.

Toughness

Toughness is the ability of a material to withstand impact and shock loading. The standard tests for toughness are the Izod impact test and the Charpy test. Both use the same testing machine in which a notched specimen is subjected to a sudden blow from a swinging pendulum (see Figure 4.22).

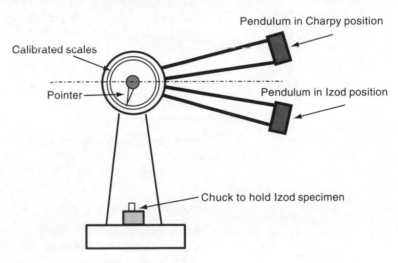

Figure 4.22 *Izod/Charpy impact tester*

With the Izod test, a notched specimen of rectangular section bar gripped in a chuck and held in a vertical position (Figure 4.23(a)). With the Charpy test, the specimen is supported horizontally at each end, so that the pendulum strikes it at its centre (see Figure 4.23(b)).

After impact with the specimen, the pendulum continues in its arc of travel and pushes the pointer around the calibrated scale. There is a separate scale for each of the two tests but both are calibrated in Joules. The pointer records the amount of energy absorbed by the specimen on impact, and this gives a measure of the toughness of the material.

An advantage of the Charpy test, which has its origins in America and Canada, is that heated specimens from a furnace

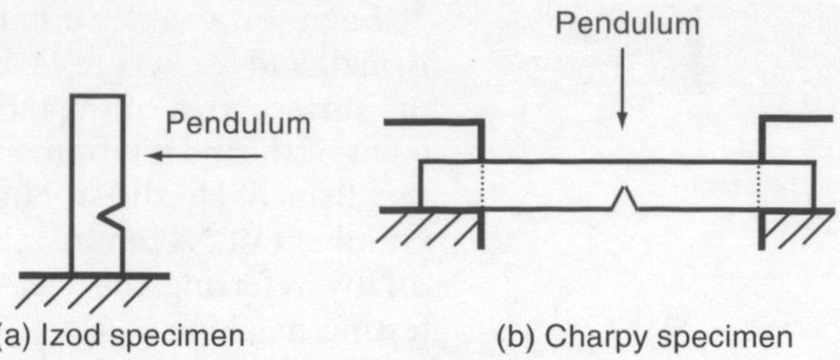

Key point

Toughness is the ability of a material to withstand impact or shock loading.

(a) Izod specimen   (b) Charpy specimen

Figure 4.23 *Izod and Charpy test specimens*

Test your knowledge 4.9

1. What is meant by the UTS of a material?
2. What is the difference between ductility and malleability in materials?
3. How do the Brinell, Vickers Pyramid and Rockwell hardness tests differ?
4. How can the toughness of a material be measured?

can be quickly placed in position and tested for toughness. Provided that the test is done quickly, the drop in temperature is small. The Izod test, where the specimen must be secured in a chuck, is unsuitable for heated specimens.
See Questions 4.4 on p. 274.

Activity 4.5

You are given samples of different engineering materials that are all in the form of 10 mm diameter bar. You have at your disposal an engineers' vice, a hammer and a centre punch. Devise a series of tests which makes use of these to compare the hardness, toughness and ductility of the materials.

Thermal properties

These include the ability of a material to conduct heat energy and the changes in dimension which occur with a rise or fall in temperature.

Expansivity

When metals are heated they generally expand. The same occurs with a great many polymer and ceramic materials although the effect is not so pronounced. Some thermoplastics contract when heated and use is made of this property in packaging. Shrink insulation sleeving for electrical conductors has the same property. Gentle heating causes it to shrink and provide a firm covering for the conductor.

Thermal expansivity is a measure of the effect of temperature change on the dimensions of a material when it is heated. It is defined as the increase in length, per unit of original length, per degree of temperature rise. It is given the symbol α (alpha), and its units are $°C^{-1}$.

If the original length of a component is l, and it expands by an amount x, when the temperature rises from $t_1 °C$ to $t_2 °C$, the linear expansivity of the material is given by the formula

$$\alpha = \frac{x}{l(t_2 - t_1)} \qquad (4.5)$$

The change in length x, of a component for a given change in temperature is given by transposing the formula, that is,

$$x = l\alpha(t_2 - t_1) \tag{4.6}$$

Thermal conductivity

Metals are mainly good conductors of heat energy whilst plastics and ceramics are generally good heat insulators. It is thought that this is because of the presence of free electrons in the crystal lattice structure of metals. When a metal is heated, the kinetic energy of the free electrons is increased in the locality of the heat source. This is passed on to other free electrons throughout the material and heat energy is transported much more quickly than in plastic and ceramic materials where the valence electrons are all employed in producing the strong covalent bonds. In these materials, it is thought that heat energy is passed on through increased thermal vibration of the atoms and molecules, which is a much slower process. The thermal conductivity of a material is defined as the amount of heat energy per second that will pass through a 1 m cube of the material when the difference in temperature between opposite faces is 1 °C. It is given the symbol k, and its units are $W\,m^{-1}\,°C^{-1}$.

Consider a piece of material whose cross-sectional area is A, and thickness is l, with a temperature difference $t_2 - t_1$ across its thickness, as shown in Figure 4.24. If the heat transfer rate through the material is Q, measured in Watts, then the thermal conductivity of the material will be given by the formula

$$k = \frac{Q\,l}{A(t_2 - t_1)} \tag{4.7}$$

The heat transfer rate for a given temperature difference is given by transposing the formula, that is,

$$Q = \frac{kA}{l}(t_2 - t_1) \tag{4.8}$$

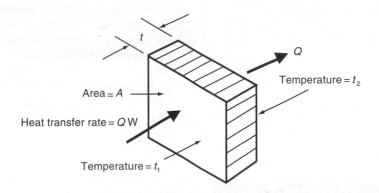

Figure 4.24 *Conduction of heat energy*

Electrical and magnetic properties

These include the ability of a material to allow or prevent the passage of electric current and the behaviour of a material in the presence of electromagnetic fields.

Resistivity

Metals are generally good conductors of electricity. As with thermal conductivity, this is due to the presence of free electrons in the crystal lattice structure. When an electrical potential difference is applied between the ends of a metal conductor, the negatively charged free electrons drift towards the positive potential. It is this flow which constitutes an electric current.

The flow of free electrons is impeded by the fixed atoms and the resistivity of a material is defined as the resistance of a 1 m length whose cross-sectional area is 1 m. It is given the symbol ρ (rho), and its units are ohm-metres. For a conductor of length l, cross-sectional area A, and resistance R, the resistivity of the material is given by the formula

$$\rho = \frac{RA}{l} \tag{4.9}$$

Alternatively, the resistance of the conductor can be found using the formula

$$R = \frac{\rho l}{A} \tag{4.10}$$

Temperature coefficient of resistance

The atoms and molecules of a material are thought to be in a state of vibration whose amplitude increases with temperature rise. This partly accounts for the expansion of the material and also the increase in electrical resistance since the free electron flow is further impeded by the increased vibration.

The resistance of some materials is not much affected by temperature change. Carbon is a common example of these, which are said to be *ohmic* materials. With most metals, the resistance increases uniformly with temperature and they are said to be *non-ohmic*. The temperature coefficient of resistance of a material is a measure of the effect. It is defined as the increase in resistance per unit of its resistance at $0\,°C$, per degree of temperature rise. It is given the symbol α (alpha), and its units are $°C^{-1}$. It should not be confused with linear expansivity which has the same units and symbol.

The temperature coefficient of resistance of a material whose resistance at $0\,°C$ is R_0 and whose resistance at some higher temperature $t\,°C$ is R, is given by the formula

$$\alpha = \frac{(R - R_0)}{R_0 t} \tag{4.11}$$

Alternatively, the resistance of a material at some temperature $t\,°C$ is given by the formula

$$R = R_0(1 + \alpha t) \tag{4.12}$$

With semiconductor materials such as silicon and germanium, the resistance falls with temperature rise and they are said to have a negative temperature coefficient of resistance. The fall in resistance is not uniform however, which means that α is not a constant. As a result, the above formulae cannot be used with semiconductor materials.

Permeability

Electromagnets, transformers, motors and generators all contain current carrying coils around which a magnetic field is produced. The core material, on which the coils are wound, can greatly affect the intensity and strength of the magnetic field. Certain metals, which are known as ferromagnetic materials, have the greatest effect. They include iron, nickel, cobalt and a number of specially developed alloys which contain these metals.

Permeability is a measure of a material's ability to intensify the magnetic field produced by a current carrying coil. A comparison is made with the magnetic field produced by a coil which has nothing whatsoever in its core. The relative permeability μ_r of a core material indicates the increased intensity of the field produced compared to that of the same empty coil. Materials other than those listed above have little or no effect on field intensity and so have a relative permeability of $\mu_r = 1$. Soft low-carbon steel on the other hand can have a relative permeability of $\mu_r = 100\,000$.

Ferromagnetic materials can be sub-divided into 'soft' and 'hard' categories. The soft materials are the low-carbon steels used for the cores of transformers and electromagnets. These have a high permeability but soon loose their magnetism when the current is switched off. Hard magnetic materials are generally made from hard alloy steels containing nickel. They are used to make permanent magnets such as those used in audio speakers. Hard magnetic materials have a lower permeability but once magnetised, they retain their magnetism for long periods of time.

Permittivity

Permittivity is a measure of a material's ability to intensify an electric field such as that produced between the plates of a capacitor. The plates of a capacitor are separated by an insulating material known as a 'dielectric'. Depending on the use and type of capacitor, air, waxed paper, plastics and ceramic materials are used as dielectrics. When a potential difference is applied across the plates of a capacitor, electrical charge is stored on them and an electric field is set up between them.

The intensity of the electric field, and the amount of charge which can be stored in a capacitor, depends on the permittivity of the dielectric material. As with magnetism, a comparison is made with the intensity of the field produced by the same potential difference when the plates of a capacitor are separated by vacuum. The relative permittivity ε_r gives a measure of the increased field intensity when a dielectric is present. For an air space $\varepsilon_r = 1.0006$, for paper $\varepsilon_r = 2$ to 2.5, for mica, which is a ceramic, $\varepsilon_r = 3$ to 7.

Durability

There is a tendency for all engineering materials to deteriorate over a period of time. This may be due to corrosion, attack by chemical solvents or degradation due to electromagnetic radiation. The durability of engineering components can be maximised by choosing materials which are best suited to their service conditions.

Key point

Permeability is a measure of the ability of a material to intensify the magnetic field produced by a current carrying coil when it is used as a core material.

Key point

Permittivity is a measure of the ability of a material to intensify an electric field such as that set up between the plates of a charged capacitor.

Test your knowledge 4.10

1. Why are most metals good conductors of both heat and electricity?
2. How is the 'resistivity' of a material defined?
3. What is an 'ohmic' material?
4. Which materials have a high magnetic permeability?
5. What is meant by the relative permittivity of a dielectric material?

Surface protection and shielding from the above kinds of attack can also prolong service life.

Corrosion resistance

All metals, and particularly ferrous metals, are subject to corrosion. This is the result of a chemical reaction that takes place between the metal and some other element in its service environment. The other element is very often oxygen, which is present in the atmosphere and in water. Non-ferrous metals usually have a higher resistance to corrosion than those containing iron. They react with oxygen from the atmosphere but the oxide film that forms on the surface is more dense. Once formed, it protects the metal against further attack.

Ferrous metals are affected by two kinds of corrosion. Wet corrosion, which occurs at normal temperatures in the presence of moisture, and dry corrosion which occurs when the metal is heated to high temperatures. Both processes will be discussed in detail later in this chapter when considering the ways in which materials can fail in service. It is sufficient here to say that wet corrosion is an electro-chemical process where the moisture acts as an electrolyte and different areas of the metal become anodes and cathodes. Corrosion always occurs at the anodes.

The oxides that form on the surface of ferrous metals are rust, at normal temperatures, and black millscale, at high temperatures. Both are loose and flakey. They do not protect the metal from further attack but there are a number of ways in which it can be protected at normal temperatures. The most common methods are painting, plating with a corrosion resistant metal and coating with a polymer material. The choice of paint depends on the service environment but oil paints which contain bitumen and enamel paints containing polymers, give good surface protection against rusting.

The most common metals used for the surface protection of steel are zinc, tin and cadmium. Plating with zinc is known as *galvanising*. It is particularly effective because, if the surface is scratched to expose the steel beneath, the steel will not rust. In the presence of moisture it is the zinc which becomes an anode and which will corrode. This is called *sacrificial* protection. The corrosion of the zinc takes place at a very slow rate and so it gives long-term protection.

Tin and cadmium plating give good protection provided that the coating is not damaged to expose the steel beneath. In this event, the steel becomes an anode and rusting will occur in the presence of moisture. The following list of metals is part of a much longer list that is called the *electrochemical series*. When two dissimilar metals are joined together there is a danger of electrolytic corrosion and it is the metal lower down in the series which becomes the anode and which corrodes.

Gold

Platinum

Silver

Copper

Lead

Key point

Wet corrosion is an electro-chemical process which occurs when moisture is present. Dry corrosion is a direct chemical reaction between a material and oxygen in the atmosphere.

Material processing

Tin

Nickel

Cadmium

Iron

Chromium

Zinc

Aluminium

Magnesium

You will note that zinc is below iron and that tin and cadmium are above it. This accounts for the degree of protection that they give to steel. Rubber and other polymer materials are also used to protect steel. Rubber compounds are sometimes used for tank linings. Plastic material is very often used to protect steel lamp standards and the posts that support road traffic signs.

Solvent resistance

There are certain chemical substances that attack plastics and rubbers. They are known generally as solvents. The liquids and gases used in many industrial processes can have such a degrading effect. Petrol, fuel oil, lubricating oils and greases can also act as solvents. Thermosetting plastics tend to have a high resistance to solvents and it is generally thermoplastics and rubbers which are most vulnerable. Checks must be made on their solvent resistance when selecting these materials for service conditions where chemical substances are present. Metals and ceramics usually have a high resistance to the solvents that attack polymers but they can be attacked by other substances, particularly acids.
See Questions 4.5 on p. 274.

Engineering components often pass through a number of forming and finishing processes during the production cycle. These may be sub-divided into primary and secondary processes. The primary processes include moulding, casting, forging, drawing, rolling and extrusion. They are performed on the raw material to give it form and shape. Sometimes, as is the case with polymers and die-cast metals, the finished product can be made in one primary forming process. More often, a component requires secondary forming and finishing by machining, heat treatment and surface protection.

The properties of a material affect the choice of production process and conversely, a production process may affect the properties of a material. In the case of metals, it is the primary forming processes and subsequent heat treatment which have the greatest effect on the final properties of a component.

Effects of forming

When metal components are cast to shape, the rate at which they are allowed to cool affects the size of the grains. Slow cooling tends to produce large grains, which may leave the material weak. Fast cooling tends to produce smaller grains and a stronger material but if the cooling is uneven there may be a variation in grain size. This can result in internal stress concentrations, which may lead to distortion and cracking.

Hot and cold forming by forging, rolling, drawing and extrusion distorts the grain structure of a metal and causes it to flow in a particular direction. This can have a strengthening effect on the material but in the case of cold forming, it can also produce work hardness and brittleness.

Primary forming is often followed by heat treatment, the objective of which is to refine the grain size and remove internal stress concentrations. The material can then proceed to secondary forming by machining after which it may undergo further heat treatment to give it the required degree of hardness and toughness. The most common heat treatment processes are annealing, normalising, quench hardening, tempering, case hardening and precipitation hardening. Some ceramics can also have their properties modified by heat treatment.

Annealing

The purpose of annealing is to restore the ductility and malleability of work hardened material after cold working. The material is heated in a furnace to what is known as the recrystallisation temperature. At this temperature, new crystals or grains start to form and grow in the regions where the old grains are most distorted. The process continues until the new grains have completely replaced the old deformed structure. When recrystallisation is complete the material is cooled. If the annealing process is carried on for too long a time, the new grains will grow by feeding off each other to give a very course grain structure. This can make the material too soft and weak (Figure 4.25).

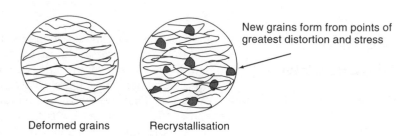

Deformed grains   Recrystallisation

Figure 4.25 *Recrystallisation*

Carbon steels, copper brass and aluminium can all have their grain structure reformed by annealing. Figure 4.26 shows the range of temperatures at which annealing is carried out for plain carbon steels. After annealing for a sufficient length of time, steel components are cooled slowly in the dying furnace. Pure aluminium is annealed between 500 °C and 550 °C, cold worked brass between 600 °C and 650 °C and copper between 650 °C and 750 °C. Unlike steels, these materials can be quench-cooled after annealing.

Normalising

Normalising is a process which is mainly carried out to refine the grain structure and relieve internal stress concentrations in

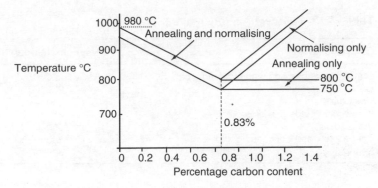

Figure 4.26 *Annealing temperature range for plain carbon steels*

components which have been hot formed to shape. Components formed by hot forging and pressing are very often normalised prior to machining. The process for steel components is similar to annealing except for carbon contents above 0.83%. Here the normalising temperature is higher than that used for annealing as shown in Figure 4.26. Normalising also differs from annealing in the rate at which components are cooled after recrystallisation. The usual practice is to allow steel components to cool more quickly in still air.

Quench hardening

Medium and high carbon steels which have a carbon content of above 0.3% can be hardened by heating them to within the same temperature band as for annealing and then quenching them in water or oil. Structural changes take place at these high temperatures. The iron atoms rearrange themselves from a body-centred to a face-centred cubic structure, which is known as *austenite*, and all of the carbon atoms are taken into solution.

In this condition the steel becomes very malleable and ductile which is why hot working is done at these high temperatures. It must be remembered that the change of structure takes place with the metal still in the solid state. When the steel is quenched, it does not have time to revert to its original grain structure. The result is a new grain formation called *martensite*, which consists of hard needle like crystals.

Quenching in water gives the fastest rate of cooling and the maximum hardness. The violence of the cooling can however cause cracking, particularly with high carbon steels. Oil quenching is slower and less violent. The steel is slightly less hard but cracking is less likely to occur. High carbon steels should always be oil quenched. Mild steel with a carbon content below 0.3% does not respond to quench hardening. It can, however, be case hardened on its surface, as will be described.

Tempering

Quench hardened components are generally too hard and brittle for direct use. The tempering process removes some of the hardness and toughens the steel.

Table 4.13 *Tempering temperatures*

Components	Tempering temperature	Oxide colour film
Craft knife blades and woodworking tools	220 °C	Pale straw
Lathe cutting tools	230 °C	Medium straw
Twist drills	240 °C	Dark straw
Screwcutting taps and dies	250 °C	Brown
Press tools	260 °C	Brown to purple
Cold chisels and punches	280 °C	Purple
Springs	300 °C	Blue

Tempering is achieved by re-heating it to temperatures between 200 °C and 600 °C and quenching again in water or oil. The temperature to which steel components are re-heated depends on their final use as shown in Table 4.13.

Large batches of components are hardened and tempered in special temperature-controlled furnaces. Small single items may be hardened in the workshop by heating in a gas flame to a bright cherry red colour and quenching. The surface is then polished and they are then gently re-heated until oxide colour films start to spread over the surface. When the colour film that corresponds to the required tempering temperature starts to appear, the components are quenched.

Key point

Quench hardened components are tempered to remove some of the hardness and increase the toughness of the material.

Case hardening

Mild steel does not respond to quench hardening because of its low carbon content. Case hardening increases the surface hardness of the material whilst leaving the core in its soft and tough condition. The first part of the process is known as carburising where the components are 'soaked' for a period of time at high temperature in a carbon bearing material. The traditional method is to pack them in cast iron boxes with a carbon rich powder. This may be purchased under a variety of trade names or made up from a mixture of charcoal and bone meal. The carbon slowly soaks into the steel to give an outer case with a high carbon content. The depth of the case depends on the time of soaking.

The second part of the process is to re-heat the components to refine the grain size in the core and then to quench harden and temper the outer case. Case hardened component have a hard and wear resistant outer case and a tough impact resistant core. This is an ideal combination of properties for many engineering components.

Case hardening is not confined to mild steels. Medium and high carbon steels in the normalised condition can be case hardened by rapidly heating and quenching the outer surface. Induction hardening is such a process, where an induction coil induces a high frequency electric current into the component as it passes through the coil. This has a rapid heating effect after which the component passes through a water or oil jet to quench the surface.

Key point

Case hardening leaves a component with a hard wear resistant surface whilst retaining a tough, impact resistant core.

Key point

Precipitation or age hardening can only be carried out on aluminium alloys which contain copper or magnesium and silicon.

Test your knowledge 4.12

1. What takes place at the recrystallisation temperature in a work hardened material?
2. What is the purpose of the annealing and normalising processes?
3. How should high carbon steels be quench hardened?
4. At what temperature should springs be tempered?
5. What happens in aluminium alloys during precipitation hardening?

Selection of engineering materials

Precipitation hardening

Aluminium alloys which contain copper or small amounts of magnesium and silicon among their constituents can be hardened by this process. After hot forming followed by slow cooling the material is quite soft and its properties are similar to those of pure aluminium. The first part of the process is to heat the material to around 500 °C at which temperature, the copper atoms become fully absorbed in the aluminium crystal lattice structure. The material is then quenched to retain the new formation and is then found to be harder and tougher than before.

The new structure is, however, unstable and over a period of time tiny particles of $CuAl_2$ start to precipitate out of the solid solution. These are evenly scattered throughout the structure and have the effect of making the material harder still. The effect is known as 'age hardening' and over a period of several days the material becomes noticeably harder and stronger.

The precipitation process can be speeded up by re-heating the material to between 120 °C and 160 °C for a period of around 10 hours. The final hardness and strength is then greater than that achieved by age hardening at room temperature. Aluminium alloys which contain silicon and magnesium also respond to precipitation hardening. Here it is the intermetallic compound Mg_2Si which precipitates. See Table 4.8 for aluminium alloys which can be heat treated.

See Questions 4.6 on p. 275.

Activity 4.6

A cold chisel that has been in use for some time has become blunted and has a heavily mushroomed head. Write an account of how you would completely recondition the chisel to make it serviceable and safe for use.

The success of an engineering company depends on the quality of its products. A quality product is one that is fit for its purpose and which will have an acceptably long and trouble free service life. Furthermore its material, production and maintenance costs must be kept within reasonable limits so that the company can make an acceptable profit and remain competitive. The selection of engineering materials is therefore an important part of the design process in which the following considerations have to be taken into account.

- What exactly is the function of the product and what will be its service conditions?
- What, if any, are the legal requirements of the product and are there any National or International Standards or Design Codes which must be adhered to?
- What are the properties required of the materials?
- What are the available processing facilities and labour skills?
- Which are the most suitable materials, are they readily available and what will be their cost?

Design considerations

One has only to think of the components which go into the production of a mountain bike to realise the importance of material selection. The designers must be fully aware of the mechanical properties required. The frame, forks and brake cables must have sufficient tensile strength. They must also be impact resistant as must the wheels and pedals. The chain, sprockets and bearings must be tough and wear resistant. The springs in the suspension must be tough and elastic and the tyres must be durable.

The designers must also be aware of the production processes that will be required. The frame parts will need to be formed to shape and joined by welding or hard soldering. Other components will need to be machined and heat treated. Before final assembly, all of the parts will need to be given some form of surface protection to guard against corrosion. Many of the components that go to make up a mountain bike will of course be bought-in complete from specialist suppliers. They will have followed the same material selection procedures to ensure that their products are of a high quality.

It is important to know the loads which engineering components are likely to carry. This often involves complex calculations, particularly in the design of aircraft, motor vehicles, and manufacturing machinery. Materials with appropriate load-bearing properties can then be selected and the component dimensions calculated to ensure that the working stresses are below the strength limits of the materials by a safe margin. Prototype components are often subjected to tests that simulate service conditions. Component failure or excessive wear might lead to the choice of alternative materials with better mechanical properties. New and untried materials should not be selected for their novelty value. There are many instances where unforeseen deterioration has occurred in service. New materials should be rigorously tested to confirm the properties claimed for them and information sought from other users as to their success.

Processability is an important factor that must be taken into account when selecting materials for engineered products. Materials with the required mechanical properties might be difficult and expensive to machine to the required dimensional tolerances and surface finish. The materials selected for castings must have good fluidity when molten. Steel has many good mechanical properties but it is difficult to cast into complex shapes. The materials for forging or pressing must have a sufficient degree of malleability and ductility and some materials, such as aluminium, are difficult to weld without specialised skill and equipment. These are all factors which must be considered in material selection to make the best use of existing plant, equipment and skills.

Key point

The materials selected for engineered products must be fit for their intended purpose, be suitable for the available process plant and be readily available at a reasonable cost.

Costs

Design engineers must always be aware of material and processing costs. There have been many examples of products which have been 'over engineered'. That is to say that the design has been too complex and the use of materials too extravagant. Many firms have gone out of business as a result. The material costs of a product are often a significant part of the overall manufacturing cost. The cost of a material depends on its scarcity as a natural resource, the amount of processing needed to convert it from its raw state into a usable form and the quantity and regularity with which it is purchased.

It may be that the material which appears to be the most suitable in terms of its properties, is the most expensive. Cheaper alternatives may be available with similar properties but these could prove to be the more expensive in the long term. A judgement must be made of the comparative processing costs and also the cost of replacement under warrant if the alternatives prove to be unreliable in service.

Wherever possible, materials such as bar stock, rolled, drawn and extruded sections should be purchased in standard sizes and with the required physical properties and surface finish. The cost of processing can then be kept to a minimum. Discount can usually be obtained by buying in bulk or placing orders for regular delivery. Bulk purchases might, however, require a considerable financial outlay and then take up storage space for a period of time. Overall, the cost of this might cancel out any initial savings. There are also pitfalls in placing too much reliance on a single supplier. Labour, manufacturing and transport problems might interrupt the supply which might again cancel out any cost savings.

Key point

Discount from the bulk buying of materials can be cancelled out by the cost of storing and lack of return on the money which has been invested in them.

Forms of supply

Engineering materials are available in a variety of forms. Some of the most common are shown in Table 4.14. The availability of different materials varies with material type and the form of supply. Engineering metals such as carbon steels, copper, brasses and aluminium alloys may be readily obtained in the form of bar, tube, sheet etc. from a number of alternative suppliers in standard sizes.

Castings, forgings and pressings are not so easily obtained. Patterns and dies must be made before supply can commence. These are expensive to produce and the cost may only be justified

Table 4.14 *Forms of supply*

Metals	Polymers	Ceramics	Composites
Ingots	Granules	Mouldings	Mouldings
Castings	Powders	Sheet	Sheet
Forgings	Resins	Plate	Plate
Pressings	Mouldings	Cements	
Barstock	Sheet		
Sheet	Pipe and tube		
Plate	Extrusions		
Rolled sections			
Pipe and tube			
Wire			
Extrusions			

if large quantities are required. Careful forward planning and close liaison with the supplier is required if supply is to commence on time. As has already been stated, problems may then arise from placing too much reliance on a single supplier. The cost of duplicate tooling and the quantities required might however dictate that this is a gamble that has to be taken.

The most common raw polymer materials in the form of granules, powders, liquid resins and hardeners are also readily available from a number of suppliers. As with metal castings and pressings, moulded polymers require the production of moulds and dies, which might take a considerable time to produce. The same applies to ceramics. Sheet glass and building materials are readily available but the supply of specially moulded components requires careful forward planning.

Newly developed alloys, polymers and composites might appear to have superior properties but they may only be available in limited quantities from a single supplier. Many firms now operate on a low stock or 'just in time' system to minimise storage costs. Costly production stoppages can result from an interruption in deliveries and wherever possible, it is good practice to have a back-up supplier. When selecting materials and liasing with suppliers, care should also be taken to ensure that supplies will be available to meet possible increases in demand.

Key point

Components that are to be cast, moulded, forged or pressed to shape will require expensive patterns, moulds and dies. The cost of these needs to be justified by the quantities that will be required.

Information sources

Product safety is often a concern when selecting engineering materials. This is particularly true for items such as pressure vessels, steam plant, structural supports and electrical equipment. Reference often needs to be made to British Standards Specifications (BS) and International Standards Organisation (ISO) specifications for particular sectors of engineering. Typical examples are BS 5500 for the design of pressure vessels, BSEN 60335 which lays down the durability and safety specifications for domestic products and BSEN 10002 for the tensile testing of materials. Particular grades of material are also covered by British Standards. For example, BS 970 covers the different grades of plain carbon and alloy steels, BS 1449 for steel plate sheet and strip, BS 2875 for copper and copper alloy plate and BS 2870 which specifies the composition of brasses.

A great deal of information which is of help when selecting materials is to be found in manufacturers' and stockholders' catalogues and data sheets. They are particularly useful for selecting standard forms of supply such as barstock, sheet and rolled structural section. They may be obtained in hard copy and in many cases they may be accessed via the Internet. Trade directories which contain the names, addresses and websites of material suppliers are often held in the reference section of public libraries and the resource centres.

The selection of materials has been made easier in recent years by the availability of computer databases that are compiled and updated by manufacturers and research bodies. Two such packages are the Cambridge Materials Selector, compiled by the Cambridge University Engineering Department, and MAT.DB which is compiled by the American Society for Metals.

They lead the user to a choice of suitable alternative materials through a structured question pathway. Access costs are involved but the time and money saved often make it well worth the outlay. They are particularly useful in introducing designers to new materials.

Trade association brochures and journals published by the engineering institutes often contain information on new materials. Much information can also be obtained from specialist data books such as the following:

Metals Reference Book – R.J. Smithells (Butterworth)
Newnes Engineering Materials Pocket Book – W. Bolton (Heinemann-Newnes)
Metals Data Book – C. Robb (Institute of Metals)
Handbook of Plastic and Elastomers – edited by C.A. Harper (McGraw-Hill)

Text books with titles such as Metallurgy, Polymer Science, Engineering Materials etc. are useful for providing the underpinning knowledge and theory of materials engineering. These are to be found under the Dewey Codes 620 and 621 in college and public libraries. Some, from the many available, are as follows:

Properties of Engineering Materials – R.A. Higgins (Butterworth-Heinemann)
Materials for Engineering – W. Bolton (Newnes)
Materials Science – E. Ramsden (Stanley Thornes)
Engineering Materials 1 & 2 – M. Ashby and R.H. Jones (Butterworth-Heinemann)

Test your knowledge 4.13

1. Why do design engineers have to consider processability when selecting materials for engineered products?
2. What might be the advantages and disadvantages of purchasing material from a single supplier?
3. What form does the raw material take which is used for producing moulded polymers?
4. Why must the purchase orders for castings be placed well in advance of the time when they will be required?
5. What might be the most appropriate information source on the availability of engineering materials in standard sizes?

Activity 4.7

Select any of the following products and identify the materials from which it is made. Explain why you think they were selected and what, if any, are the possible alternatives. Describe the form in which the materials might have been supplied and how they might have been processed. Identify any standards which are relevant to their design and manufacture: (a) a bicycle pump, (b) a woodworking plane, (c) a mains extension lead, (d) a 250 mm (10 inch) hacksaw, (e) a motorcycling-crash helmet.

Modes of failure

When the load on a ductile material exceeds the elastic limit, it becomes permanently deformed and *elastic failure* is said to have occurred. The material may still be intact but it is likely that the component from which it is made will no longer be fit for its intended purpose. Brittle materials such as cast iron, very often fail in the elastic range with the brittle types of fracture shown in Figure 4.27. Brittle fracture, which is also known as cleavage fracture, is more prevalent in materials with BCC and CPH crystal lattice structures. Under certain conditions, ductile materials can also fail with a brittle type of fracture, as will be explained.

Both kinds of failure are to be avoided by incorporating a suitable *factor of safety* into the design of engineering components. As a general rule, factor of safety of at least 2 should be employed

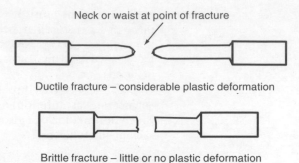

Ductile fracture – considerable plastic deformation

Brittle fracture – little or no plastic deformation

Figure 4.27 *Fracture characteristics*

on static structures. With this in place, the working stress in the material should always be less than half of that which will cause failure. That is,

$$\text{factor of safety} = \frac{\text{stress at which failure occurs}}{\text{safe working stress}}$$

In spite of the best intentions of design engineers, components sometimes fail in service. Static loads can be hard to predict and dynamic loads on the component parts of machinery, motor vehicles and aircraft are very difficult to analyse. Combinations of direct loading, shearing, bending and twisting are very often present. A complex stress system is then said to exist, the resultant of which may exceed the predicted working stress and lead to failure.

An additional danger is the presence of stress concentrations in a component. These can occur at sharp internal corners, holes, fixing points and welds. They are known as *stress raisers*, where the stress may exceed that at which failure occurs. Under certain conditions, cracks can spread from these points, which eventually lead to failure. These kinds of failure are usually detected at the prototype stage and the design modified to prevent them occurring. Material faults such as the presence of cavities, impurities, large grain size and inappropriate heat treatment can also contribute to failure if not detected by quality control procedures.

Under certain circumstances, materials can fail at comparatively low stress levels that would normally be considered to be quite safe. The main reasons for this are changes in temperature, which can affect the properties of a material, and cyclic loading. Low temperatures can cause brittleness and loss of strength. High temperatures can cause the material to *creep*, and eventually fail, under loads that are well below the normal elastic limit. A material is subjected to cyclic loading when it is repeatedly being loaded and unloaded. The loads may be well below that which would be expected to cause failure, but over a period of time, failure can occur due to *metal fatigue*. Some of these failure modes will now be described in detail.

Key point

Materials can fail due to metal fatigue, creep and brittle fracture at stress levels which would normally be considered safe.

Brittle fracture

The plastic deformation which precedes a ductile fracture takes a finite amount of time to take place. If a load in excess of that which will cause fracture is suddenly applied, as with an impact load, there will be insufficient time for plastic deformation to take place and a brittle form of fracture may occur. This can be observed during an Izod or Charpy impact test where an otherwise ductile material is suddenly fractured by an impact load.

Brittle, or cleavage fractures usually have a granular appearance due to the reflection of light from the individual grains. Too large a grain size can affect the strength of a material and make it brittle. Grain growth can occur when materials are operating at high temperatures for long periods of time. Here the grains feed off each other in cannibal fashion, reducing the strength of the material and increasing the likelihood of brittle fracture.

Some metals which exhibit ductile behaviour under normal conditions become very brittle at low temperatures. The temperature at which the change occurs is called the *transition temperature*. Mild steel becomes brittle at around 0 °C. As can be seen from Figure 4.28 the transition temperature is judged to be that at which the fracture surface of an Izod or Charpy specimen is 50% granular and brittle and 50% smooth and ductile. It is generally the metals with BCC and CPH crystal lattice structures that are affected in this way by low temperatures. The ferrite grains in steel, which are almost pure iron, have a BCC structure at normal temperatures. Chromium, tungsten and molybdenum are also BCC and suffer from low temperature brittleness in the same way as iron.

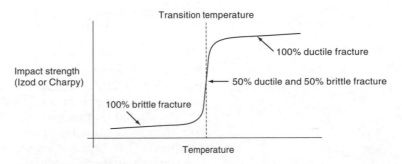

Figure 4.28 *Transition from ductile to brittle material*

The cracking and sinking of the all-welded liberty ships in the Second World War was attributed to brittle fracture. The cold north Atlantic temperatures produced brittleness in the steel hulls. Cracks appeared in places of stress concentration and these quickly spread with disastrous effects. The transition temperature in mild steel is raised by the presence of phosphorous and lowered by the addition of manganese and nickel. A relatively high level of phosphorous was present in the steel of the liberty ships and all ship hulls are now made from steel which is low in phosphorous.

Key point

At a transition temperatures of around 0 °C mild steel with a high phosphorous content becomes brittle. The presence of nickel and manganese lowers the transition temperature.

Creep

Creep is a form of plastic deformation which takes place over a period of time at stress levels which may be well below the yield stress of a material. It is temperature related and as a general rule, there will be little or no creep at temperatures below $0.4 \times T$, where

T, is the melting point of the material measured on the Kelvin scale. For mild steel, $T = 1500\,°C$ which is 1773 K and so there should be very little creep below $0.4 \times 1773 = 709$ K which is $436\,°C$. It should be stressed that this is only a general rule and that some of the softer low-melting point metals such as lead, will creep under load at normal temperatures.

With the more common engineering metals, creep is a problem encountered at sustained high temperatures such as those found in steam and gas turbine plant. Under extreme conditions it can eventually lead to failure. A typical graph of creep deformation against time is shown in Figure 4.29.

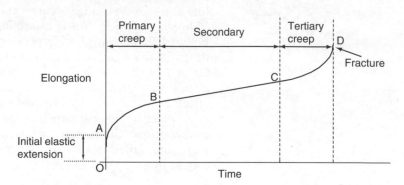

Figure 4.29 *Creep v. time graph*

Figure 4.29 shows the behaviour of a material which is above the threshold temperature at which creep is likely to occur. When it is initially loaded, the elastic extension OA is produced. If the stress in the material is below a level called the *limiting creep stress* or *creep limit* at that temperature, there will be no further extension. If, however, the stress is above this level, primary creep AB, commences. This begins at a rapid rate, as indicated by the slope of the graph, and then decreases as work hardening sets in. It is followed by secondary creep BC, which takes place over a comparatively long period at a steady rate. The final stage is tertiary creep CD, where the deformation rate increases. Necking becomes apparent at this stage, leading finally to fracture at D. Creep in polymer materials below the glass transition temperature is found to proceed in the same way.

Increases in temperature and/or increases in stress have the effect shown in Figure 4.30. The rate at which the three stages of creep take place is increased and failure occurs in a shorter time. Study of the nature of creep suggests that the plastic deformation is partly due to slipping of the planes of atoms in the grains and partly due to viscous flow at the grain boundaries. The atoms tend to pile up in an irregular fashion at the grain boundaries which would normally lead to work hardening. The high temperatures, however, have a relieving effect, and the smaller the grains the greater is the viscous flow at the grain boundaries.

Creep resistance can be increased in two ways. The first is to introduce alloying elements which reduce slipping within the grains. The second is to have as course a grain structure as possible, bearing in mind that this can lead to increased brittleness at

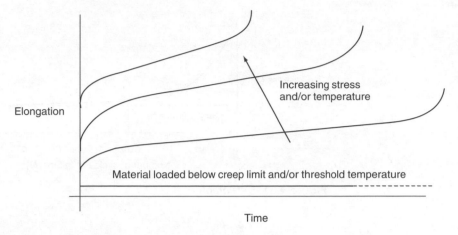

Figure 4.30 *Effect of increasing the load and temperature*

normal temperatures. Many creep resistant materials have been developed over the last 50 years, in particular the nimonic series of alloys which have been widely used in gas turbines.

Fatigue

Fatigue failure is a phenomena which can occur in components which are subjected to cyclic loading. That is to say that they are repeatedly subjected to fluctuating or alternating stresses. Typical examples are the suspension units on motor vehicles and the connecting rods and crankshafts in internal combustion engines. The forces and vibrations set up by out-of-balance rotating parts can also produce cyclic loading. The alternating stresses may be well below the elastic limit stress, and the material would be able to carry a static load of the same magnitude indefinitely. Failure usually starts with a small crack which grows steadily with time. Eventually the remaining cross-sectional area of the component becomes too small to carry the repeated loads and the material fractures.

It is found that in ferrous metals, there is a certain stress level below which fatigue failure will not occur no matter how many stress reversals take place. This is called the *fatigue limit* and is given the symbol S_D. As a general rule for steels, the fatigue limit is about one half of the UTS of the material. The higher the stress above this value, the fewer will be the number of reversals or stress cycles before failure occurs. A typical graph of stress level against the number of cycles leading to failure for a material such as steel is shown in Figure 4.31. The graph is often referred to as an *S–N* curve.

Fatigue cracks are observed to spread from points of stress concentration. Cyclic loading at stress levels above S_D produces slip in the planes of atoms in the grains of a material. This results in the appearance of small extrusions and intrusions on the surface of an otherwise smooth material, as shown in Figure 4.32. Although the intrusions are very small, they act as stress raisers from which a fatigue crack can spread. If other stress raisers are

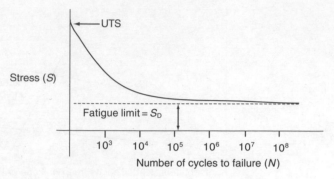

Figure 4.31 *S–N graph*

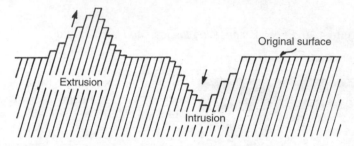

Figure 4.32 *Intrusions and extrusions due to local slip*
Source: *From Properties of Engineering Materials by R.A. Higgins. Reprinted by permission of Elsevier Ltd*

present such as sharp internal corners, tool marks, and quench cracks from heat treatment, the process can be accelerated and these should be guarded against.

The fracture surfaces of a fatigue failure have a characteristic appearance as shown in Figure 4.33. As the fatigue crack spreads, its two sides rub together under the action of the cyclic loading. This gives them a burnished, mother-of-pearl appearance. Eventually, the material can no longer carry the load and fractures. The remainder of the surface, where fracture has occurred, has a crystalline or granular appearance.

There are many non-ferrous metals which do not have a fatigue limit and which will eventually fail even at very low levels of cyclic loading. Some steels, when operating in corrosive conditions, exhibit these characteristics. The *S–N* graph for such materials is

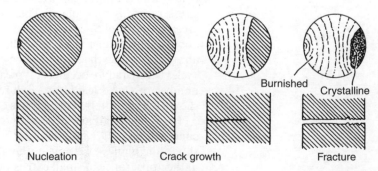

Figure 4.33 *Fatigue fracture appearance*
Source: *From Properties of Engineering Materials by R.A. Higgins. Reprinted by permission of Elsevier Ltd*

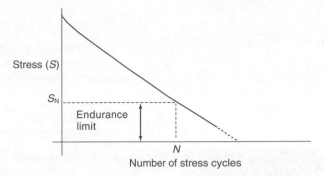

Figure 4.34 *Endurance limit*

shown in Figure 4.34. Instead of a fatigue limit, they are quoted as having an *endurance limit* which is given the symbol S_N. The endurance limit is defined as the stress which can be sustained for a given number of loading cycles. Components made from these materials should be closely monitored, especially when used in aircraft, and replaced at a safe time before the specified number of cycles N, has been reached.

Degradation

Ferrous metals are affected by two kinds of corrosion. Low temperature or 'wet' corrosion is due to the presence of moisture and results in the formation of red rust. This, as you well know, is very loose and porous. Red rust is an iron oxide formed by electrochemical action, in which the moisture acts as an electrolyte. Adjacent areas of the metal, which have a different composition, such as the alternate layers of ferrite and cementite in the pearlite grains, become the anodes and cathodes. Corrosion occurs at the anode areas resulting in rust formation.

Figure 4.35 shows that the ferrite layers, which are almost pure iron, become anodes and corrode to form $FeOH_3$ which is red rust. The same kind of electrolytic action can occur between adjacent areas which have been cold worked to a different extent. Figure 4.36 shows a fold in a sheet of metal which is more highly stressed than the surrounding areas. In the presence of moisture, the region in the fold becomes an anode and corrodes. This kind of electrolytic action is called *stress corrosion*. It is the form from which motor vehicle panels can suffer if they are not properly protected.

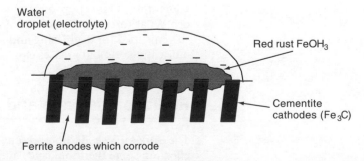

Figure 4.35 *Wet corrosion of pearlite grain*

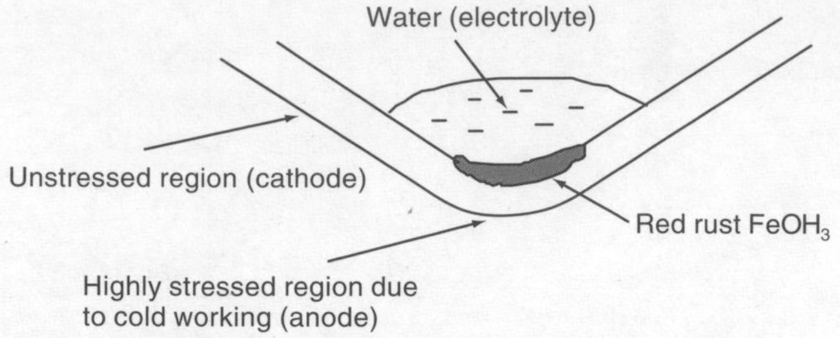

Water (electrolyte)

Unstressed region (cathode)

Red rust FeOH₃

Highly stressed region due
to cold working (anode)

Figure 4.36 *Stress corrosion*

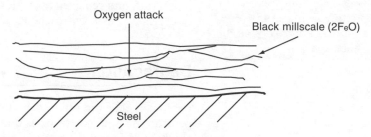

High temperature or 'dry' corrosion occurs due to a direct chemical reaction between the metal and oxygen of the atmosphere. It results in the formation of black millscale when the metal is heated for forging or for heat treatment. Millscale is another form of loose and porous iron oxide whose chemical formula is FeO (Figure 4.37).

Oxygen attack

Black millscale (2F$_e$O)

Steel

Figure 4.37 *High temperature corrosion*

As has been mentioned, the oxide films that form on the surface of non-ferrous metals and alloys are generally quite dense. Polished copper, brass and silver very soon become tarnished but once a thin oxide film has formed, it protects the metal from further attack. Sometimes the oxide film is artificially thickened by an electrolytic process known as *anodising*. Aluminium alloys for outside uses such as door and window frames, are treated in this way.

Solvent attack

Thermosetting plastics tend to have a high resistance to solvents and it is generally thermoplastics and rubbers which are most vulnerable. The action of the solvent is to break down the Van der Waal forces and take the polymers into solution. Industrial solvents used for degreasing and for paint thinners, petrol, fuel oil, lubricating oils and greases can have this effect on some polymers.

Radiation damage and ageing

The ultra-violet radiation present in sunlight can have a degrading effect on some thermoplastics and rubbers. It progressively causes oxygen atom cross-links to form between the polymers. These cause the material to become brittle and can also lead to discolouration. Ultra-violet lamps and X-rays used in industrial processes can also cause this kind of degradation. Colouring pigments are often added during the polymer forming process, and this reduces the effect. The darker colours are the most effective, black being the best of all.

Deterioration of ceramics

The ceramic tiles, bricks, cements and natural stone used for building degrade with time due to moisture and pollutant gases in the atmosphere. The absorption of rain-water into the surface pores can cause deterioration in winter. When the moisture freezes, it expands and over a period of time it can cause cracking and flaking. Sulphur from flue and exhaust gases combines with moisture in the atmosphere to form sulphurous acid which falls as acid rain. This attacks many types of ceramic building material and in particular, natural stone. The refractory ceramics used to line furnaces, and the ladles for carrying molten metal, can suffer from thermal shock if heated too quickly. Because they are poor conductors of heat, there can be a very large temperature difference between the heated surface and the material beneath it. As a result, the expansion of the surface layer can cause flaking or *spalling*.

Refractory linings can also be attacked at high temperatures by the slag which rises to the surface of molten metal. There are two types of slag which form, depending on the impurities present in the metal. One is acidic and the other is known as basic slag. Linings of silica brick are resistant to acidic slag whilst linings of *dolomite*, which contains calcium and magnesium carbonates, and *magnesite*, which contains magnesium oxide, are resistant to basic slag.

See Questions 4.7 on p. 276.

Test your knowledge 4.14

1. Define what is meant by a *factor of safety*.
2. How might a brittle type of fracture occur in what is normally a ductile material?
3. How does the rate of material creep vary with stress and working temperature?
4. How can fatigue failure be identified from the appearance of the fracture surfaces?
5. What are the mechanisms which lead to the formation of red rust on steel components?

Activity 4.8

You are required to design a test rig to investigate the creep characteristics of lead specimens at normal temperature when subjected to different loads. A requirement of the apparatus is that it must be able to supply an electrical output signal which is proportional to the extension in order to obtain graphs of creep v. time from a *y–t* plotter.

or

You are required to design a fatigue testing machine to investigate the behaviour of mild steel test specimens when subjected to different values of cyclic stress. A requirement of the apparatus is that it must run unattended, record the number of cycles to failure and automatically switch off the power supply when failure occurs.

Questions 4.1

1. Medium carbon steel has a carbon content of:

   (a) 0.1% to 0.15%
   (b) 0.15% to 0.3%
   (c) 0.3% to 0.8%
   (d) 0.8 to 1.4%.

2. A property of grey cast iron is that:

   (a) it is strong in tension
   (b) it is easy to machine
   (c) it is ductile and malleable
   (d) it is weak in compression.

3. The main constituents of brass are:

   (a) copper and tin
   (b) copper and nickel
   (c) copper and lead
   (d) copper and zinc.

4. The tin–bronze used for bearings and gears contains a small amount of:

   (a) phosphorous
   (b) iron
   (c) nickel
   (d) manganese.

5. Precipitation hardening is a heat treatment process which is carried out on:

   (a) medium and high carbon steels
   (b) certain aluminium alloys
   (c) aluminium bronzes
   (d) grey cast iron.

Answers: (c), (b), (d), (a), (b)

Questions 4.2

1. Polychloroethene is better known as:

   (a) perspex
   (b) polythene
   (c) PVC
   (d) polystyrene.

2. At the glass transition temperature:

   (a) cross-linking occurs in thermosetting plastics
   (b) thermoplastics become liquid
   (c) thermosetting plastics become hard and brittle
   (d) thermoplastics become soft and flexible.

3. The polyamide group of thermoplastics contains

   (a) nylon
   (b) PTFE

(c) terylene

(d) perspex.

4. Bakelite is used for

(a) insulation for electrical wiring
(b) drinks cups
(c) boat hulls
(d) vehicle distributor caps.

5. The synthetic rubber which best retains its flexibility over a wide temperature range is:

(a) natural rubber
(b) silicone rubber
(c) butyl rubber
(d) styrene rubber.

Answers: (c), (d), (a), (d), (b)

Questions 4.3

1. The main constituent of glass is:

(a) aluminium oxide
(b) silicon oxide
(c) beryllium oxide
(d) calcium oxide.

2. Amorphous ceramics are a group of materials which include the different types of:

(a) cement
(b) whiteware
(c) abrasive
(d) glass.

3. Bonding of the crystals in clay ceramics occurs at the:

(a) glass transition temperature
(b) vitrification temperature
(c) melting point
(d) tempering temperature.

4. A common property of ceramic materials is that they are:

(a) good conductors of heat
(b) impact resistant
(c) good electrical insulators
(d) strong in tension.

5. A cermet such as tungsten carbide, may be classified as a:

(a) fibrous composite
(b) laminate
(c) amorphous ceramic
(d) particulate composite.

Answers: (b), (d), (b), (c), (d)

Questions 4.4

1. A tensile test specimen of initial diameter 10 mm which fractures at a tensile load of 31 kN will have a UTS value of:

   (a) 550 MPa
   (b) 395 MPa
   (c) 210 MPa
   (d) 430 Mpa.

2. A bend test may be carried out on a material specimen to investigate its:

   (a) toughness
   (b) malleability
   (c) elasticity
   (d) ductility.

3. Malleability is a property of a material which enables it to:

   (a) withstand impact loading
   (b) be drawn out by tensile forces
   (c) be deformed by compressive forces
   (d) return to its original shape when unloaded.

4. The toughness of a material may be investigated by carrying out a:

   (a) tensile test
   (b) Brinell test
   (c) Charpy test
   (d) bend test.

5. At the point of fracture the diameter of a tensile test specimen is 9.5 mm. If the initial diameter was 12 mm, its ductility measured as percentage reduction in area will be:

   (a) 20.8%
   (b) 27.6%
   (c) 32.9%
   (d) 37.3%.

Answers: (b), (d), (c), (c), (d)

Questions 4.5

1. A bar of a material is 500 mm in length at a temperature of 20 °C. If its length at 150 °C is 501.04 mm, the linear expansivity of the material will be:

   (a) $12 \times 10^{-6} \,°C^{-1}$
   (b) $13 \times 10^{-6} \,°C^{-1}$
   (c) $16 \times 10^{-6} \,°C^{-1}$
   (d) $19 \times 10^{-6} \,°C^{-1}$.

2. The temperature difference across a tile of surface area $22.5 \times 10^{-3} \, m^2$ and thickness 10 mm is 100 °C. If the rate of heat transfer through the plate is 200 W, the thermal conductivity of the material will be:

   (a) $1.51 \, W \, m^{-1} \, °C^{-1}$
   (b) $0.48 \, W \, m^{-1} \, °C^{-1}$

(c) $2.33\,\mathrm{W\,m^{-1}\,{}^{\circ}C^{-1}}$
(d) $0.89\,\mathrm{W\,m^{-1}\,{}^{\circ}C^{-1}}$.

3. If a 10 m length of resistance wire of diameter 2 mm has a resistance of $1.55\,\Omega$, its resistivity will be:

(a) $48.7 \times 10^{-9}\,\Omega\,\mathrm{m}$
(b) $36.6 \times 10^{-9}\,\Omega\,\mathrm{m}$
(c) $53.6 \times 10^{-9}\,\Omega\,\mathrm{m}$
(d) $21.8 \times 10^{-9}\,\Omega\,\mathrm{m}$.

4. The ability of a material to intensify an electric field is indicated by the value of its:

(a) permittivity
(b) conductivity
(c) permeability
(d) resistivity.

5. Sacrificial protection occurs when steel is coated with

(a) cadmium
(b) tin
(c) nickel
(d) zinc.

Answers: (c), (d), (a), (a), (d)

Questions 4.6

1. Annealing is a heat treatment process carried out to:

(a) harden and toughen the material
(b) increase the carbon content of the material
(c) restore ductility and malleability after cold working
(d) induction harden the material.

2. In the normalising process, steel components are cooled:

(a) by quenching in water
(b) by exposing them in still air
(c) by leaving them in the dying furnace
(d) by quenching in oil.

3. Springs are tempered at a temperature of around:

(a) $220\,^{\circ}\mathrm{C}$
(b) $240\,^{\circ}\mathrm{C}$
(c) $260\,^{\circ}\mathrm{C}$
(d) $300\,^{\circ}\mathrm{C}$.

4. Carburising is a process carried out on components made from

(a) mild steel
(b) aluminium alloys
(c) cast iron
(d) high carbon steel.

5. Age hardening in certain aluminium alloys results from the precipitation of:

(a) spheroidal graphite
(b) new grains or crystals

(c) stress concentrations
(d) intermetallic compounds.

Answers: (c), (b), (d), (a), (d)

Questions 4.7

1. The transition temperature, below which steels become brittle, can be lowered by:

    (a) the addition of magnesium and silicon
    (b) increasing the phosphorous content
    (c) the addition of manganese and nickel
    (d) reducing the carbon content.

2. Creep in materials is more likely to occur:

    (a) at high working temperatures
    (b) when cyclic loading is present
    (c) when there is a coarse grain structure
    (d) in a moist atmosphere.

3. The cyclic loading of a component can result in:

    (a) brittle fracture
    (b) creep failure
    (c) stress corrosion
    (d) material fatigue.

4. The chemical compound $FeOH_3$ is better known as

    (a) millscale
    (b) cementite
    (c) red rust
    (d) silica.

5. Ultra-voilet radiation can cause degradation in some:

    (a) cermets
    (b) thermoplastics
    (c) refractory ceramics
    (d) ferrous alloys.

Answers: (c), (a), (d), (c), (b)

Chapter 5 Fluid mechanics

Summary

By definition, a fluid may be a liquid or a gas. In hydraulic and pneumatic control systems, engineers use fluids as a means of transmitting force. The hydraulic and pneumatic braking systems on motor cars, buses and trucks are typical examples which you meet everyday. Sometimes fluids need to be transported. Water and gas are supplied to your home through pipes. Rainwater and sewage is transported away and a steady flow of fresh air is maintained in workshop and office ventilation systems. Design engineers need to be aware of the forces required to maintain a steady flow so that they can install sufficiently powerful pumps and fans. Engineers also need to be aware of the forces which act on solid bodies as they move through a fluid. This is essential to the efficient design of ships, aircraft and motor vehicles.

Physical properties of fluids

The physical properties of fluids, whose values need to be defined and measured, include density, specific weight, relative density, dynamic and kinematic viscosity, and surface tension. You will find that some of these, particularly density and dynamic viscosity, recur quite often in fluid mechanics calculations.

Density

The density of a substance is defined as its mass per unit volume. In other words, it is the number of kilograms that go to make up a cubic metre of a substance. It is denoted by the Greek letter ρ (rho) and its units are kilograms per cubic metre (kg m^{-3}). If you know the mass m (kg) of a liquid or a gas and the volume V (m^3) in which it is contained, its density can be calculated using the following formula (see Table 5.1 for some typical values):

$$\text{density} = \frac{\text{mass}}{\text{volume}}$$

$$\rho = \frac{m}{V} \tag{5.1}$$

Specific weight

This is closely related to density. It is defined as the weight per cubic metre of a substance and is usually denoted by the symbol w. Weight is of course the mass multiplied by g, the acceleration due to gravity. Specific weight is given by the formula

$$\text{specific weight} = \frac{\text{weight}}{\text{volume}}$$

$$w = \frac{mg}{V} \tag{5.2}$$

or

$$w = \rho g \tag{5.3}$$

The units of specific weight is newtons per cubic metre, or more usually kilonewtons per cubic metre ($kN\,m^{-3}$). You might note that the density of a substance under the same conditions of temperature and pressure will be the same on the Earth, the Moon, Mars or anywhere out in space. The specific weight will however be different, as it depends on the force of gravity.

Relative density

The relative density of a substance is the ratio of its density to the density of water. Sometimes, it is also called *specific gravity* and since it is a ratio, it does not have any unit. You will note from Table 5.1 that the density of water is $1000\,kg\,m^{-3}$.

$$\textbf{relative density} = \frac{\textbf{density of substance}}{\textbf{density of water}} \tag{5.4}$$

> **Key point**
>
> The density of a substance is the same anywhere in the universe for given conditions of temperature and pressure, whereas specific weight changes with the force of gravity.

> **Key point**
>
> The density of water in normal conditions of temperature and pressure is taken to be exactly $1000\,kg\,m^{-3}$.

Table 5.1 *Density, specific weight and relative density values*

Substance	Density ($kg\,m^{-3}$)	Specific weight ($kN\,m^{-3}$)	Relative density
Water	1000	9.81	1
Alcohol	810	7.95	0.81
Mercury	13 600	13.34	13.6
Steel	7870	7.72	7.87
Aluminium	2700	2.65	2.7
Copper	8500	8.34	8.5

Example 5.1

The cylindrical fuel tank on a lorry is 0.5 m in diameter and 1.5 m in length. What will be the weight of fuel contained when it is full? The relative density of diesel oil is 0.86.

Finding volume of tank:

$$V = \frac{\pi d^2}{4} l$$

$$V = \frac{\pi \times 0.5^2}{4} \times 1.5$$

$$\boldsymbol{V = 0.295\,m^3}$$

Finding density of diesel oil:

ρ = relative density $\times$ density of water

$\rho = 0.86 \times 1000$

$\boldsymbol{\rho = 860\,kg\,m^{-3}}$

Finding specific weight of diesel oil:

$w = \rho g = 860 \times 9.81$

$\boldsymbol{w = 8.44 \times 10^3\,N\,m^{-3}}$ or $\boldsymbol{8.44\,kN\,m^{-3}}$

Finding weight of fuel W in tank:

W = volume $\times$ specific weight

$W = Vw$

$W = 0.295 \times 8.44 \times 10^3$

$\boldsymbol{W = 2.49 \times 10^{-3}\,N}$ or $\boldsymbol{2.49\,kN}$

Viscosity

The force needed to change the shape of a fluid is very small when compared with the force needed to deform a solid body. A fluid assumes the shape of its container which may be a tank, a gas cylinder, a pipe, a duct, etc. Fluids do however offer resistance when they are being stirred, poured or made to flow. It is more difficult to do this with some fluids than with others, particularly those that stick to the walls of a container. This resistance is due to the forces of attraction between the fluid molecules themselves, and between the fluid molecules and the container.

Viscosity is a measure of the resistance to shape deformation. It should not be confused with density. Lubricating oil is more viscous than water but it is less dense, and mercury is very dense but has a low viscosity. You will remember that shear stress and strain in solid materials were investigated in Chapter 1. It is a similar shearing action that takes place in a fluid when it is being stirred, poured or pumped along a pipe. The fluid adjacent to the containing surfaces is slowed down and has a much lower velocity than that in the centre of the flow. Consider now a small quantity of fluid which is being subjected to shearing forces as shown in Figure 5.1.

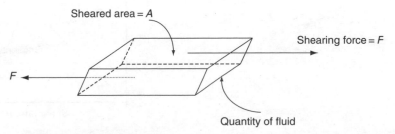

Figure 5.1 *Shearing in a fluid*

As with solid materials, the shear stress τ in the fluid is given by

$$\text{shear stress} = \frac{\text{shear force}}{\text{sheared area}}$$

$$\tau = \frac{F}{A} \ \textbf{(Pa)} \tag{5.5}$$

The fluid can be considered to be made up of an infinite number of layers or *laminae*, rather like the pages of a book. Each one moves a little faster than the one below it, as shown by the velocity profile in Figure 5.2, under the action of the shearing forces. The shearing forces set up a *velocity gradient* across the fluid whose value is given by the symbol μ.

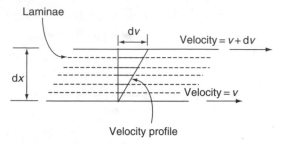

Figure 5.2 *Velocity profile*

$$\text{velocity gradient} = \frac{\text{change in laminar velocity}}{\text{distance between laminae}}$$

$$\mu = \frac{dv}{dx} \ (\text{s}^{-1}) \tag{5.6}$$

The unit of the velocity gradient is metres per second divided by metres, which reduces to s^{-1}. This is very closely related to the shear strain in solid materials. The difference is that in solids the shear strain is static and here it is continuous, i.e. it is the shear strain which occurs per second. For this reason, the velocity gradient μ is also called the *shear rate*.

With liquids such as water and lubricating oils, the shear rate is proportional to the shear stress which is causing it. If you double the shear stress, double the shear rate, etc., a graph plotted of shear stress τ against shear rate μ will be a straight line as shown in Figure 5.3. Liquids that behave in this way are called *Newtonian fluids*. The gradient of the graph is known as

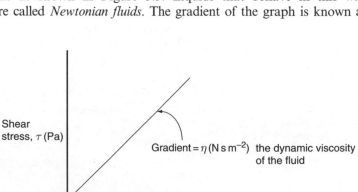

Figure 5.3 *Graph of shear stress v. shear rate*

the *dynamic viscosity* of the fluid. It is denoted by the Greek letter η (eta) and has the units $\mathrm{N\,s\,m^{-2}}$. The value for water is $1.145 \times 10^{-3}\,\mathrm{N\,s\,m^{-2}}$ and the values for a typical lubricating oil is $0.15\,\mathrm{N\,s\,m^{-2}}$.

$$\text{dynamic viscosity} = \frac{\text{shear stress}}{\text{shear rate}}$$

$$\eta = \frac{\tau}{\mu} \ (\mathbf{N\,s\,m^{-2}}) \tag{5.7}$$

You may have noticed that the dynamic viscosity of a fluid is closely related to the shear modulus of a solid. The only difference is that the shear strain is replaced by shear rate in the above expression. Substituting the values of τ and μ from equations (5.5) and (5.6) in equation (5.7) respectively, gives

$$\eta = \frac{\frac{F}{A}}{\frac{\mathrm{d}v}{\mathrm{d}x}}$$

$$F = \eta A \frac{\mathrm{d}v}{\mathrm{d}x}$$

This expression can be used to calculate the viscous resistance between lubricated surfaces in sliding contact, such as machine slides and plane bearings. If the thickness of the lubrication film is x metres and the sliding velocity is $v\,\mathrm{m\,s^{-1}}$, the above formula can be written as

$$F = \eta A \frac{v}{x} \tag{5.8}$$

In such applications, the power loss due to viscous resistance is given by

$$\mathbf{power\ loss} = \mathbf{\mathit{Fv}} \tag{5.9}$$

> ## Key point
>
> A Newtonian fluid is one whose dynamic viscosity at a given temperature is constant irrespective of the shearing which it subjected to.

Example 5.2

A machinework table rests on slides with a contact area of $0.35\,\mathrm{m^2}$. The slides are lubricated with oil of dynamic viscosity $0.15\,\mathrm{N\,s\,m^{-2}}$ and the thickness of the oil film is $0.5\,\mathrm{mm}$. What will be the viscous resistance to motion and the power required to overcome it when the worktable is moving at a speed of $2\,\mathrm{m\,s^{-1}}$?

Finding viscous resistance to motion:

$$F = \eta A \frac{v}{x}$$

$$F = \frac{0.15 \times 0.35 \times 2}{0.5 \times 10^{-3}}$$

$$F = 210\,\mathbf{N}$$

Finding power to overcome viscous resistance:

$$\text{power} = Fv$$

$$\text{power} = 210 \times 2$$

$$\mathbf{power} = 420\,\mathbf{W}$$

Example 5.3

A shaft 60 mm in diameter rotates in a plane bearing of length 80 mm with a radial clearance of 0.15 mm. The shaft is lubricated by oil of dynamic viscosity $0.2 \, N \, s \, m^{-2}$ and rotates at a speed of 2500 rpm. Assuming the shaft and bearing to be concentric, determine (a) the torque required to overcome viscous resistance, (b) the power loss in the bearing.

Finding lubricated surface area:

$$A = \pi dl = \pi \times 0.06 \times 0.08$$

$$A = 0.0151 \, m^2$$

Finding angular velocity of shaft:

$$\omega = \frac{2\pi N}{60} = \frac{2\pi}{60} \times 2500$$

$$\omega = 262 \, rad \, s^{-1}$$

Finding surface speed of shaft:

$$v = \omega r = 262 \times 0.03$$

$$v = 7.86 \, m \, s^{-1}$$

Finding tangential viscous resistance:

$$F = \eta A \frac{v}{x} = \frac{0.2 \times 0.0151 \times 7.86}{0.15 \times 10^{-3}}$$

$$F = 158 \, N$$

Finding torque required to overcome viscous resistance:

$$T = Fr = 158 \times 0.03$$

$$T = 4.74 \, N \, m$$

Finding power loss in bearing:

$$\text{power loss} = Fv = 158 \times 7.86$$

$$\textbf{power loss} = 1242 \, W \quad \text{or} \quad 1.24 \, kW$$

The viscosity of a substance is sometimes defined in a slightly different way for use in advanced fluid mechanics calculations. It is known as *kinematic viscosity* which is the ratio of dynamic viscosity and density. Kinematic viscosity is denoted by the Greek letter ν (nu) and has the unit $m^2 \, s^{-1}$.

$$\text{kinematic viscosity} = \frac{\text{dynamic viscosity}}{\text{density}}$$

$$\nu = \frac{\eta}{\rho} \, (m^2 \, s^{-1}) \tag{5.10}$$

Non-Newtonian fluids

As can be seen in Figure 5.3, the graph of shear stress against shear rate for a Newtonian fluid is a straight line which passes through the origin. This indicates that its viscosity does not change, no matter how quickly it is sheared by stirring or pumping. There are, however, many fluids which do not behave in this way and they may be broadly divided into two classes:

- *Thixotropic fluids*, in which the viscosity falls as the rate of shearing increases. These are also known as *shear thinning* fluids.
- *Rheopectic fluids*, in which the viscosity rises as the rate of shearing increases. These are also known as *shear thickening* fluids.

The graphs of shear stress τ against shear rate μ for these fluids are as follows. They are known as *rheograms*. The ratio of shear stress to shear rate at any point on the curve is known as the *apparent dynamic viscosity* η_a of the fluid. Another way of defining apparent viscosity is to say that it is the gradient of a line drawn from the origin to that point on the curve. There are three common types of thixotropic, or shear thickening, fluid. They are called Bingham plastic, pseudoplastic and Casson plastic.

1. *Bingham plastic* Here the stirrer or pump must first apply an initial yield stress τ_0, before shearing can begin. Afterwards the graph is linear as shown in Figure 5.4. Margerine, cooking fats, greases, chocolate mixtures, toothpaste, some soap and detergent slurries, and some paper pulps exhibit this kind of behaviour.

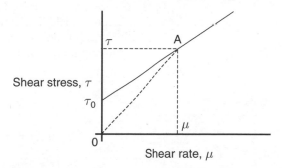

Figure 5.4 *Bingham plastic rheogram*

The rheogram curve follows a law of the form

$$\tau = \tau_0 + K_p\mu \tag{5.11}$$

where K_p is a constant for the fluid known as the *coefficient of rigidity*. The apparent viscosity η_a at the point A on the curve is the gradient of the line OA.

$$\eta_a = \frac{\tau}{\mu} = \frac{\tau_0 + K_p\mu}{\mu}$$

$$\eta_a = \frac{\tau_0}{\mu} + K_p \tag{5.12}$$

2. *Pseudoplastic* Here the fluid becomes progressively less viscous as the shear rate increases due to intermolecular bonds being broken. The graph follows a curve as shown in Figure 5.5. Rubber solutions, adhesives, polymer solutions and mayonnaise exhibit this kind of behaviour.

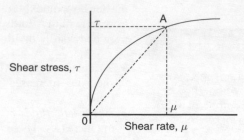

Figure 5.5 *Pseudoplastic rheogram*

The rheogram curve follows a power law of the form

$$\tau = K\mu^n \tag{5.13}$$

where K and n are constants for the fluid. The constant K is known as the *consistency coefficient* of the fluid. The index n is known as the *behaviour flow index* whose value is always less than 1. The apparent viscosity at the point A on the curve is the gradient of the line OA.

$$\eta_a = \frac{\tau}{\mu} = \frac{K\mu^n}{\mu}$$

$$\eta_a = K\mu^{n-1} \tag{5.14}$$

3. *Casson plastic*   Here the stirrer or pump must apply to induce an initial yield stress τ_0, before shearing can begin. Afterwards the graph follows a curve as shown in Figure 5.6. Printing ink, non-drip paint, tomato ketchup and blood behave in this way.

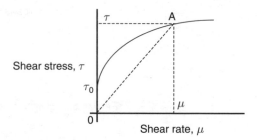

Figure 5.6 *Casson plastic rheogram*

The rheogram curve follows a law of the form

$$\tau = \tau_0 + K_c\,\mu^n \tag{5.15}$$

where K_c and n are constants for the fluid. The constant K_c is known as the *Casson viscosity* of the fluid. As with pseudo-plastic fluids, the behaviour flow index n has values which are always less than 1. The apparent viscosity at the point A on the curve is the gradient of the line OA.

$$\eta_a = \frac{\tau}{\mu} = \frac{\tau_0 + K_c\mu^n}{\mu}$$

$$\eta_a = \frac{\tau_0}{\mu} + K_c\,\mu^{n-1} \tag{5.16}$$

Rheopectic or shear thickening fluids are very often fluids which contain solid particles in suspension. They are also referred to as *dilatent fluids*. Starch solutions, quicksand, some cornflower and sugar solutions, and iron powder dispersed in low-viscosity liquids exhibit this kind of behaviour. The apparent viscosity increases with shear rate as shown in Figure 5.7.

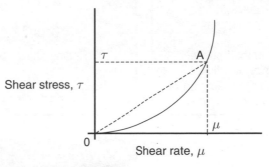

Figure 5.7 *Dilatent fluid rheogram*

As with the pseudoplastic rheogram, the curve follows a power law of the form

$$\tau = K\mu^n \tag{5.17}$$

where K is again the *consistency coefficient* of the fluid and the behaviour flow index n has values which are always greater than 1. The apparent viscosity at the point A on the curve is the gradient of the line OA.

$$\eta_a = \frac{\tau}{\mu} = \frac{K\mu^n}{\mu}$$

$$\eta_a = K\mu^{n-1} \tag{5.18}$$

Key point

Apparent viscosity of a non-Newtonian fluid at any instant of time is the ratio of shear stress to shear rate.

Example 5.4

(a) A pseudoplastic fluid obeys the power law $\tau = 0.25\mu^{0.6}$. Determine its apparent viscosity when subjected to a shear rate of $80\,\text{s}^{-1}$.

(b) A Bingham plastic fluid has a rheogram which obeys the law $\tau = 4.3 + 0.015\mu$. Determine the conditions of shear rate and shear stress when its apparent viscosity is $0.05\,\text{N s m}^{-2}$.

(a) Finding shear stress in pseudoplastic fluid:

$$\tau = 0.25\mu^{0.6} = 0.25 \times 80^{0.6}$$

$$\tau = 3.47\,\text{Pa}$$

Finding apparent viscosity under these conditions:

$$\eta = \frac{\tau}{\mu} = \frac{3.47}{80}$$

$$\eta = 0.0434\,\text{N s m}^{-2}$$

Test your knowledge 5.1

1. If a liquid has a relative density of 0.9 what will be its specific weight?
2. Define dynamic viscosity.
3. What is a Newtonian fluid?
4. What is the Kinematic viscosity of a fluid?
5. Sketch the rheogram of a Bingham plastic fluid.

(b) Finding shear rate in Bingham plastic fluid using equation (5.12):

$$\eta_a = \frac{\tau_0}{\mu} + K_p \quad \text{where } \tau_0 = 4.3\,\text{Pa} \quad \text{and} \quad K_p = 0.015\,\text{N s m}^{-2}$$

$$\mu = \frac{\tau_0}{\eta_a - K_p} = \frac{4.3}{0.05 - 0.015}$$

$$\mu = 123\,\text{s}^{-1}$$

Finding shear stress:

$$\tau = 4.3 + 0.015\mu = 4.3 + (0.015 \times 123)$$

$$\tau = 6.15\,\text{Pa}$$

Activity 5.1

In a printing machine, two parallel rectangular plates measuring $1.5\,\text{m} \times 0.5\,\text{m}$ are separated by a distance of 2 mm. This space is filled with printing ink which behaves as a Casson plastic fluid. The apparent viscosity of the fluid is known to be $0.15\,\text{N s m}^{-2}$ at a shear rate of $150\,\text{s}^{-1}$ and the shear stress in the ink as the surfaces move over each other is given by the expression

$$\tau = 3.9 + K_c\mu^{0.5}$$

Determine the viscous resistance to motion when the plates move over each other at a speed of $1.5\,\text{m s}^{-1}$ and the power dissipated.

Surface tension

The surface of a liquid acts like an elastic skin which is in a state of tension. As you probably know, small insects can walk across the surface of water and with care, it is possible to make a dry steel needle float on the surface. You will also have noticed that water droplets try to assume a spherical shape as though surrounded by an elastic skin. It is thought that the molecules at the surface of a liquid have more energy than those beneath which may account for the phenomena.

The *surface tension* of a liquid, which is also sometimes called *the coefficient of surface tension*, is denoted by the Greek letter γ (gamma). It is defined as the force per unit length acting on the surface at right angles to one side on a line drawn on the surface. The unit of γ is thus newtons per metre. Its value for water at 20 °C in contact with air, is around $70 \times 10^{-3}\,\text{N m}^{-1}$ and its value for mercury under the same conditions is $465 \times 10^{-2}\,\text{N m}^{-1}$.

At the interface between a liquid and a solid surface, some liquids will adhere to the solid and others will not. If the forces of *adhesion* between the liquid molecules and those of the solid are greater than the forces of *cohesion* between the liquid molecules themselves, the liquid will 'wet' the solid and vice versa. Water and alcohol will readily adhere to clean glass whereas mercury will not.

Key point

Surface tension is due to the force of attraction between the molecules in the surface of a liquid.

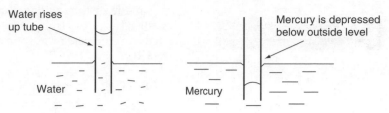

Figure 5.8 *Capillary action*

The forces of adhesion and cohesion between molecules gives rise to the phenomenon of *capillary action*. When one end of a clean glass capillary tube is immersed vertically in water, the liquid rises a short distance up the tube. The smaller the internal diameter of the tube, the higher the water will rise. When the same tube is immersed in mercury however, the opposite occurs and the liquid in the tube is depressed below the outside level. This is shown in Figure 5.8. The surface tension for water in contact with glass can be found by measuring the rise in a capillary tube of known internal diameter as shown in Figure 5.9.

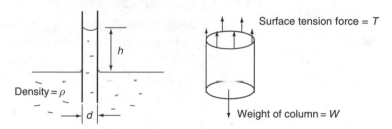

Figure 5.9 *Rise of water in a capillary tube*

The weight of the column is found as follows:

volume of column, $V = \dfrac{\pi d^2}{4} h$

mass of column, $m = \rho V = \rho \dfrac{\pi d^2}{4} h$

weight of column, $W = mg = \rho g \dfrac{\pi d^2}{4} h$

If γ is the surface tension force per metre, the upward surface tension force is given by

surface tension, $T = \pi d \gamma$

Equating these two expressions enables γ to be found.

surface tension force, $T = $ Weight, W

$$\pi d \gamma = \rho g \frac{\pi d^2}{4} h$$

$$\boldsymbol{\gamma = \rho g \frac{d}{4} h} \qquad (5.19)$$

In deriving the above formula, it is assumed that the meniscus at the top of the column meets the glass vertically around its edge. This is in fact very close to being the case for water and alcohol provided that the glass is clean.

Use is made of the capillary action phenomenon in wick-feed lubrication. Here, oil from a reservoir rises up a wick to be deposited on the slide-ways or rotating parts of a machine (Figure 5.10).

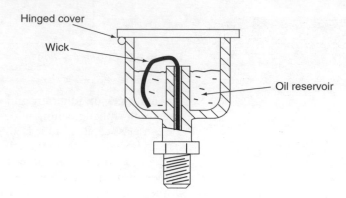

Hinged cover

Wick

Oil reservoir

Figure 5.10 *Wick-feed lubrication*

Example 5.5

Water is observed to rise through a height of 70 mm in a clean glass capillary tube of internal diameter 0.4 mm. If the density of water is 1000 kg m^{-3}, determine its surface tension. To what height would the same water rise in a capillary tube of internal diameter 1 mm?

Finding γ from equation (5.19):

$$\gamma = \rho g \frac{d}{4} h$$

$$\gamma = 1000 \times 9.81 \times \frac{0.4 \times 10^{-3}}{4} \times 70 \times 10^{-3}$$

$$\boldsymbol{\gamma = 68.7 \times 10^{-3}\,N\,m^{-1}}$$

Finding height in 1 mm diameter capillary tube:

$$\gamma = \rho g \frac{d}{4} h$$

$$h = \frac{4\gamma}{\rho g d}$$

$$h = \frac{4 \times 68.7 \times 10^{-3}}{1000 \times 9.81 \times 1 \times 10^{-3}}$$

$$\boldsymbol{h = 0.028\,m} \quad \text{or} \quad \boldsymbol{28\,mm}$$

Laminar and turbulent flow

It has already been stated that a fluid can be considered to be made up of an infinite number of layers or *laminae*. If these are all flowing smoothly in the same direction, the flow is said to be *laminar* or *streamlined*. Streamlines are imaginary lines in the direction of flow which can be simulated by injecting die or smoke into the fluid stream from a small diameter tube with holes drilled at intervals along its length (Figure 5.11).

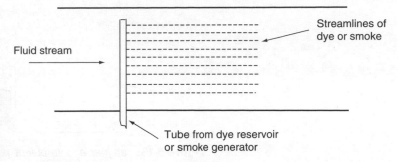

Figure 5.11 *Simulation of streamlines*

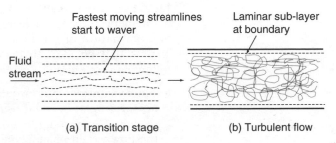

(a) Transition stage   (b) Turbulent flow

Figure 5.12 *Transition to turbulent flow*

Laminar flow can take place only at relatively low flow velocities. If the velocity is increased in any flow system such as a pipeline, open channel or wind tunnel, a stage is reached very soon where the laminar flow becomes unstable. This is called the *transition stage*. The flow near the boundary surfaces is slowed down due to viscous resistance. Shearing forces are thus present and the fastest moving streamlines are in the centre of the flow. It is these that start to waver with the onset of transition as shown in Figure 5.12(a).

As the flow velocity increases beyond the transition stage, the streamlines disintegrate and the flow then consists of a random eddying motion. It is then said to be fully turbulent as shown in Figure 5.12(b). Even with turbulent flow, however, there remains a thin laminar sub-layer adjacent to the system boundary where the flow velocity is retarded by viscous resistance. It is across this sub-layer that most of the shearing action in the fluid takes place. A body placed in the fluid stream experiences this as a drag force. To be more precise, it is known as *skin friction drag* which will be described in more detail later.

Although the practice is not to be encouraged, the transition from laminar to turbulent flow can be observed in the smoke rising from a lit cigarette as shown in Figure 5.13. The smoke contains hot gases which accelerate as they rise. The stationary air surrounding them has a retarding effect which produces a shearing action. Initially the smoke rises with straight vertical streamlines but eventually it starts to waver with the onset of transition. The flow then becomes fully turbulent with a characteristic random eddying motion.

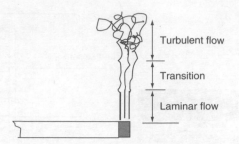

Figure 5.13 *Laminar and turbulent flow in cigarette smoke*

Behaviour of an ideal fluid

Fluid flow systems can be very difficult to analyse and model using mathematics because of the different properties involved and the random nature of the flow. To simplify matters, it is usual to assume that an ideal fluid exists. Such a fluid would have the following properties:

- The dynamic viscosity would be zero. As a result there would be no shearing at boundary surfaces or around bodies in the fluid stream. There would be no retarding forces or drag forces and the flow would remain laminar at all velocities.
- The effects of cohesion and adhesion would be absent, i.e. there would be no surface tension in the fluid and no capillary action.
- In liquid form the fluid would be incompressible.
- In gaseous form the fluid would obey the gas laws at all temperatures and pressures.

Having made these assumptions it is possible to predict the behaviour of fluid systems using mathematics. Very often there is some variance between the theoretical and actual behaviour and this becomes apparent when the predictions are tested experimentally. Nevertheless, the assumption of ideal fluid behaviour gives us a starting point from which we can develop formulae. Correction factors can then be applied in the light of experimental evidence to make the theoretical predictions correlate more nearly with the actual behaviour.

Fluid flow calculations contain many of these experimentally derived correction factors. They are sometimes disparagingly known as 'fudge' factors, but more correctly they are known as correction coefficients. Typical examples are velocity coefficients, discharge coefficients and drag coefficients. Fortunately, water and air behave surprisingly close to the ideal fluid model in many respects and there is not too much of a difference between their actual and predicted behaviour.

Key point

The effects of viscosity and surface tension would be absent in an ideal fluid.

Test your knowledge 5.2

1. How is the coefficient of surface tension defined?
2. What is capilliary action?
3. What are streamlines?
4. What name is given to the change from laminar to turbulent flow?
5. What would be the properties of an ideal fluid?

Activity 5.2

A glass microscope slide of length 60 mm and thickness 2 mm is suspended from a sensitive balance with its lower edge touching the surface of a beaker of water. The beaker is slowly lowered and at the point where the slide breaks free of the water the balance reading is 11.13 g. The balance reading then falls to 10.25 g, which is the mass of the slide. What is the surface tension coefficient for the water?

Problems 5.1

1. A storage tank of diameter 5 m and height 4 m is 2/3 full of crude oil. Load cells on which the tank is mounted record the mass of the oil to be 45 tonnes. Determine (a) the density of the oil, (b) its specific weight, (c) its relative density.
[859 kg m^{-3}, 8.43 kN m^{-3}, 8.59]

2. A plane rectangular plate measuring 300 mm × 250 mm rests on a flat horizontal surface from which it is separated by a film of fluid of thickness 0.1 mm. The force necessary to slide the plate over the surface at a steady speed of 2.5 m s^{-1} is 220 N. Find (a) the shear stress in the fluid, (b) the shear rate or velocity gradient, (c) the dynamic viscosity of the fluid.
[2.93 kPa, 25 × 10^3 s^{-1}, 0.117 N s m^{-2}]

3. Oil of dynamic viscosity 0.12 N s m^{-2} is used to lubricate the slides of a machine worktable. The thickness of the oil film is 0.25 mm and the contact area is 0.15 m^2. Determine the viscous resistance and the power dissipated when the worktable is moving at a speed of 3 m s^{-1}.
[216 N, 648 W]

4. A shaft of diameter 50 mm is supported by a plane bearing of length 100 mm and diameter 50.25 mm. The shaft and bearing are concentric and the space between them is filled with oil of dynamic viscosity 0.15 N s m^{-2}. Determine the torque required to overcome viscous resistance and the power loss when the shaft is rotating at a speed of 3000 rpm.
[3.7 N m, 1.16 kW]

5. A Casson plastic fluid is subjected to a shear stress of 5.5 Pa. If the fluid obeys a law of the form $\tau = 1.5 + 0.2\mu^{0.45}$, determine the shear rate and the apparent viscosity under these conditions.
[778 s^{-1}, 7.1 × 10^{-3} N s m^{-2}]

6. A liquid is observed to rise to a height of 55 mm in a clean glass capillary tube of internal diameter 0.5 mm. If a volume of 50 cm^3 of the liquid has a mass of 45 g what will be its density and its surface tension coefficient. The top of the column can be assumed to meet the glass vertically around its edge.
[900 kg m^{-3}, 60.7 × 10^{-3} N m^{-1}]

Hydrostatic systems

Hydrostatic systems are those in which the fluid is at rest or in which the movements of the fluid are relatively small. Pressure-measuring devices such as piezometers, manometers and barometers are examples of this type of system as are floating and immersed bodies which experience up-thrust and fluid pressure. In other types of hydrostatic system, the fluid is used as a means of transmitting force. Hydraulic jacks, presses and breaking systems are examples of this category.

Hydrostatic pressure

Pressure is the intensity of the force acting on a surface and is measured in pascals (Pa). A pascal is defined as a force of 1 N which acts evenly and at right angles over a surface of area $1\,m^2$. This is rather a small unit for practical purposes and pressure is more generally measured in kPa or even MPa. Although pascal is the standard SI unit of pressure, there are others in common use. One of these is the 'bar' which is widely used for measuring high pressures in steam plant and industrial processes.

1 bar = 100 000 Pa   or   100 kPa

To convert from bar to Pa multiply by 100 000 or 10^5.
The 'bar' gets its name from barometric or atmospheric pressure which is approximately equal to 1 bar.

standard atmospheric pressure = 101 325 Pa   or

101.325 kPa   or   1.01325 bar

It should be remembered that manometers and mechanical pressure gauges do not usually measure the total or *absolute pressure* inside a vessel. Instead they measure the difference between the pressure inside and the atmospheric pressure outside. This is known as *gauge pressure*. If the absolute pressure is required, atmospheric pressure must be added to it, that is,

absolute pressure = gauge pressure + atmospheric pressure

In fluid mechanics calculations it is generally the gauge pressure that is used and unless you are told otherwise, any pressure values which you are given can be assumed to be gauge pressures.

There are three important points relating to the pressure inside a pressure vessel or any other closed hydrostatic system as illustrated in Figure 5.14.

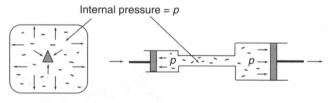

Internal pressure = p

(a) Pressure normal to surfaces   (b) Pressure transmission

Figure 5.14 *Properties of fluid pressure*

1. The pressure exerted on a body inside the fluid at rest is the same from all directions.
2. The pressure exerted on a solid surface by a fluid at rest is always at right angles to the surface, i.e. normal to the surface.
3. If the pressure is increased at any point in a fluid which is at rest in a closed system, the increase is transmitted without loss to all other points within the system.

Pressure at depth

Below the free surface of a liquid, the pressure increases with depth. Consider a tank that contains a liquid of density $\rho\,kg\,m^{-3}$ to a depth of $h\,m$. The pressure at the bottom of the tank is the

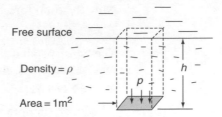

Figure 5.15 *Pressure at depth*

force acting on each square metre of the base. This is equal to the weight of the column of liquid of cross-sectional area $1\,\text{m}^2$ which is carried by each square metre (Figure 5.15).

volume of column = cross-sectional area × height = $1 \times h$

$$V = h\,(\text{m}^3)$$

mass of column = density × volume

$$m = h\rho\,(\text{kg})$$

pressure at base = weight of column

$$p = mg$$

$$p = h\rho g\,(\text{Pa}) \tag{5.20}$$

Thrust on an immersed surface

Consider a plane surface of area $A\,\text{m}^2$ whose centroid lies at a depth $\bar{x}\,\text{m}$ below the free surface of a liquid of density $\rho\,\text{kg m}^{-3}$, as shown in Figure 5.16. The pressure of the liquid will exert a force F on the surface. We will show that it is the depth of the centroid that determines this force and that it will be the same whether the immersed surface is vertical, horizontal or inclined at some angle. Consider now a thin strip of length b and thickness δx which lies at a depth x below the free surface.

The pressure p on the elemental strip is given by

$$p = x\rho g$$

The force δF acting on the elemental strip will be

$$\delta F = \text{pressure} \times \text{area}$$

$$\delta F = x\rho g b\,\delta x$$

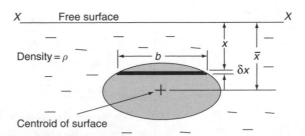

Figure 5.16 *Immersed surface*

Now the total force F will be the sum of all these small forces:

$$F = \Sigma(x\rho gb\,\delta x)$$

The density ρ and the acceleration due to gravity g are constants and can be taken outside the summation sign giving

$$F = \rho g\Sigma(xb\,\delta x)$$

Now the term in the brackets is the area of the strip multiplied by its distance below the free surface X–X. An area multiplied by a distance is known as a *first moment of area* and the above expression contains the sum of all such elemental first moments of area about the axis X–X in the free surface. This is equal to the total first moment of the surface which is also given by the product of the total area A and the depth of the centroid $\bar{x}$, that is,

total first moment of surface about axis X–$X =$

sum of elemental first moments about axis X–X

$$A\bar{x} = \Sigma(xb\,\delta x)$$

Substituting for the $\Sigma(xb\,\delta x)$ term in the expression for force gives

$$F = \rho gA\bar{x} \tag{5.21}$$

As you will see from equation (5.21), the thrust on the surface depends on the depth of the centroid and is independent of the orientation of the surface. The expression can be rearranged in the form

$$F = \rho g\bar{x}A \tag{5.22}$$

Now, $\rho g\bar{x}$ is the pressure at the depth of the centroid and another way of interpreting these expressions for thrust is to say that

thrust = pressure at centroid × immersed surface area

Key point

The thrust on the side of a tank or a retaining wall increases as the square of the depth of liquid contained.

Example 5.6

A square tank measuring 2 m × 2 m contains oil of relative density 0.88 to a depth of 3 m when full. Find (a) the pressure at the base of the tank when full, (b) the force acting on each side when the tank is full, (c) the force acting on each side when the tank is half-full. The density of water is 1000 kg m^{-3}.

(a) Finding the density of the oil:

ρ = relative density × density of water

$\rho = 0.88 \times 1000$

$\rho = 880$ kgm^{-3}

Finding pressure on base when tank is full:

$p = h\rho g = 3 \times 880 \times 9.81$

$p = 25.9 \times 10^3$ Pa   or   25.9 kPa

(b) Finding the force F_1 acting on each side when the tank is full:

$F_1 = \rho g\bar{x}_1 A_1 = 880 \times 9.81 \times 1.5 \times 3 \times 2$

$F_1 = 77.7 \times 10^3$ N   or   77.7 kN

(c) Finding the force F_2 acting on each side when the tank is half-full:

$$F_2 = \rho g \bar{x}_2 A_2 = 880 \times 9.81 \times 0.75 \times 1.5 \times 2$$

$$F = 19.4 \times 10^3 \, \text{N} \quad \text{or} \quad 19.3 \, \text{kN}$$

You should note that the thrust on the sides increases by a factor of 4 when the depth is doubled, i.e. the thrust increases as the square of the depth of liquid contained.

Pressure-measuring devices

There are certain pressure-measuring devices which operate on what may be called, the fluid column principle. These include piezometers, U-tube manometers, well-type manometers and mercury barometers. In each of these, the pressure that is being measured causes liquid to rise up a small diameter tube.

A piezometer is a clear glass or plastic tube inserted into a pipe or vessel that contains liquid under pressure, as shown in Figure 5.17(a). A U-tube manometer, as its name suggests, consists of a U-shaped tube containing a liquid. One limb of the tube is attached to the pressure source and the other is open to the atmosphere, as shown in Figure 5.17(b).

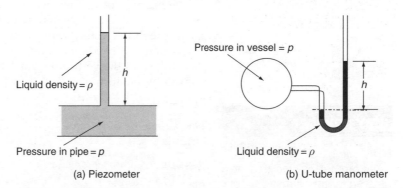

(a) Piezometer   (b) U-tube manometer

Figure 5.17 *Piezometer and U-tube manometer*

The difference between the pressure of the contained fluid and atmospheric pressure causes the liquid to rise up the piezometer tube and to be displaced in the U-tube of the manometer. This pressure difference, which is of course the gauge pressure of the fluid, is very often expressed directly in millimetres, e.g. mmH_2O if it is water, or mmHg for U-tube manometers containing mercury. Its value may also be calculated in pascals using equation (5.20), that is

$$p = h\rho g \, (\textbf{Pa})$$

Well-type manometers are closely related to U-tube manometers. The difference is that the two limbs of a well-type manometer are of different cross-section as shown in Figure 5.18.

With the reservoir at atmospheric pressure, the level in the reservoir and the small diameter glass tube is adjusted to correspond with the zero reading on the graduated scale. When

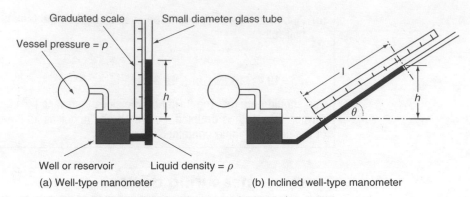

Graduated scale   Small diameter glass tube

Vessel pressure = p

h

l

h

θ

Well or reservoir   Liquid density = ρ

(a) Well-type manometer   (b) Inclined well-type manometer

Figure 5.18 *Well-type and inclined well-type manometers*

connected to a pressure source, the manometer liquid is forced up the glass tube but the change in level in the reservoir is negligible. The gauge pressure in pascals can then be calculated using the formula

$$p = h\rho g \ (\text{Pa})$$

Before the invention of reliable mechanical pressure gauges in the mid-nineteenth century, the well-type manometer filled with mercury was the only available means of measuring the pressure in steam boilers. Mercury is still used in the type with the vertical glass tube, shown in Figure 5.18(a), for the measurement of relatively high pressures.

The purpose of the inclined well-type manometer is to amplify the rise of fluid h, in the glass tube when measuring small values of gauge pressure. Here the manometer fluid is generally water or a very thin oil. The liquid moves a distance l m along the inclined glass tube and the vertical height h through which it actually rises is given by the formula

$$h = l\sin\theta$$

The gauge pressure can then be calculated by again using the formula

$$p = h\rho g$$

or

$$p = \rho g l \sin\theta \ (\text{Pa})$$

Key point

The limb of a well-type manometer is inclined to increase the sensitivity of the instrument.

Atmospheric pressure is measured by means of a barometer. The basic principle of the mercury barometer is shown in Figure 5.19(a). In days gone by it was common practice to construct one in the school physics laboratory and observe the day-to-day changes in atmospheric pressure. A small diameter glass tube of length approximately 1 m, and with one end closed, was filled with mercury. This was then upended with the thumb placed under the open end. The open end was then placed under the surface of a mercury reservoir and the thumb removed. The mercury fell down the tube to a height h, creating a vacuum above it, and balanced by atmospheric pressure on the surface of the reservoir. Atmospheric pressure could then be calculated using equation (5.20), that is,

atmospheric pressure $= h\rho g \ (\text{Pa})$

This practice is now frowned upon as the inhalation of mercury vapour and contact with mercury through the skin can have harmful effects.

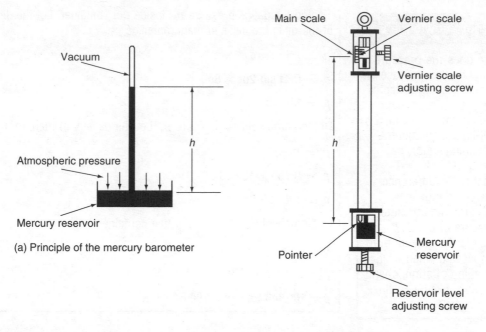

Figure 5.19 *Basic mercury barometer and Fortin barometer*

The Fortin barometer, which is shown in Figure 5.19(b), is virtually identical in its construction and principle of operation but it is totally encased to prevent the escape of mercury vapour. To take a reading, the level of the mercury reservoir is adjusted until it just makes contact with the pointer. The slide containing the vernier scale is then adjusted until its base is level with the top of the mercury column and the height h is read off. Standard atmospheric pressure is 101.325 kPa which gives a height of 760 mm of mercury on the Fortin barometer.

The portable type of barometer, which you may have in your home, is known as an *aneroid* barometer. This is a purely mechanical device containing a sealed metal capsule in which the air pressure has been reduced. Changes in atmospheric pressure cause the capsule to flex and the movement is amplified through a mechanical linkage to move the pointer round the viewing scale. Although the display is usually calibrated to predict the weather conditions, it might also contain numbers around its periphery. These are the corresponding heights of a mercury column, given in either millimetres or inches.

Example 5.7

The gas pressure inside a container is measured by means of an inclined well-type manometer containing a fluid of density 850 kg m^{-3}. The angle of the inclined limb is 20° and the fluid is observed to travel a distance of 410 mm along the inclined scale. A Fortin barometer shows that the prevailing atmospheric pressure corresponds to 755 mmHg. What is the absolute pressure of the gas inside the container? The density of mercury is 13 600 kg m^{-3}.

Finding the gauge pressure p_g, inside the container. Let the density of the oil in the inclined manometer be ρ_1:

$$p_g = l \sin \theta \, \rho_1 g$$

$$p_g = 0.41 \sin 20° \times 850$$

$$p_g = \mathbf{119 \ Pa}$$

Finding atmospheric pressure p_a. Let the density of mercury be ρ_2:

$$p_a = h \rho_2 g$$

$$p_a = 0.755 \times 13\,600 \times 9.81$$

$$p_a = \mathbf{100\,729 \ Pa}$$

Finding absolute pressure p in the container:

$$p = p_g + p_a$$

$$p = 119 + 100\,729$$

$$p = \mathbf{100\,848 \ Pa} \quad \text{or} \quad \mathbf{100.848 \ kPa}$$

Activity 5.3

The wall of a dam contains a sluice gate of mass 50 kg as shown in Figure 5.20. The gate can be raised in guides by means of the handwheel. If the limiting coefficient of friction between the gate and the guides is 0.4, determine the vertical force required to raise it. The density of the water is $1000 \ \text{kg m}^{-3}$.

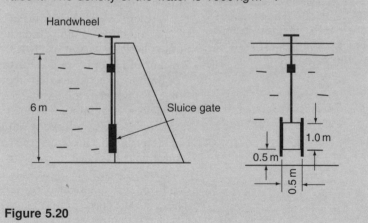

Figure 5.20

Problems 5.2

1. A rectangular tank measuring $4 \ \text{m} \times 2.5 \ \text{m}$ contains water to a depth of 2 m. (a) What will be the thrust acting on the sides. (b) What will be the thrust acting on a circular valve flange of diameter 250 mm that is bolted to the base? The density of water is $1000 \ \text{kg m}^{-3}$.

[78.5 kN, 49.1 kN, 963 N]

2. A vertical dam wall is 20 m above the floor of the dam. What will be the thrust per metre width (a) when the water is level with the top of the dam, (b) when it is half-full, (c) when it is one-third full. State how the thrust varies with the depth of water contained. The density of water is $1000\,\text{kg}\,\text{m}^{-3}$.

[1962 kN, 491 kN, 218 kN]

3. A lock gate of width 12 m contains sea water. The depth is 6 m on one side and 2.5 m on the other. What is the resultant thrust on the gate? The specific weight of sea water is $10.1\,\text{kN}\,\text{m}^{-3}$.

[1.8 MN]

4. A sluice gate in a rectangular section channel of width 2 m has water to a depth of 1.5 m to one side. The sluice gate has a mass of 500 kg and can be raised in vertical guides. The coefficient of friction in the guides is 0.25. Determine (a) the thrust on the gate, (b) the force needed to raise it.
   The density of water is $1000\,\text{kg}\,\text{m}^{-3}$.

[22.1 kN, 10.4 kN]

5. Two piezometer tubes are fitted to a water pipeline 20 m apart. If the difference in levels is 0.75 m, what is the pressure drop per metre length measured in pascals.
   The density of water is $1000\,\text{kg}\,\text{m}^{-3}$.

[368 Pa per metre length]

6. A well-type mercury manometer has its limb inclined at an angle of 20° to the horizontal. If the length of the mercury column is 200 mm and atmospheric pressure is equivalent to 760 mmHg, determine the absolute pressure which is being indicated.

[110.5 kPa]

Buoyancy

When a body is placed in a fluid, it experiences an up-thrust. The amount of up-thrust is explained by the principle of Archimedes which states that *the resultant up-thrust on a body immersed in a fluid is equal to the weight of the fluid displaced by the body*. It is said that Archimedes, who lived in ancient Greece, came upon this principle intuitively whilst taking a bath. It can, however, be verified by considering the forces exerted on an immersed body due to hydrostatic pressure.

In the case of a floating body, the up-thrust F is equal to its weight and like the weight, it is considered to act at the centre of gravity of the body. For the body shown in Figure 5.21, the pressure on the sides makes no contribution to the up-thrust. It is the pressure on the base which produces the upward force.

up-thrust = pressure on base × area of base

$$F = p \times A$$
$$F = h\rho g A$$

This can be rearranged into the form

$$F = hA\rho g$$

Now hA is the volume of the fluid displaced, or pushed aside by the floating body, $hA\rho$ the mass of liquid displaced and $hA\rho g$ the

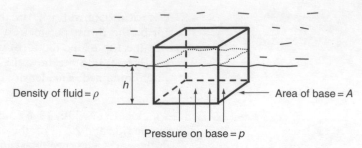

Figure 5.21 *Up-thrust on a floating body*

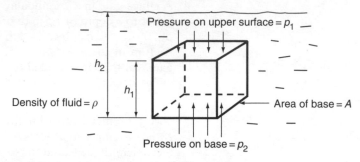

Figure 5.22 *Up-thrust on an immersed body*

weight of liquid displaced. The up-thrust F is thus equal to the weight of fluid displaced, as stated in Achimedes' principle.

For a body which is totally immersed, as shown in Figure 5.22, the up-thrust F may be greater than, equal to, or less than, the weight of the body depending on whether it is rising, stationary or sinking. However, the up-thrust is still equal to the weight of fluid displaced. Once again the pressure on the sides makes no contribution to the resultant upward force.

up-thrust = force on base − force on upper surface

$$F = p_2 A - p_1 A$$
$$F = (p_2 - p_1)A$$
$$F = (h_2 \rho g - h_1 \rho g)A$$
$$\boldsymbol{F = (h_2 - h_1)\rho g A}$$

This can be rearranged into the form

$$\boldsymbol{F = (h_2 - h_1)A\rho g}$$

Here, $(h_2 - h_1)A$ is the volume of the fluid displaced, or pushed aside by the floating body, $(h_2 - h_1)A\rho$ the mass of liquid displaced and $(h_2 - h_1)A\rho g$ the weight of the liquid displaced. Once again, the up-thrust F is thus equal to the weight of fluid displaced, as stated in Achimedes' principle.

Key point

The up-thrust on a body which is fully or partly immersed in a fluid is equal to the weight of fluid which it displaces.

Example 5.8

A steel bar of length 0.5 m and diameter 50 mm is suspended from a chain and lowered into a tank of liquid whose relative density is 0.95. Determine the tension in the chain when the bar is fully immersed. The density of water is $1000 \, \text{kg m}^{-3}$ and the density of steel is $7870 \, \text{kg m}^{-3}$.

Finding density ρ_l of liquid:

ρ_l = relative density × density of water

$\rho_l = 0.95 \times 1000$

$\rho_l = \mathbf{950\,kg\,m^{-3}}$

Finding volume of steel bar (which is also the volume of liquid displaced):

$V = \pi d^2 l = \dfrac{\pi \times 0.05^2}{4} \times 0.5$

$V = \mathbf{982 \times 10^{-6}\ m^3}$

Finding weight of steel bar:

$W = m_s g = V \rho_s g$

$W = 982 \times 10^{-6} \times 7870 \times 9.81$

$W = \mathbf{75.8\,N}$

Finding up-thrust on steel bar:

F = weight of liquid displaced

$F = m_l g = V \rho_l g$

$F = 982 \times 10^{-6} \times 950 \times 9.81$

$F = \mathbf{9.2\,N}$

Finding tension T in the chain:

$T = W - F = 75.8 - 9.2$

$T = \mathbf{66.6\,N}$

A method commonly used to find the density of a solid material is to weigh it, by means of a spring or chemical balance, in air and when totally immersed in water. The method is only suitable for solid objects which will sink, and so they must have a density greater than that of water (Figure 5.23).

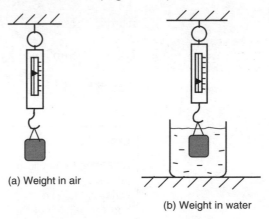

(a) Weight in air

(b) Weight in water

Figure 5.23 *Determination of density by immersion*

The difference between the two readings is equal to the up-thrust on the solid body and this, by Archimedes' principle, is equal to the weight of liquid displaced.

Let the weight reading in air be W_1 and the reading when immersed in water be W_2.
Let the mass of the solid material be m and its density be ρ.
Let the density of the water be ρ_w.
Let the volume of the solid be V, which is also the volume of water displaced.

The mass of the solid material can be found from its weight:

$$W_1 = mg$$
$$m = \frac{W_1}{g} \tag{i}$$

The up-thrust, F is the difference between the two weight readings:

$$F = W_1 - W_2$$

From Archimedes' principle, the up-thrust is also equal to the weight of water displaced:

$$W_1 - W_2 = V\rho_w g$$
$$V = \frac{W_1 - W_2}{\rho_w g} \tag{ii}$$

An expression for the density ρ of the solid material can now be obtained by dividing the mass from equation (i) by the volume from equation (ii):

$$\rho = \frac{m}{V}$$
$$\rho = \frac{\frac{W_1}{g}}{\frac{W_1 - W_2}{\rho_w g}}$$
$$\rho = \frac{W_1}{g} \times \frac{\rho_w g}{W_1 - W_2}$$
$$\rho = \frac{W_1 \rho_w}{W_1 - W_2} \tag{5.23}$$

Now the relative density of the material is given by

$$\text{relative density} = \frac{\text{density of material}}{\text{density of water}}$$
$$\text{relative density} = \frac{\rho}{\rho_w}$$

This can be obtained by transposing equation (5.23) to give

$$\text{relative density} = \frac{\rho}{\rho_w} = \frac{W_1}{W_1 - W_2} \tag{5.24}$$

Example 5.9

A sample of a solid material weighs 250 N in air, 210 N when immersed in water and 215 N when immersed in oil. If the density of water is 1000 kg m^{-3}, determine the density of the material sample and the density of the oil.

Finding density ρ_m of solid material from equation (5.23)

$$\rho_m = \frac{W_1 \rho_w}{W_1 - W_2}$$

where $W_1 = 250$ N and $W_2 = 210$ N

$$\rho_m = \frac{250 \times 1000}{250 - 210}$$

$$\rho_m = \textbf{6250 kgm}^{-3}$$

Equation (5.23) can also be applied when the solid material is weighed in oil rather than water. If ρ_0 is the density of the oil and the weight in oil is $W_3 = 215\,\text{N}$, equation (5.23) can be written as

$$\rho_m = \frac{W_1 \rho_0}{W_1 - W_3}$$

$$\rho_0 = \frac{\rho_m(W_1 - W_3)}{W_1}$$

$$\rho_0 = \frac{6250 \times (250 - 215)}{250}$$

$$\rho_0 = \textbf{875 kg m}^{-3}$$

Test your knowledge 5.4

1. State Archimedes' principle.
2. Why does an immersed body experience up-thrust?
3. How can the density of a solid material be found by weighing it in air and immersed in water?
4. Could the same method for finding density be used on the Moon or on Mars?

Activity 5.4

A ball bearing of diameter 35 mm is weighed whilst totally immersed in oil of relative density 0.85 and its apparent weight is 1.54 N. What is the density of the ball bearing material?

Problems 5.3

1. A ship of mass of 1000 tonnes floats on sea water of density $1026\,\text{kg m}^{-3}$. Calculate the volume of water which it displaces.
   [975 m³]

2. A submarine has a volume of $1200\,\text{m}^3$ and floats with 75% of this below the surface. What is the mass of the vessel and what mass of water is required to be taken into its ballast tanks for it to float completely submerged? The density of sea water is $1030\,\text{kg m}^{-3}$.
   [927 tonnes, 309 tonnes]

3. The weight of an engineering component when weighed in air is 2.7 kN. It is weighed again when immersed in water and the recorded weight is then 2.1 kN. What is the volume of the component and its density? The density of water is $1000\,\text{kg m}^{-3}$.
   [0.061 m³, 4511 kg m⁻³]

4. A cube of wood with sides 250 mm long floats on water and has a density of $700\,\text{kg m}^{-3}$. (a) What will be the depth to which the cube is submerged? (b) What mass must be placed on top of the cube so that its top face is level with the water surface? The density of water is $1000\,\text{kg m}^{-3}$.
   [175 mm, 4.32 kg]

5. A component of density $3000\,\text{kg m}^{-3}$ is weighed when immersed in water and again when immersed in a second liquid. The recorded weights are 20 N and 21 N, respectively.
   What will be the mass of the component and the density of the second liquid? The density of water is $1000\,\text{kg m}^{-3}$.
   [3.06 kg, 899 kg m⁻³]

6. A steel bar of length 1 m and diameter 150 mm is suspended by a chain with two-thirds of its volume submerged in oil of density 780 kg m^{-3}. What will be the tension in the chain? The relative density of steel is 7.86.

[1.27 kN]

Force transmission

One of the basic principles of hydraulic engineering is that if the pressure is increased at any point in a fluid which is at rest in a closed system, the increase is transmitted without loss to all other points within the system. This is the principle on which hydraulic devices such as jacks, presses and braking systems operate.

Hydraulic jacks and presses consist of a small diameter *effort* or *master cylinder* coupled to a larger diameter *load* or *slave cylinder*. Each cylinder contains a closely fitting piston and the system is filled with hydraulic oil. In the basic arrangement shown in Figure 5.24, the pressure p is created by the input effort E as it moves through a distance x. The pressure is transmitted instantaneously to the piston in the load cylinder which moves through a smaller distance y.

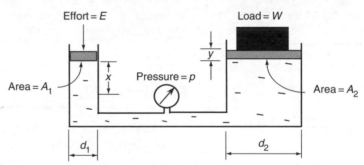

Figure 5.24 *Principle of hydraulic jack*

In the master cylinder,

effort = pressure × area of master cylinder piston

$$E = pA_1$$ (i)

$$p = \frac{E}{A_1}$$

In the slave cylinder,

load = pressure × area of slave cylinder piston

$$W = p A_2$$

$$p = \frac{W}{A_2}$$ (ii)

Equating (i) and (ii) gives

$$\frac{E}{A_1} = \frac{W}{A_2}$$

$$\frac{W}{E} = \frac{A_2}{A_1}$$

$$\frac{W}{E} = \frac{\frac{\pi d_2^2}{4}}{\frac{\pi d_2^1}{4}} = \frac{d_2^2}{d_2^1}$$

$$\frac{W}{E} = \left(\frac{d_2}{d_1}\right)^2 \tag{5.25}$$

Hydraulic jacks and presses are simple machines which perform similar tasks to the mechanical devices described in Chapter 1, i.e. they can be used to raise heavy loads and transmit force. Because the pistons are well lubricated by the hydraulic oil, the friction losses are quite small, and if they are neglected, equation (5.25) gives the *mechanical advantage* or *force ratio* of the system.

As the master cylinder piston moves through a distance x, it displaces a volume of oil which is given by

$$\text{volume displaced} = \frac{\text{area of master}}{\text{cylinder piston}} \times \frac{\text{distance moved}}{\text{by piston}}$$

$$V = \frac{\pi d_1^2}{4} x \tag{iii}$$

The same displaced volume enters the slave cylinder causing the piston to move through the distance y.

$$\text{volume displaced} = \frac{\text{area of slave}}{\text{cylinder piston}} \times \frac{\text{distance moved}}{\text{by piston}}$$

$$V = \frac{\pi d_2^2}{4} y \tag{iv}$$

Equating (iii) and (iv) gives,

$$\frac{\pi d_1^2}{4} x = \frac{\pi d_2^2}{4} y$$

$$d_1^2 x = d_2^2 y$$

$$\frac{x}{y} = \frac{d_2^2}{d_1^2}$$

$$\frac{x}{y} = \left(\frac{d_2}{d_1}\right)^2 \tag{5.26}$$

This is the *velocity ratio* or *movement ratio* of the system. It follows that when the effects of friction are small enough to be neglected, the two can be equated giving

$$\frac{W}{E} = \frac{x}{y} = \left(\frac{d_2}{d_1}\right)^2 \tag{5.27}$$

Key point

If friction losses are negligible, the mechanical advantage of a hydraulic jack or press is equal to its velocity ratio.

Example 5.10

A hydraulic jack is required to raise a load of 10 tonnes. The master and slave cylinder diameters are 75 mm, and 250 mm, respectively. Calculate (a) the effort required to raise the load, (b) the distance moved by the master cylinder piston when the load is raised through 50 mm, (c) the pressure in the hydraulic fluid.

(a) Finding effort required:

$$\frac{W}{E} = \left(\frac{d_2}{d_1}\right)^2$$

$$E = W\left(\frac{d_1}{d_2}\right)^2 = 10 \times 10^3 \times 9.81 \times \left(\frac{75}{250}\right)^2$$

$$E = 8.83 \times 10^3 \, \text{N} \quad \text{or} \quad 8.83 \, \text{kN}$$

(b) Finding distance moved by effort:

$$\frac{x}{y} = \left(\frac{d_2}{d_1}\right)^2$$

$$x = y\left(\frac{d_2}{d_1}\right)^2 = 50 \times \left(\frac{250}{75}\right)^2$$

$$x = 556 \, \text{mm}$$

(c) Finding pressure in hydraulic oil:

$$\text{pressure} = \frac{\text{load}}{\text{area of slave cylinder piston}}$$

$$p = \frac{W}{\frac{\pi d_2^2}{4}} = \frac{4\,W}{\pi d_2^2}$$

$$p = \frac{4 \times 10 \times 10^3}{\pi \times 0.25^2}$$

$$p = 204 \times 10^3 \, \text{Pa} \quad \text{or} \quad 204 \, \text{kPa}$$

Hydraulic braking systems operate on the same principle as hydraulic jacks and presses. The difference is that there is more than one slave cylinder and that the slave cylinders on the front and rear brakes may be of different diameters. Figure 5.25 shows a simple braking system for a vehicle in which the brake on each wheel is operated by a single slave cylinder. A modern disc braking system on a car or motor cycle will have at least two slave cylinders per disc.

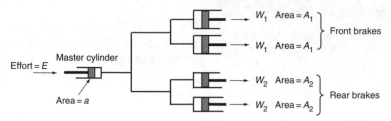

Figure 5.25 *Braking system*

In Figure 5.25, the pressure created in the master cylinder is transmitted equally to the slave cylinders.

In the master cylinder,

$$p = \frac{E}{a}$$

In the front slave cylinders,

$$p = \frac{W_1}{A_1}$$

In the rear slave cylinders,

$$p = \frac{W_2}{A_2}$$

Equating these gives

$$\frac{E}{a} = \frac{W_1}{A_1} = \frac{W_2}{A_2} \qquad (5.28)$$

It should be noted that the effort to the master cylinder piston is not applied directly by a car driver's foot. The brake pedal is a lever which magnifies the force applied by the foot and some additional force is also applied by servo-assistance from the engine.

Example 5.11

In the hydraulic braking system of a vehicle, a total force of 1.2 kN is applied to the master cylinder piston which has a diameter of 20 mm and moves through a distance of 15 mm. There are eight slave cylinders, four of diameter 30 mm and four of diameter 25 mm. Assuming no friction losses and that the hydraulic fluid is incompressible, determine (a) the pressure in the hydraulic fluid, (b) the force exerted by each of the different slave cylinders, (c) the distance moved by each slave cylinder piston if they all have the same brake clearance.

(a) Finding area a of master cylinder piston:

$$a = \frac{\pi d^2}{4} = \frac{\pi \times 0.02^2}{4}$$

$$a = 314 \times 10^{-6} \, \mathbf{m^2}$$

Finding pressure in hydraulic fluid:

$$p = \frac{E}{a} = \frac{1.2 \times 10^3}{314 \times 10^{-6}}$$

$$p = 3.82 \times 10^6 \, \mathbf{Pa} \quad \text{or} \quad 3.82 \, \mathbf{MPa}$$

(b) Finding force W_1 exerted by the slave cylinder pistons whose diameter D_1 is 30 mm area is A_1:

$$W_1 = p A_1 = p \frac{\pi d_1^2}{4}$$

$$W_1 = 3.82 \times 10^6 \times \frac{\pi \times 0.03^2}{4}$$

$$W_1 = 2.7 \times 10^3 \, \mathbf{N} \quad \text{or} \quad 2.7 \, \mathbf{kN}$$

Finding force W_2 exerted by the slave cylinder pistons whose diameter D_2 is 25 mm area is A_1:

$$W_2 = p A_2 = p \frac{\pi d_2^2}{4}$$

$$W_2 = 3.82 \times 10^6 \times \frac{\pi \times 0.025^2}{4}$$

$$W_2 = 1.88 \times 10^3 \, \mathbf{N} \quad \text{or} \quad 1.88 \, \mathbf{kN}$$

Test your knowledge 5.5

1. How may the velocity ratio of a hydraulic jack be calculated from the master and slave cylinder diameters?
2. What assumptions are made as to the compressibility of the hydraulic fluid when doing this calculation?
3. What assumption is made when the mechanical advantage and the velocity ratio of a hydraulic jack are taken to be equal?
4. What is the essential difference between the hydraulic braking system of a motor vehicle and basic hydraulic jack or press?

(c) Finding distance y moved by slave cylinder pistons when master cylinder piston moves through a distance $x = 15$ mm:

$$\frac{\text{volume displaced in}}{\text{master cylinder}} = \frac{\text{total volume displaced in}}{\text{slave cylinders}}$$

$$\frac{\pi d^2}{4} x = \frac{2\pi D_1^2}{4} y + 2\frac{\pi D_2^2}{4} y$$

$$d^2 x = 2D_1^2 y + 2D_2^2 y$$

$$d^2 x = 2(D_1^2 y + D_2^2) y$$

$$y = \frac{d^2 x}{2(D_1^2 + D_2^2)}$$

$$y = \frac{0.02^2 \times 0.015}{2(0.03^2 + 0.025^2)}$$

$$\boldsymbol{y = 1.97 \times 10^{-3} \, \textbf{m} \quad \text{or} \quad 1.97 \, \textbf{mm}}$$

Activity 5.5

Figure 5.26 is a schematic diagram for a hydraulic tensile testing machine. The spindle moves the master cylinder piston whose diameter is 25 mm. The twin output slave cylinders are each 60 mm in diameter and are connected to a rigid crossbeam. This in turn transmits the load to the tensile test specimen. Frictional resistance in the hydraulic cylinders can be neglected and the hydraulic fluid can be considered to be incompressible. The steel test specimen is of length 75 mm and diameter 3 mm. Its elastic limit stress is 250 MPa and its modulus of elasticity is 200 GPa.

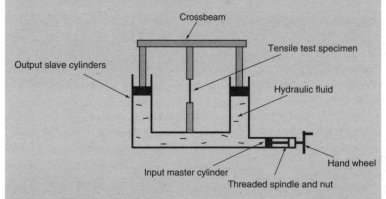

Figure 5.26

What will be the input effort to the master cylinder as the specimen reaches its elastic limit and how far will its piston have moved?

Problems 5.4

1. A hydraulic jack is required to raise a load of 1.5 tonnes. The area of the load piston is 200 mm. The area of the input piston

is 25 mm with a stroke length of 150 mm. Calculate the effort required to raise the load and the distance moved by the load per input stroke. Neglect friction losses.

[230 N, 2.34 mm]

2. A hydraulic press is required to deliver a load of 5 kN to a die which moves through a distance of 12 mm during a forming operation. The load cylinder diameter is 80 mm and the effort piston moves through a distance 110 mm during the operation. Determine (a) the diameter of the effort cylinder, (b) the effort required assuming negligible friction losses.

[26.4 mm, 545 N]

3. A hydraulic tensile testing machine has twin load cylinders, each of diameter 50 mm and a single input cylinder of diameter 20 mm. An alloy specimen 3 mm diameter has a UTS of 300 MPa. Determine the load required to fracture the specimen and the effort.

[2.12 kN, 170 N]

4. A disc brake on a machine shaft is operated by two slave cylinders, each of diameter 25 mm at a radius of 150 mm from the centre of the disc. The master cylinder has a diameter of 15 mm and the input effort is 120 N. What will be the output braking torque?

[99.9 N m]

5. A hydraulic disc braking system has a master cylinder of diameter 18 mm with a stroke of 20 mm. There are four slave cylinders of diameter 32 mm operating on the front brake discs and four of diameter 24 mm operating on the rear brake discs. If the force exerted by each of the front slave cylinders is 2.5 kN determine (a) the pressure in the hydraulic fluid, (b) the force exerted by each of the rear slave cylinders, (c) the input effort to the master cylinder, (d) the distance moved by the slave cylinder pistons assuming that they had same initial brake clearance.

[3.11 MPa, 1.41 kN, 791 N, 2.03 mm]

Fluids in motion

When considering fluids in motion it is usual to assume ideal conditions. You may remember that an ideal fluid is one whose dynamic viscosity and surface tension are zero. As a result it would not experience internal shearing forces, viscous drag would be absent and its flow would always be laminar. Furthermore, an ideal liquid would be incompressible and would never vaporise or freeze. These assumptions enable steady fluid flow systems to be modelled mathematically and formulae to be derived for calculating flow velocities, flow rates, pressures and forces. In many cases they accurately predict the behaviour of the flow systems. In cases where there is some variance, adjustments can often be made to the formulae, based on experimental evidence, to give closer agreement between theory and practice.

A steady flow system is one in which the flow velocity is constant at any cross-section. The flow may be through a pipe, an open channel or a duct. Alternatively, it may be in the form of a jet issuing at a steady rate from a nozzle or an orifice. There are two basic equations which can be applied to a steady flow process.

They are (i) the continuity equation, and (ii) the steady flow energy equation.

The continuity equation

Consider the steady flow of a fluid at velocity $v\,\mathrm{m\,s^{-1}}$ along a parallel pipe or duct of cross-sectional area $A\,\mathrm{m^2}$, as in Figure 5.27. Let the volume flow rate be $V\,\mathrm{m^3\,s^{-1}}$.

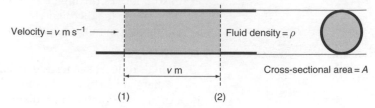

Velocity $= v\,\mathrm{m\,s^{-1}}$ Fluid density $= \rho$

$v\,\mathrm{m}$ Cross-sectional area $= A$

(1)   (2)

Figure 5.27 *Steady flow in a parallel pipe*

The volume between sections (1) and (2) in Figure 5.27 has a length equal to the flow velocity. In 1 s this volume will pass through section (2), and so the volume flow rate, measured in $\mathrm{m^3\,s^{-1}}$, is given by the formula

$$V = Av\ (\mathrm{m^3\,s^{-1}}) \tag{5.29}$$

The mass flowing per second, or mass flow rate, is the volume flow rate multiplied by the fluid density

$$m = \rho V$$

$$m = \rho Av\ (\mathrm{kg\,s^{-1}}) \tag{5.30}$$

Consider now a pipe whose cross-section changes between two stations (1) and (2) as in Figure 5.28. If the fluid is a gas it is quite possible that its density will change especially if there is a change of temperature and pressure between sections (1) and (2). Irrespective of whether the fluid is a liquid or a gas, the mass per second passing section (1) will be the same as that passing section (2) if the velocities v_1 and v_2 are steady. There is said to be *continuity of mass* which is given by the equation

$$\rho_1 A_1 v_1 = \rho_2 A_2 v_2 = m\ (\mathrm{kg\,s^{-1}}) \tag{5.31}$$

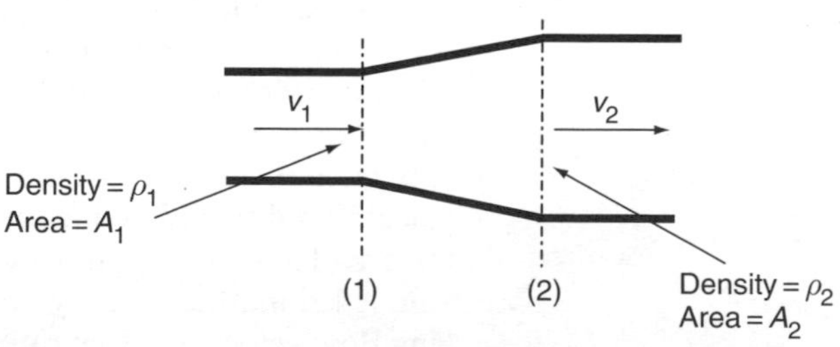

v_1 v_2

Density $= \rho_1$
Area $= A_1$

Density $= \rho_2$
Area $= A_2$

(1)   (2)

Figure 5.28 *Steady flow in a tapering pipe*

If the fluid is an incompressible liquid whose temperature is constant, then the density will be constant and $\rho_1 = \rho_2$. Equation (5.31) then becomes

$$A_1 v_1 = A_2 v_2 = V\ (\mathrm{m^3\,s^{-1}}) \tag{5.32}$$

The volume per second passing section (1) is now equal to the volume per second passing section (2) and there is then said to be *continuity of volume*. Substituting the expressions for cross-sectional area in place of A_1 and A_2 gives

$$\frac{\cancel{\pi} d_1^2}{\cancel{4}} v_1 = \frac{\cancel{\pi} d_2^2}{\cancel{4}} v^2$$

$$d_1^2 v_1 = d_2^2 v_2$$

$$\frac{v_1}{v_2} = \frac{d_2^2}{d_1^2}$$

$$\frac{v_1}{v_2} = \left(\frac{d_2}{d_1}\right)^2 \tag{5.33}$$

From this it will be seen that the ratio of the velocities at the two sections is equal to the inverse ratio of the diameters squared, i.e. if d_2 is twice the value of d_1, then the velocity v_1 will be four times bigger than v_2, etc.

Key point

The flow velocities at two sections through the steady flow of an incompressible fluid are in the inverse ratio of the cross-sectional areas.

Example 5.12

Figure 5.29 conveys oil of density $875\,\text{kg m}^{-3}$. The diameters at sections (1), (2) and (3) are 100 mm, 50 mm and 150 mm, respectively. If the flow velocity is $8\,\text{m s}^{-1}$ at section (2), determine (a) the volume and mass flow rates, (b) the velocities at sections (1) and (3).

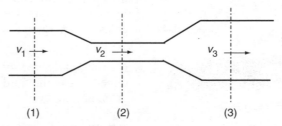

Figure 5.29

(a) Finding cross-sectional area at (2):

$$A_2 = \frac{\pi d_2^2}{4} = \frac{\pi \times 0.05^2}{4}$$

$$A_2 = 1.96 \times 10^{-3}\ \text{m}^2$$

Finding volume flow rate:

$$V = A_2 v_2 = 1.96 \times 10^{-3} \times 8$$

$$V = 0.0157\ \text{m}^3\,\text{s}^{-1} \quad \text{or} \quad 15.7\ \text{l s}^{-1}$$

Finding mass flow rate:

$$m = \rho V = 875 \times 0.0157$$

$$m = 13.7\ \text{kg s}^{-1}$$

(b) Finding velocity at (1):

$$V = A_1 v_1 = \frac{\pi d_1^2}{4} v_1$$

$$v_1 = \frac{4 V}{\pi d_1^2} = \frac{4 \times 0.0157}{\pi \times 0.1^2}$$

$$\boldsymbol{v_1 = 2.0 \ \mathbf{m \, s^{-1}}}$$

Finding velocity at (3):

$$V = A_3 v_3 = \frac{\pi d_3^2}{4} v_3$$

$$v_3 = \frac{4 V}{\pi d_3^2} = \frac{4 \times 0.0157}{\pi \times 0.15^2}$$

$$\boldsymbol{v_3 = 0.884 \ \mathbf{m \, s^{-1}}}$$

It will be noted that the velocity at (2) is four times bigger than the velocity at (1) and nine times bigger than the velocity at (3), as predicted by equation (5.33), i.e. the ratio of the velocities is equal to the inverse ratio of the cross-sectional areas and of the diameters squared.

Energy of a fluid

A moving fluid can contain energy in a number of different forms. If the mass flow rate is $m \ \mathrm{kg \, s^{-1}}$, the values of these energy forms, given as the energy per second passing a particular cross-section, are as follows:

1. *Gravitational potential energy*   This is the work which has been done to raise the fluid to its height, z m, measured above some given datum level.

    potential energy $= m \, g \, z \ (\mathbf{J \, s^{-1}}$ or $\mathbf{W})$ (5.34)

2. *Kinetic energy*   This is the work which has been done in accelerating the fluid from rest up to some particular velocity $v \, \mathrm{m \, s^{-1}}$.

    kinetic energy $= \dfrac{1}{2} m v^2 \ (\mathbf{J \, s^{-1}}$ or $\mathbf{W})$ (5.35)

3. *Internal energy*   This is the energy contained in a fluid by virtue of the absolute temperature, T K, to which it has been raised. It is the sum of the individual kinetic energies of the molecules of the fluid. These are in a state of random motion and their kinetic energy is directly proportional to the absolute temperature of the fluid. It is separate from the above linear kinetic energy of flow. A fluid at a particular temperature has the same internal energy when stationary and when it is in motion. It is difficult to calculate internal energy although the changes of internal energy which accompany temperature change can be determined and will be considered in Chapter 6. Internal energy is generally expressed as

    internal energy $= \mathbf{U} \ (\mathbf{J \, s^{-1}}$ or $\mathbf{W})$ (5.36)

4. *Pressure-flow energy* This is also called *flow work* and *work of introduction*. It is the energy of the fluid by virtue its pressure, p Pa. This is the pressure to which it has been raised by a pump, fan or compressor to make it flow into or through a system against the prevailing back-pressure. It is analogous to the work that must be done against friction to keep a solid body in motion.

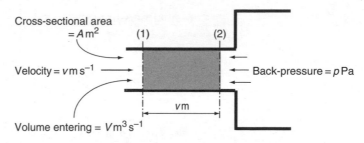

Figure 5.30 *Steady flow against back-pressure*

Consider the steady flow of fluid entering a system as shown in Figure 5.30. The volume entering per second, V m^3, is the volume between sections (1) and (2) whose length is v m. The pressure-flow energy received by the fluid is the work done per second in pushing it past section (2).

pressure-flow energy = force applied × distance moved per second

pressure-flow energy $= p A v$

But $A v = V$, the volume entering per second.

pressure-flow energy $= p V$ (J s^{-1} or W) (5.37)

The steady flow energy equation

When a fluid is flowing through a device such as a pump, compressor or turbine, the forms of energy listed above may all undergo a change. Furthermore, there may be a flow of energy into or out of the system in the form of heat and work. Such a system is shown in Figure 5.31.

Key point

Although energy changes might occur in a fluid system, the principle of conservation of energy tells us that the total energy of the system and its surrounding is constant.

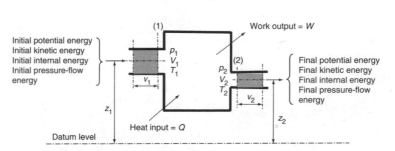

Figure 5.31 *Steady flow process*

For steady flow conditions, the mass per second flowing into the system is equal to the mass per second leaving. Also the energy per second entering the system must equal the energy per second leaving. It follows that for the system shown in Figure 5.31,

$$
\left.\begin{array}{c}
\text{initial potential energy} \\
+ \\
\text{initial kinetic energy} \\
+ \\
\text{initial internal energy} \\
+ \\
\text{initial pressure-flow energy} \\
+ \\
\text{heat energy input}
\end{array}\right\} = \left\{\begin{array}{c}
\text{final potential energy} \\
+ \\
\text{final kinetic energy} \\
+ \\
\text{final internal energy} \\
+ \\
\text{final pressure-flow energy} \\
+ \\
\text{work output}
\end{array}\right.
$$

Substituting the expressions which have been derived for these terms gives

$$
\begin{aligned}
mg z_1 + \tfrac{1}{2}mv_1{}^2 &= m g z_2 + \tfrac{1}{2}m v_2^2 \\
+ U_1 + p_1 V_1 + Q &\quad + U_2 + p_2 V_2 + W
\end{aligned} \tag{5.38}
$$

This is known as the *full steady flow energy equation* (*FSFEE*) which we shall make use of in Chapter 6 in the study of thermodynamics. It is also very useful in the study of fluid mechanics but here we change it into a more usable form which is known *as Bernoulli's equation.*

Bernoulli's equation

In the study of liquids flowing steadily through a pipe or channel, or issuing from a nozzle in the form of a jet, it is assumed that temperature changes are so small as to be negligible. This means that there is little or no change in internal energy, i.e. $U_1 = U_2$, and that these terms can be eliminated from the steady flow energy equation. It is also assumed that no heat transfer takes place and that no external work is done, i.e. $Q = 0$ and $W = 0$, and these terms can also be eliminated. The steady flow energy equation thus reduces to

$$
mg z_1 + \tfrac{1}{2} mv_1^2 + p_1 V_1 = m g z_2 + \tfrac{1}{2}m v_2^2 + p_2 V_2
$$

Dividing throughout by the product mg gives

$$
z_1 + \frac{v_1^2}{2g} + \frac{p_1 V_1}{m g} = z_1 + \frac{v_2^2}{2g} + \frac{p_2 V_2}{m g}
$$

Now for an incompressible liquid, $\frac{V_1}{m} = \frac{V_2}{m} = \frac{1}{\rho}$, where ρ is the density of the liquid.
Substituting and rearranging gives

$$
\frac{p_1}{\rho g} + \frac{v_1^2}{2g} + z_1 = \frac{p_2}{\rho g} + \frac{v_2^2}{2g} + z_2 \tag{5.39}
$$

This is *Bernoulli's equation* in which the units of each term are metres. Dividing throughout by mg has changed the units from Joules per second or Watts to the equivalent potential energy height, or *head*, measured in metres:

The $\frac{p}{\rho g}$ terms are called *pressure heads*.

The $\frac{v^2}{2g}$ terms are called *velocity heads*.

The z terms are called *potential heads*.

The terms on each side of Bernoulli's equation give the total head at section (1) of a steady flow system which is equal to the total head at section (2). In practice there is often some energy loss, z_f due to viscous resistance and turbulence, particularly in long pipelines. This can sometimes be estimated and added to the right-hand side of Bernoulli's equation to give

$$\frac{p_1}{\rho g} + \frac{v_1^2}{2g} + z_1 = \frac{p_1}{\rho g} + \frac{v_2^2}{2g} + z_2 + z_f \tag{5.40}$$

When used together, the continuity equation (5.32) and Bernoulli's equation (5.40) can solve many problems concerned with steady flow. Sometimes they have to be solved as simultaneous equations as in Example 5.13. This provides an opportunity for you to put into practice some of the mathematical techniques which you have learned.

Example 5.13

Water is pumped upwards through a height of 8.8 m in a vertical pipe whose diameter diverges gradually from 75 mm to 125 mm over part of its length. The pressure at the lower end is 234 kPa and the pressure at the upper end is 145 kPa (Figure 5.32). The flow is steady and there is a pipe friction loss equivalent to a head of 1.2 m. Determine the flow velocity in the two sections of the pipe and the volume flow rate.

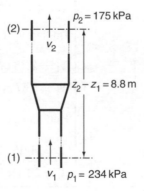

Figure 5.32

Finding the cross-sectional areas at (1) and (2)

$$A_1 = \frac{\pi d_1^2}{4} = \frac{\pi \times 0.075^2}{4}$$

$$A_1 = 4.42 \times 10^{-3} \text{ m}^2$$

$$A_2 = \frac{\pi d_2^2}{5} = \frac{\pi \times 0.125^2}{4}$$

$$A_2 = 12.27 \times 10^{-3} \text{ m}^2$$

Applying continuity equation to sections (1) and (2)

$$A_1 v_1 = A_2 v_2$$

$$v_1 = \frac{A_2 v_2}{A_1} = \frac{12.27 \times 10^{-3}}{4.42 \times 10^{-3}} \times v_2$$

$$\mathbf{v_1 = 2.78\, v_2} \tag{i}$$

Applying Bernoulli's equation to sections (1) and (2)

$$\frac{p_1}{\rho g} + \frac{v_1^2}{2g} + z_1 = \frac{p_2}{\rho g} + \frac{v_2^2}{2g} + z_2 + z_f$$

$$\frac{v_1^2}{2g} - \frac{v_2^2}{2g} = \frac{p_2}{\rho g} - \frac{p_1}{\rho g} + z_2 - z_1 + z_f$$

$$v_1^2 - v_2^2 = 2g\left(\frac{p_2 - p_1}{\rho g} + z_2 - z_1 + z_f\right)$$

$$v_1^2 - v_2^2 = 2 \times 9.81\left[\frac{(145 - 234) \times 10^3}{1000 \times 9.81} + 8.8 + 1.2\right]$$

$$\mathbf{v_1{}^2 - v_2{}^2 = 18.2} \tag{ii}$$

Substitute for v_1 from equation (i)

$$2.78^2\, v_2^2 - v_2^2 = 18.2$$

$$v_2^2\,(7.73 - 1) = 18.2$$

$$6.73\, v_2^2 = 18.2$$

$$v_2 = \sqrt{\frac{18.2}{6.73}}$$

$$\mathbf{v_2 = 1.64\, m\,s^{-1}}$$

Finding v_1 from equation (i):

$$v_1 = 2.78\, v_2 = 2.78 \times 1.64$$

$$\mathbf{v_1 = 4.57\, m\,s^{-1}}$$

Finding volume flow rate:

$$V = A_2 v_2 = 12.27 \times 10^{-3} \times 1.64$$

$$\mathbf{V = 20.1 \times 10^{-3}\, m^3\,s^{-1}} \quad \text{or} \quad \mathbf{20.1\,l\,s^{-1}}$$

Flow measurement

A variety of instruments have been developed for measuring the rate of fluid flow. Three in particular which are within the scope of this unit are the *venturi meter*, the *orifice meter* and the *pitot-static tube*. They are known as differential pressure devices because they make use of the pressure difference between two points in the flow to give an indication of the flow rate. The continuity equation and Bernoulli's equation are applied to obtain expressions for flow rate and sometimes they are said to work on the Bernoulli principle.

The venturi and orifice meters

A venturi meter consists of a section of pipe which converges to a throat section and then diverges back to the original pipe diameter as shown in Figure 5.33. The piezometers at sections (1) and (2) record the pressure heads h_1 and h_2. The difference in the pressure heads h varies with the rate of flow but it is not a linear relationship, as will be shown.

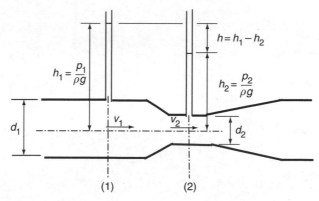

Figure 5.33 *Venturi meter*

The ratio of the flow velocities at sections (1) and (2) is given by equation (5.24):

$$\frac{v_1}{v_2} = \left(\frac{d_2}{d_1}\right)^2$$

or

$$v_2 = v_1 \left(\frac{d_1}{d_2}\right)^2 \tag{i}$$

Applying Bernoulli's equation to sections (1) and (2), we get

$$\frac{p_1}{\rho g} + \frac{v_1^2}{2g} + z_1 = \frac{p_2}{\rho g} + \frac{v_2^2}{2g} + z_2$$

Now $z_1 = z_2$. Cancelling and rearranging gives,

$$\frac{p_1}{\rho g} - \frac{p_2}{\rho g} = \frac{v_2^2}{2g} - \frac{v_1^2}{2g}$$

But $\frac{p_1}{\rho g} = h_1$ and $\frac{p_2}{\rho g} = h_2$

$$h_1 - h_2 = \frac{v_2^2 - v_1^2}{2g}$$

Also, $h_1 - h_2 = h$, the difference in the pressure heads at (1) and (2)

$$h = \frac{v_2^2 - v_1^2}{2g}$$

or

$$v_2^2 - v_1^2 = 2gh \tag{ii}$$

Substituting for v_2 from equations (i) gives

$$\left[v_1 \left(\frac{d_1}{d_2} \right)^2 \right]^2 - v_1^2 = 2gh$$

$$v_1^2 \left(\frac{d_1}{d_2} \right)^4 - v_1^2 = 2gh$$

$$v_1^2 \left[\left(\frac{d_1}{d_2} \right)^4 - 1 \right] = 2gh$$

$$v_1^2 = \frac{2gh}{\left(\frac{d_1}{d_2} \right)^4 - 1}$$

$$v_1 = \sqrt{\frac{2gh}{\left(\frac{d_1}{d_2} \right)^4 - 1}} \tag{5.41}$$

The volume flow rate can now be found:

$$V = A_1 v_1$$

$$V = A_1 \sqrt{\frac{2gh}{\left(\frac{d_1}{d_2} \right)^4 - 1}} \tag{5.42}$$

This is of course, the theoretical volume flow rate. The actual volume flow rate V' is always slightly less than this due to viscous drag and turbulence. It is found by collecting the actual discharge over a period of time in a weighing tank. The ratio of the actual to theoretical flow rate is given by the discharge coefficient c_d for the meter. Its value changes a little with the flow rate but is usually around 0.95–0.97.

$$\text{discharge coefficient} = \frac{\text{actual flow rate}}{\text{theoretical flow rate}}$$

$$c_d = \frac{V'}{V}$$

or

$$V' = c_d V$$

$$V' = c_d A_1 \sqrt{\frac{2gh}{\left(\frac{d_1}{d_2} \right)^4 - 1}} \tag{5.43}$$

You will note that all except the difference in pressure head h are constants and the expression may be written as

$$V' = c_d A_1 \sqrt{\frac{2g}{\left(\frac{d_1}{d_2} \right)^4 - 1}} \ \sqrt{h}$$

or

$$V' = k\sqrt{h} \tag{5.44}$$

where k is the *meter constant*, whose value can be calculated from a knowledge of the pipe and throat diameters and the discharge coefficient. As has been stated, the relationship between flow rate and head difference is not linear and this can now be seen clearly from equation (5.44). Plotting volume flow rate V' against pressure head difference h gives a curve, as shown in Figure 5.34, and plotting V' against the square root of h gives a straight line graph whose gradient is the meter constant k.

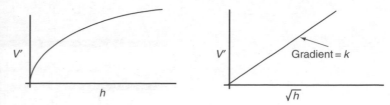

Figure 5.34 *Venturi meter characteristics*

An orifice meter operates on exactly the same principle as the venturi meter. In its basic form, it consists of a sharp-edged orifice plate which is inserted between the connecting flanges of a pipeline. Piezometers measure the pressure difference between the upstream and downstream pressures as shown in Figure 5.35.

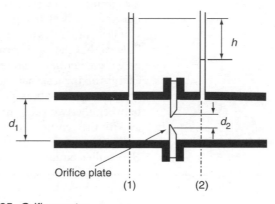

Figure 5.35 *Orifice meter*

As with the venturi meter, the actual flow rate V' is given by equations (5.43) and (5.44)

$$V' = c_d A_1 \sqrt{\frac{2gh}{\left(\frac{d_1}{d_2}\right)^4 - 1}}$$

or

$$V' = k\sqrt{h}$$

The advantage of the orifice meter is that it is easy to install and is relatively inexpensive when compared with the venturi meter. Its main disadvantage is that it impedes the flow to a much greater extent as is evidenced by a much lower discharge coefficient. This ranges from 0.6 to 0.65 for most orifice meters. The piezometer tappings should be positioned one pipe diameter upstream of the orifice and half a pipe diameter downstream to give maximum value of pressure difference. For industrial purposes, however, it is often more convenient to drill through the pipe flanges where there is a greater depth of metal for the tappings.

The pitot-static tube

The pitot-static tube is used for measuring the speed of gas flow in ducts and wind tunnels. It is also used in aircraft for measuring air speed. In its basic form, the instrument consists of two tubes as shown in Figure 5.36. The static tube records the static pressure in

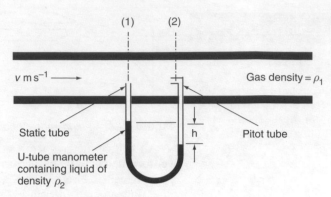

Figure 5.36 *Pitot and static tubes*

the gas flow. The name is a little confusing because the gas is moving, but imagine it as being the pressure which you would experience if you were moving with the gas. The pitot tube is angled to face the gas flow so that the gas is brought to rest inside its mouth. The pressure here is greater than that recorded in the static tube because the kinetic energy of the gas has been transformed into pressure head. It is known as stagnation pressure. The difference in pressure head h between the two tubes increases with the speed of the gas flow but, as with the venturi and orifices meters, it is not a linear relationship.

To obtain an expression for the free stream velocity v, apply Bernoulli's equation to station (1), which is in the free stream and to station (2), which is in the mouth of the pitot tube.

$$\frac{p_1}{\rho_1 g} + \frac{v_1^2}{2g} + z_1 = \frac{p_2}{\rho_1 g} + \frac{v_2^2}{2g} + z_2$$

Now $v_1 = v$, the free stream velocity and $v_2 = 0$. Also $z_1 = z_2$. Substituting and eliminating leaves

$$\frac{p_1}{\rho_1 g} + \frac{v^2}{2g} = \frac{p_2}{\rho_1 g}$$

$$\frac{v^2}{2g} = \frac{p_2}{\rho_1 g} - \frac{p_1}{\rho_1 g}$$

$$\frac{v^2}{2g} = \frac{p_2 - p_1}{\rho_1 g}$$

But $p_2 - p_1 = h\rho_2 g$, where ρ_2 is the density of the liquid in the manometer.

$$\frac{v^2}{2g} = \frac{h\rho_2 \cancel{g}}{\rho_1 \cancel{g}}$$

$$v^2 = 2gh\frac{\rho_2}{\rho_1}$$

$$\boldsymbol{v = \sqrt{2gh\frac{\rho_2}{\rho_1}}} \qquad (5.45)$$

Key point

Pitot-static tubes are not sensitive at low flow velocities.

The pitot-static tube is not very sensitive at gas velocities less than around $15\,\mathrm{m\,s}^{-1}$. The use of an inclined well-type manometer will increase its sensitivity but it is essentially a high-speed measuring device. In practical applications, the two tubes are incorporated into a single device as shown in Figure 5.37.

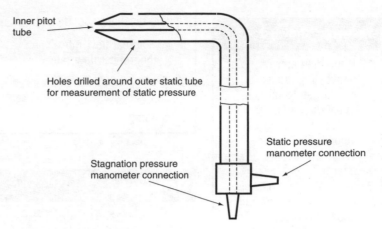

Figure 5.37 *Pitot-static tube*

Example 5.14

In a test on an orifice meter, 280 kg of water was collected in a weighing tank over a 5 min period. The pipe containing the orifice was 50 mm in diameter and the orifice was 30 mm in diameter. The pressure difference across the orifice plate during the test was recorded to be 200 mmH₂O. Determine the discharge coefficient in operation. The density of water is 1000 kg m⁻³.

Finding actual mass flow rate:

$$m' = \frac{\text{mass collected}}{\text{time taken}} = \frac{280}{5 \times 60}$$

$$m' = 0.933 \text{ kg s}^{-1}$$

Finding actual volume flow rate:

$$V' = \frac{m}{\rho} = \frac{0.933}{1000}$$

$$V' = 0.933 \times 10^{-3} \text{ m}^3 \text{ s}^{-1} \quad \text{or} \quad 0.933 \text{ l s}^{-1}$$

Finding theoretical volume flow rate using equation (5.42)

$$V = A_1 \sqrt{\frac{2gh}{\left(\frac{d_1}{d_2}\right)^4 - 1}} = \frac{\pi d_1^2}{4} \sqrt{\frac{2gh}{\left(\frac{d_1}{d_2}\right)^4 - 1}}$$

$$V = \frac{\pi \times 0.05^2}{4} \sqrt{\frac{2 \times 9.81 \times 0.2}{\left(\frac{0.05}{0.03}\right)^4 - 1}}$$

$$V = 1.5 \times 10^{-3} \text{m}^3 \text{ s}^{-1} \quad \text{or} \quad 1.5 \text{ l s}^{-1}$$

Test your knowledge 5.6

1. If the diameter of a pipe conveying liquid converges to half its initial diameter, by what factor will the flow velocity increase?
2. What is pressure-flow energy?
3. What are the units of each term in Bernoulli's equation?
4. What is the relationship between flow rate and pressure difference between the pipe and throat sections of a venturi meter?
5. For what kinds of application is a pitot-static tube used?

Finding discharge coefficient:

$$c_d = \frac{\text{actual flow rate}}{\text{theoretical flow rate}}$$

$$c_d = \frac{V'}{V} = \frac{0.933}{1.5}$$

$$c_d = 0.622$$

Activity 5.6

In Figure 5.38, which conveys water, the pipe diameter at (1) is 100 mm and the diameter at (2), the venturi throat section, is 60 mm. The pressure in the pipe at (1) is 750 mmH$_2$O and the venturi meter has a discharge coefficient of 0.97. The pipe diameter at section (3) is 40 mm. Determine (a) the volume flow rate, (b) the flow velocity at (3), (c) the pressure at (3). The density of water is 1000 kg m^{-3}.

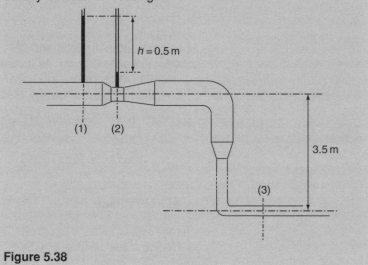

Figure 5.38

Problems 5.5

1. Oil of density 875 kg m^{-3} flows at the rate of 400 litres per second along a pipe of diameter 400 mm. If the pipe converges to a diameter of 200 mm and then diverges to a diameter of 600 mm, determine the flow velocity at each section and also the mass flow rate.

$$[3.18 \, \text{m s}^{-1}, 12.73 \, \text{m}^{-1}, 1.41 \, \text{m s}^{-1}, 350 \, \text{kg s}^{-1}]$$

2. A horizontal pipe which runs full of water diverges gradually in the direction of the flow from 150 mm to 300 mm in diameter. If the mass flow rate is 110 tonnes per hour, determine (a) the flow velocities at the two sections, (b) the pressure difference between the two sections. The density of water is 1000 kg m^{-3}.

$$[1.73 \, \text{m s}^{-1}, 0.43 \, \text{m s}^{-1}, 1.4 \, \text{kPa}]$$

3. Oil of density $800\,kg\,m^{-3}$ flows along a horizontal pipe which varies in section from 100 mm to 150 mm diameter in the direction of flow. The pressure at the smaller section is 126 kPa and that at the larger section is 140 kPa. Determine (a) the flow velocity at the two sections, (b) the volume flow rate, (c) the mass flow rate.

$$[6.62\,m\,s^{-1},\,2.94\,m\,s^{-1},\,521\,s^{-1},\,41.6\,kg\,s^{-1}]$$

4. The diameter of a pipe conveying water tapers gradually in the direction of flow from 200 mm to 100 mm in diameter whilst falling through a vertical height of 9 m. The pressure at the upper section is 210 kPa and the volume flow rate is $72\,l\,s^{-1}$. Calculate (a) the flow velocity at each section, (b) the pressure at the upper section. The density of water is $1000\,kg\,m^{-3}$.

$$[2.29\,m\,s^{-1},\,9.17\,m\,s^{-1},\,259\,kPa]$$

5. The flow velocity along a horizontal pipeline of 120 mm diameter is measured by means of an orifice meter. The orifice diameter is 50 mm and the recorded pressure difference across it is 12 kPa. If the discharge coefficient is 0.64, determine the mass flow rate in the pipe. The density of water is $1000\,kg\,m^{-3}$.

$$[6.16\,kg\,s^{-1}]$$

6. A pitot-static tube is used to measure the velocity of the air in a duct. If the difference between the static and dynamic pressures is $30\,mmH_2O$, determine the air velocity. The density of water is $1000\,kg\,m^{-3}$ and the density of the air is $1.2\,kg\,m^{-3}$.

$$[22.1\,m\,s^{-1}]$$

Force exerted by a jet of fluid

A jet of fluid contains energy and will exert a force on anything which lies in its path. In water, steam and gas turbines, the jet impinges on the rotor blades causing them to rotate. Some of the energy in the jet is thus changed into rotational kinetic energy. Turbine output and efficiency calculations can be quite complex and will be left for study at a higher level. Here we will consider only the force exerted by a jet of fluid on stationary objects, which will lay the foundation for future work.

Force exerted by a jet of fluid on a stationary flat plate

Consider a parallel jet of fluid of density $\rho\,kg\,m^{-3}$ and cross-sectional area $A\,m^2$ which strikes a stationary flat plate at right angles with a velocity of $v\,m\,s^{-1}$. On striking the plate, the jet is deflected at right angles in all directions so that its final velocity in the original x-direction is zero, as shown in Figure 5.39. The mass of fluid striking the plate per second is given by equation (5.30), that is,

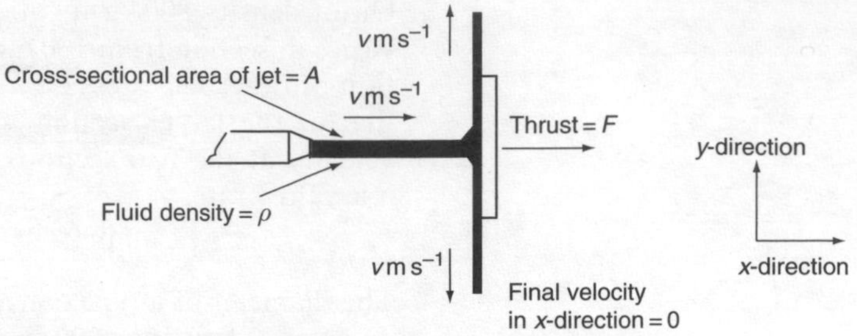

Figure 5.39 *Impact of a jet on a stationary flat plate*

mass flow rate = density × volume flow rate

$$m = \rho A v \qquad \text{(i)}$$

As you are well aware by now, velocity is a vector quantity. It has both magnitude and direction, and if either of should change, a change in velocity will occur. Here the jet direction is changed by the plate and as a result, its velocity changes from $v\,\mathrm{m\,s^{-1}}$ to zero in the x-direction. Momentum is the product of mass and velocity, and so there will also have been a change in momentum in the x-direction. From Newton's second law of motion, the force acting on the plate is given by

force = rate of change of momentum

$$\text{force} = \frac{\text{mass} \times \text{change in velocity}}{\text{time taken}}$$

or

$$\textbf{force} = \frac{\textbf{mass}}{\textbf{time taken}} \times \textbf{change in velocity} \qquad \text{(ii)}$$

But the mass divided by the time taken is the mass flow rate given by equation (i) above, and the change in velocity is from $v\,\mathrm{m\,s^{-1}}$ to zero. Substituting in equation (ii) we get

$$F = m(v - 0)$$
$$F = \rho A v(v - 0)$$
$$F = \rho A v^2 \qquad \text{(5.46)}$$

From equation (5.46), it can be seen that the thrust on the plate is proportional to the density of the fluid, the cross-sectional area of the jet and the square of the jet velocity. This means that if you double the velocity of the jet, the force on the plate will increase by a factor of 4. Similarly, if you double the diameter of the jet, its cross-sectional area will increase by a factor of 4 and once again, there will be four times as much force on the plate.

> **Key point**
>
> It is convenient to apply Newton's second law of motion to fluid systems in the form
>
> **force = mass flow rate × change in velocity**

Force exerted by a jet of fluid on a stationary hemispherical cup

Consider now the effect of replacing the flat plate by a hemispherical cup as shown in Figure 5.40.

The mass striking the cup per second is the same as for the flat plate, that is,

$$m = \rho A v \qquad \text{(i)}$$

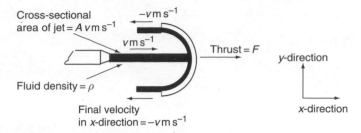

Figure 5.40 *Impact of a jet on a hemispherical cup*

From Newton's second law of motion, the force exerted on the cup is again given by

$$\text{force} = \frac{\text{mass}}{\text{time taken}} \times \text{change in velocity} \qquad \text{(ii)}$$

The difference now is that the jet is deflected through 180° and the change in velocity in the x-direction is from $v\,\text{m s}^{-1}$ to $-v\,\text{m s}^{-1}$. Substituting in equation (ii) for these values, we obtain

$$F = m[v - (-v)]$$
$$F = \rho A v[v - (-v)]$$
$$\mathbf{F = 2\rho A v^2} \qquad \text{(5.47)}$$

In reversing the velocity of the fluid, the change of velocity, and so also the change of momentum is double that for the flat plate. As a result there is twice as much force exerted by the jet.

Reaction of a convergent nozzle

When a jet of fluid issues from a convergent nozzle as shown in Figure 5.41, the nozzle experiences a reactive force in the opposite direction to the jet. An accelerating force causes the fluid to increase its velocity from v_1 in the supply pipe to velocity v_2 as it emerges from the nozzle. The nozzle experiences an equal and opposite reaction as predicted by Newton's third law of motion. You may have experienced the same backward force when holding a water hose pipe. Applying Newton's second law of motion enables the force to be found.

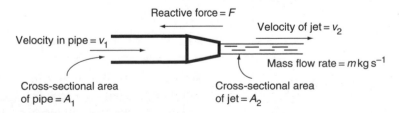

Figure 5.41 *Reaction of a nozzle*

The mass flow rate in the pipe and issuing from the nozzle is again given by

$$m = \rho A_1 v_1 = \rho A_2 v_2 \qquad \text{(i)}$$

> ### Key point
>
> The thrust exerted by a jet of fluid on a hemispherical cup is double the normal force exerted by the same jet on a flat plate.

From Newton's second law of motion, reaction of the nozzle is given by

force = rate of change of momentum

$$\text{force} = \frac{\text{mass} \times \text{change in velocity}}{\text{time taken}}$$

or

$$\textbf{force} = \frac{\textbf{mass}}{\textbf{time taken}} \times \textbf{change in velocity} \tag{ii}$$

Substituting for these gives

$$\left.\begin{aligned}
F &= m(v_2 - v_1) \\
F &= \rho A_1 v_1 (v_2 - v_1) \\
\text{or} & \\
F &= \rho A_2 v_2 (v_2 - v_1)
\end{aligned}\right\} \tag{5.48}$$

Power of a jet

The power delivered by a jet of fluid is the rate at which kinetic energy emerges from the nozzle. For a jet of cross-sectional area $A\,\text{m}^2$, density $\rho\,\text{kg m}^{-3}$ and velocity $v\,\text{m s}^{-1}$ and mass flow rate $m\,\text{kg s}^{-1}$ this is given by,

power = rate of supply of KE

$$\text{power} = \frac{1}{2}mv^2$$

Now $m = \rho A v$,

$$\text{power} = \frac{1}{2}\rho A v v^2$$

$$\textbf{power} = \frac{1}{2}\rho A v^3 \tag{5.49}$$

> **Key point**
>
> The power of a jet of fluid increases as the cube of its velocity.

As can be seen from equation (5.49), the power of a jet of fluid increases as the cube of its velocity. In other words, if you double the velocity of a jet you will have $2^3 = 8$ times more power available.

Force exerted on a pipe bend

The change of momentum can also be used to determine the force exerted by a fluid on a pipe bend.

The pipe shown in Figure 5.42 turns through an angle θ from the original x-direction. As a result there is a change of velocity in the x-direction from v to $v \cos \theta$.

The mass of fluid passing per second is again given by equation (5.30), that is,

$$m = \rho A v \tag{i}$$

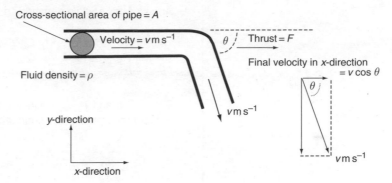

Figure 5.42 *Force on a pipe bend*

From Newton's second law of motion, the force exerted on the pipe bend is again given by equation

$$\text{force} = \frac{\text{mass}}{\text{time taken}} \times \text{change in velocity} \qquad (ii)$$

Substituting in equation (ii), we obtain

$$F = m(v - v\cos\theta)$$
$$F = \rho A v(v - v\cos\theta)$$
$$F = \rho A v^2(1 - \cos\theta) \qquad (5.50)$$

It will be noted that when $\theta = 90°$, the expression for the force on the pipe bend is the same as for the flat plate and when $\theta = 180°$, it is the same as for the hemispherical cup.

Example 5.15

A nozzle of diameter 50 mm is fed by a pipe of diameter 90 mm from a supply tank. The free surface of the water in the tank is 30 m above the nozzle. The jet emerges horizontally from the nozzle and strikes a hemispherical cup (Figure 5.43). Determine (a) the reaction of the nozzle, (b) the force on the hemispherical cup, (c) the power available in the jet. The density of water is 1000 kg m^{-3}. Ignore pipe friction losses.

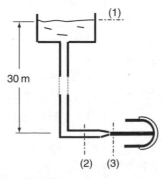

Figure 5.43

(a) Applying Bernoulli's equation to stations (1) and (3) to find v_3:

$$\frac{p_1}{\rho g} + \frac{v_1^2}{2g} + z_1 = \frac{p_3}{\rho g} + \frac{v_3^2}{2g} + z_3$$

But $v_1 = 0$ and $p_1 = p_3$ since both stations (1) and (3) are at atmospheric pressure. Eliminating these leaves

$$z_1 = \frac{v_3^2}{2g} + z_3$$

$$z_1 - z_3 = \frac{v_3^2}{2g}$$

$$v_3 = \sqrt{2g(z_1 - z_3)} = \sqrt{2 \times 9.81 \times 30}$$

$$\boldsymbol{v_3 = 24.3\,\text{m s}^{-1}}$$

Finding velocity v_2 in the nozzle feeder pipe. Applying equation (5.33) to stations (2) and (3), we get

$$\frac{v_2}{v_3} = \left(\frac{d_3}{d_2}\right)^2$$

$$v_2 = v_3 \left(\frac{d_3}{d_2}\right)^2 = 24.3 \times \left(\frac{50}{90}\right)^2$$

$$\boldsymbol{v_2 = 7.5\,\text{m s}^{-1}}$$

Finding reaction of nozzle using equation (5.48):

$$F = \rho A_3 v_3 (v_3 - v_2)$$

$$F = 1000 \times \frac{\pi \times 0.05^2}{4} \times 24.3 \times (24.3 - 7.5)$$

$$\boldsymbol{F = 802\,\text{N}}$$

(b) Finding force on hemispherical cup using equation (5.47).

$$F = 2\rho A_3 v_3^2 = 2 \times 1000 \times \frac{\pi \times 0.05^2}{4} \times 24.3^2$$

$$\boldsymbol{F = 2.32 \times 10^3\,\text{N} \quad \text{or} \quad 2.32\,\text{kN}}$$

(c) Finding power in jet using equation (5.49):

$$\text{power} = \frac{1}{2}\rho A_3 v_3^3 = \frac{1}{2} \times 1000 \times \frac{\pi \times 0.05^2}{4} \times 24.3^3$$

$$\boldsymbol{\text{power} = 580\,\text{W}}$$

Test your knowledge 5.7

1. How does the thrust of a fluid jet vary with jet velocity?
2. Why is the thrust exerted by a jet on a hemispherical cup double that exerted on a flat plate?
3. How does the power of a jet of fluid vary with jet velocity?
4. Why does a nozzle experience a force in the opposite direction to an emerging jet of fluid?

Activity 5.7

In Figure 5.44, water issues from a nozzle to strike a flat plate at right angles. The diameter of the supply pipe is 60 mm, the nozzle diameter is 30 mm and the volume flow rate is $15\,l\,s^{-1}$. Determine (a) the force on the pipe bend, (b) the reaction of the nozzle, (c) the force on the flat plate.

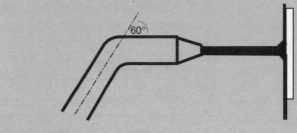

Figure 5.44

Problems 5.6

1. A jet of water issues from a horizontal nozzle of diameter 25 mm to strike a flat plate at right angles. If the volume flow rate is $7.5\,l\,s^{-1}$, determine (a) the exit velocity of the jet, (b) the thrust on the plate, (c) the power of the jet. The density of water is $1000\,kg\,m^{-3}$.

   $[15.3\,m\,s^{-1},\ 115\,N,\ 878\,W]$

2. A horizontal jet nozzle of diameter 20 mm is fed from a supply pipe of diameter 80 mm with mass flow rate is $500\,kg\,min^{-1}$. The emerging jet impinges normally on a flat plate and is deflected through 90°. Calculate (a) the flow velocity in the supply pipe and the velocity of the jet, (b) the reaction of the nozzle, (c) the thrust on the flat plate. The density of water is $1000\,kg\,m^{-3}$.

   $[1.66\,m\,s^{-1},\ 26.5\,m\,s^{-1},\ 207\,N,\ 221\,N]$

3. A jet of water strikes a hemispherical cup and is deflected through an angle of 180°. The diameter of the jet is 30 mm and the mass flow rate is $9.5\,kg\,s^{-1}$. Determine (a) the jet velocity, (b) the thrust on the cup, (c) the power of the jet. The density of water is $1000\,kg\,m^{-3}$.

   $[13.4\,m\,s^{-1},\ 255\,N,\ 853\,W]$

4. A pipe turns through an angle of 60° relative to its original direction. The diameter of the pipe is 150 mm and it contains oil of relative density 0.87. Determine the force exerted on the pipe bend if the volume flow rate is $5.3\,m^3\,min^{-1}$. The density of water is $1000\,kg\,m^{-3}$.

   $[192\,N]$

5. A nozzle of diameter 50 mm is fed by a pipe of diameter 90 mm from a water supply head of 30 m. The jet emerges horizontally and strikes a hemispherical cup. Ignoring pipe friction losses, determine (a) the reaction of the nozzle, (b) the thrust on the hemispherical cup. The density of water is $1000\,kg\,m^{-3}$.

   $[802\,N,\ 2.32\,kN]$

Aerodynamics

A body moving through the air experiences resistance and a stationary body in an air stream experiences wind force. The forces are due to the shearing action which is produced when the air flows over a surface and are generally known as drag forces. Motor vehicle and aeronautical engineers are continually seeking to reduce these forces in a quest for improved efficiency. Architects and structural engineers must also take account of wind force in the design of buildings and bridges. Aerodynamics is the study of air flow and its effects, in which the wind tunnel testing of models plays a major role.

Drag force

The total drag force acting on a body in a fluid stream is made up of two components. They are known as *form drag* and *skin friction drag*. On streamlined objects, the form drag and skin friction drag are roughly of the same magnitude. Skin friction drag is the viscous resistance caused by the shearing action which takes place across the boundary layer adjacent to the surface of a body. It is dependant on the surface area and the surface texture. The magnitude of skin friction drag is extremely difficult to predict except on the most simple shapes. It can, however, be kept to a minimum by making surfaces as smooth as possible.

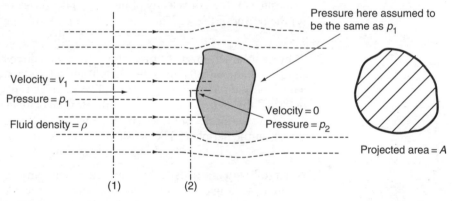

Figure 5.45 *Flow round a bluff body*

Form drag depends on the projected area which a body presents to the fluid stream. With an irregularly shaped or *bluff body*, the form drag will be high and with a streamlined body it will be comparatively low. Figure 5.45 shows a bluff body in a fluid stream. Station (1) is in the free stream where the velocity is v_1 and the pressure is p_1. It is assumed that the pressure behind the body is also p_1. Station (2) is on the front surface of the body where *stagnation conditions* are assumed to exist, i.e. where the fluid has been brought to rest. The pressure, p_2 will thus be greater than p_1. Applying Bernoulli's equation to stations (1) and (2), we obtain

$$\frac{p_1}{\rho g} + \frac{v_1^2}{2g} + z_1 = \frac{p_2}{\rho g} + \frac{v_2^2}{2g} + z_2$$

But $v_2 = 0$ *and* $z_1 = z_2$. Eliminating these, we get

$$\frac{p_1}{\rho g} + \frac{v_1^2}{2g} = \frac{p_2}{\rho g}$$

$$\frac{v_1^2}{2g} = \frac{p_2}{\rho g} - \frac{p_1}{\rho g}$$

$$\frac{v_1^2}{2} = \frac{p_2 - p_1}{\rho}$$

$$p_2 - p_1 = \frac{1}{2}\rho v_1^2 \tag{5.51}$$

This is known as the *dynamic pressure* acting on the body as a result of the fluid being brought to rest. It is assumed to act uniformly over the whole projected area A which the body presents to the fluid stream. The theoretical form drag is thus given by

theoretical drag = dynamic pressure × projected area

$$D = \frac{1}{2}\rho a v_1^2 \tag{5.52}$$

Key point

Dynamic pressure is the pressure increase that results when a fluid is brought to rest on a surface in a fluid stream.

In practice of course, the dynamic pressure does not act uniformly over the projected area and in addition to the pressure build-up in front of the body, there is often a pressure decrease in the wake behind it. There is also skin friction drag present. Nevertheless, the above theoretical drag force is used as a standard against which the actual measured drag force can be compared. The actual drag force is obtained from exhaustive wind tunnel tests where it is measured on sensitive balances. The theoretical and measured drag forces can then be used to calculate the *drag coefficient* c_D for the body.

$$\text{drag coefficient} = \frac{\text{measured drag force}}{\text{theoretical drag force}}$$

$$c_D = \frac{\textbf{measured drag force}}{\frac{1}{2}\rho A v^2} \tag{5.53}$$

Key point

The drag coefficient gives a comparison of the measured drag force on a body and the theoretical drag force that is calculated from the dynamic pressure.

where v is the free stream velocity.

The value of c_D for a flat plat is around 1.15 and for a cylindrical object it is around 0.9. For modern cars it is around 0.3 and for a fully streamlined teardrop shape it is around 0.05.

Example 5.16

In a wind tunnel test on a model car of projected area $0.03\,\text{m}^2$, the drag force was measured to be $3.5\,\text{N}$. A pitot-static tube recorded a difference of $35\,\text{mmH}_2\text{O}$ between the static and stagnation pressures in the air stream. Determine (a) the free stream velocity, (b) the theoretical drag force on the model, (c) the drag coefficient.

Take the density of the air to be 1.23 kg m^{-3} and the density of water to be 1000 kg m^{-3}.

(a) Finding velocity of air stream using equation (5.45) for the pitot-static tube:

$$v = \sqrt{2gh\frac{\rho_2}{\rho_1}}$$

where $\rho_2 = 1000$ kg m^{-3}, the density of water, and $\rho_1 = 1.23$ kg m^{-3}, the density of the air.

$$v = \sqrt{2 \times 9.81 \times 0.035 \times \frac{1000}{1.23}}$$

$$v = \mathbf{23.6\ m\ s^{-1}}$$

(b) Finding theoretical drag force on model using equation (5.52):

$$D = \frac{1}{2}\rho A v_1^2$$

$$D = \frac{1}{2} \times 1.23 \times 0.03 \times 23.6^2$$

$$D = \mathbf{10.3\ N}$$

(c) Finding drag coefficient:

$$\text{drag coefficient} = \frac{\text{measured drag force}}{\text{theoretical drag force}}$$

$$c_D = \frac{3.5}{10.3}$$

$$c_D = \mathbf{0.34}$$

Aerofoils

An *aerofoil* is a body so shaped as to produce aerodynamic reaction normal to its motion through a fluid without producing excessive drag. Aerofoils are found in various applications, e.g. aircraft wings and control surfaces, propeller blades and helicopter rotor blades. Their precise shape for different applications is determined by the use of high-level maths and associated software packages.

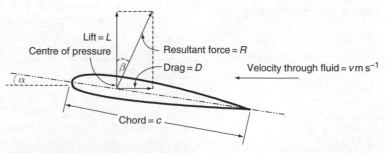

Figure 5.46 *Forces on an aerofoil*

Figure 5.46 shows the forces that act on an aerofoil as it passes through a fluid. The normal reaction to the motion of an aerofoil is known as lift force, denoted by L. The drag force, which opposes the motion, is denoted by D. The span of an aerofoil is denoted by b and its chord length by c. The plan area is denoted by S, and this is the area which is used to calculate the lift and drag forces. The angle of incidence of the chord line to the direction of motion is denoted by α, which is also called the angle of attack. The angle that the resultant R, of the lift and drag forces makes with the normal to the direction of motion is denoted by β.

The lift, drag, their resultant and the angle β are related by the following formulae:

$$\left.\begin{array}{ll} L = R\cos\beta & \text{and} \quad D = R\sin\theta \\ \tan\beta = \dfrac{D}{L} & \text{and} \quad R = \sqrt{L^2 + D^2} \end{array}\right\} \tag{5.54}$$

The flow pattern around the aerofoil is shown in Figure 5.47. The flow produces a pressure increase on the lower surface and a pressure decrease on the upper surface. The pressure difference creates lift but as can be seen in Figure 5.48, it is the pressure decrease on the upper surface that makes the greater contribution to the lift force.

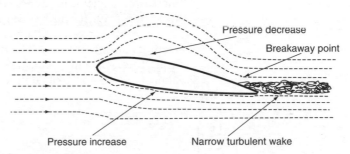

Figure 5.47 *Flow around an aerofoil*

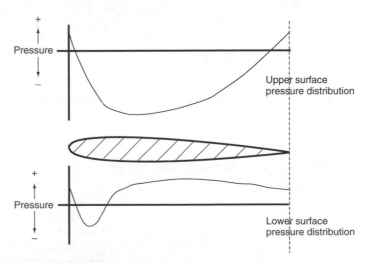

Figure 5.48 *Pressure distribution around an aerofoil*

It is assumed that for a given angle of incidence α, the lift and drag forces that act on an aerofoil are proportional to the dynamic pressure and the plan area S. The lift force is thus given by

$$L = \text{constant} \times \text{dynamic pressure} \times \text{plan area}$$

$$L = \text{constant} \times \frac{1}{2}\rho v^2 \times S$$

The constant is known as the *lift coefficient* c_L whose values for a given angle of incidence is found by wind tunnel testing, this gives

$$L = c_L \frac{1}{2}\rho v^2 S \tag{5.55}$$

In the same way, the drag force is given by

$$D = \text{constant} \times \text{dynamic pressure} \times \text{plan area}$$

$$D = \text{constant} \times \frac{1}{2}\rho v^2 \times S$$

Here the constant is the *drag coefficient* c_D whose value for a given angle of incidence is also found by wind tunnel testing, giving

$$D = c_D \frac{1}{2}\rho v^2 S \tag{5.56}$$

The variation in the values of the lift and drag coefficients with increasing angle of incidence for a typical plane aerofoil is shown in Figure 5.49. It will be seen that the lift coefficient, and hence also the lift force, rises almost linearly as the angle α increases from its no-lift value α_0. The drag coefficient, and hence also the drag force, is fairly constant over the same range, i.e. from almost $-2°$ to approximately $12°$.

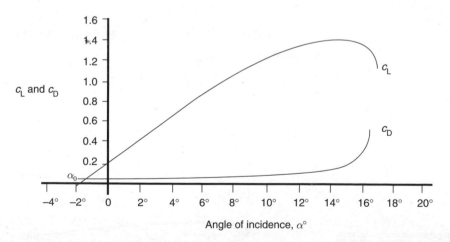

Figure 5.49 *Variation of lift and drag coefficients*

When the lift coefficient reaches its maximum value, stalling occurs. As can be seen from Figure 5.50, the breakaway point of the boundary layer on the upper surface moves towards the leading edge. This results in a pressure buildup on the upper surface and the production of a broad eddying wake. Under these conditions, the lift force rapidly decreases and the drag force increases.

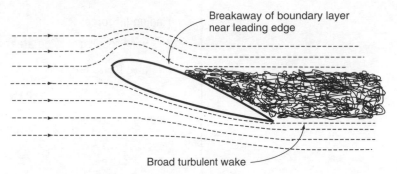

Figure 5.50 *Aerofoil in stalled condition*

An aircraft can stall when the angle of attack is increased to maintain lift at low airspeed. It can cause the aircraft to descend in a spin, often with fatal results. Various configurations of adjustable wing flaps and slots have been developed which enable aircraft to maintain lift at relatively low take-off and landing speeds. If you are able to secure an airline seat with a view of the wing, you will be able to see these in operation.

When an aircraft is travelling at a steady speed in level flight, the thrust of the engines must be equal and opposite to the drag force and the lift must be equal and opposite to its weight. The power required from the engines of a jet aircraft is the product of the thrust, F newtons, and the airspeed, $v\,\mathrm{m\,s^{-1}}$, that is

power = thrust × airspeed

But the thrust is equal and opposite to the drag force, D.

$$\textbf{power} = \boldsymbol{Dv} \tag{5.57}$$

In the case of propeller-driven aircraft, the propeller efficiency η has to be taken into account and the required shaft output power is given by

$$\textbf{power} = \frac{\boldsymbol{Dv}}{\eta} \tag{5.58}$$

Example 5.17

The lift surfaces of an aircraft have a plan area of $60\,\mathrm{m^2}$ and the wing loading is $2\,\mathrm{kN\,m^{-2}}$. The propellers are 75% efficient and the lift and drag coefficients are in the ratio 11:1. The density of the air is $1.27\,\mathrm{kg\,m^{-3}}$. Determine (a) the lift and drag coefficients and (b) the engine power required to propel the aircraft in level flight at a steady airspeed of $500\,\mathrm{kmh^{-1}}$.

(a) Finding airspeed in $\mathrm{m\,s^{-1}}$:

$$v = 500 \times \frac{1000}{60^2}$$

$$\boldsymbol{v = 139\ \textbf{m s}^{-1}}$$

Finding lift force:

$L =$ wing loading $\times$ wing area

$L = 2 \times 10^3 \times 60$

$L = 120 \times 10^3$ N   or   **120 kN**

Finding lift coefficient:

$$L = c_L \frac{1}{2}\rho v^2 S$$

$$c_L = \frac{2L}{\rho v^2 S} = \frac{2 \times 120 \times 10^3}{1.27 \times 139^2 \times 60}$$

$c_L = 0.163$

Finding drag coefficient:

$$\frac{c_L}{c_D} = 11$$

$$c_D = \frac{c_L}{11} = \frac{0.163}{11}$$

$c_D = 0.0148$

(b) Finding drag force:

$$\frac{c_L}{c_D} = \frac{L}{D} = 11$$

$$D = \frac{L}{11} = \frac{120 \times 10^3}{11}$$

$D = 10.9 \times 10^3$ N   or   **10.9 kN**

Finding engine power required:

$$\text{power} = \frac{Dv}{\eta}$$

$$\text{power} = \frac{10.9 \times 10^3 \times 139}{0.75}$$

power $= 2.02 \times 10^6$ W   or   **2.02 MW**

Test your knowledge 5.8

1. Distinguish between skin friction drag and form drag.
2. How is the drag coefficient defined for a body in a fluid stream?
3. How is the lift force on an aerofoil produced?
4. What occurs during stalling?
5. How is engine power calculated for an aircraft travelling at a steady speed in level flight?

Activity 5.8

A rectangular aerofoil of chord 300 mm and span 1.5 m is tested in a wind tunnel in which the airspeed is 60 ms^{-1} and the air density is 1.3 kg m^{-3}. The resultant force on the aerofoil is measured to be 850 N at an angle of 4° to the vertical axis of lift. Determine the lift and drag coefficients.

A propeller-driven aircraft of mass 10 tonnes has lift surfaces with the same characteristics. What will be the engine power required to maintain an airspeed of 450 km h^{-1} in level flight if the propeller efficiency is 80%?

Problems 5.7

1. In a wind tunnel test on a model car of projected cross-sectional area 0.03 m^2, the drag force was measured to be 3.1 N and the free stream velocity was indicated to be 22.1 m s^{-1}. The density of the air was 1.2 kg m^{-3}. Determine (a) the theoretical drag force, (b) the drag coefficient.

[8.79 N, 0.353]

2. A drag force of 2.1 N was recorded on a model of circular cross-section, 100 mm in diameter, when tested in a wind tunnel. The pressure difference between the free stream static pressure and the stagnation pressure was recorded as 35 mmH$_2$O on a pitot-static tube. Calculate (a) the dynamic pressure in pascals, (b) the velocity of the airstream, (c) the drag coefficent of the model. The density of the air is 1.2 kg m^{-3} and the density of water is 1000 kg m^{-3}.

[343 Pa, 23.7 m s^{-1}, 0.78]

3. A rectangular aerofoil of 300 mm chord and 1.5 m span is tested in a wind tunnel in which the airspeed is 60 m s^{-1}. The resultant force on the aerofoil is measured to be 850 N acting at an angle of 4° to the axis of lift. If the air density is 1.3 kg m^{-3}, determine the lift and drag coefficients.

[0.801, 0.056]

4. The lift surfaces of an aircraft have an equivalent plan area of 90 m^2 and a drag coefficient of 0.05. Determine the power of the engines required to propel it at a velocity of 450 km h^{-1} in level flight through air of density 1.3 kg m^{-3}. Assume a propeller efficiency of 75%.

[7.61 MW]

5. An aircraft has a mass of 12.2 tonnes and the ratio of the lift and drag coefficients is 11:1. Assuming the propellers to be 75% efficient determine the engine power to propel it at a speed of 500 km h^{-1} in level flight. [2.0 MW]

6. The lift surfaces of an aircraft have a plan area of 60 m^2. At a speed of 450 km h^{-1} in level flight the lift coefficient is 0.8 and the resultant thrust on the lift surfaces is inclined at an angle of 4.5° to the axis of lift. If the air density is 1.2 kg m^{-3}, determine (a) the lift force, (b) the drag force, (c) the engine power developed if the propeller efficiency is 75%.

[450 kN, 35.4 kN, 5.9 MW]

Chapter 6 Thermodynamics

Summary

We take it for granted that electricity will be available on demand and that our cars will start at the turn of the ignition key. The greater part of our electricity is generated from fossil and nuclear fuel and it is fossil fuel that powers our cars. The energy stored in the fuels is liberated in the form of heat, some of which is transformed into mechanical and electrical energy. Unfortunately, a great deal of the energy stored in the fuels is wasted and engineers are continually searching for ways to make the transformation processes more efficient.

Applied thermodynamics may be described as a study of the properties of matter and the relationship between heat energy and work. It might equally be described as the study of energy management and the efficient use of natural resources. The aim of this unit is to provide you with a basic knowledge of thermodynamics. You will investigate the combustion of fuels and the properties of air and steam. These are the working substances most widely used in heat- and work-transfer processes. You will also be introduced to the concept of closed and open thermodynamic systems and investigate practical examples of each type.

Expansion and compression of gases

Air, steam and other working substances used in thermodynamic systems pass through a cycle of processes as heat energy is changed into work and vice versa. Expansion and compression processes often form an essential part of the cycle and we will now consider the laws, principles and properties associated with them.

Property measurement

One of the basic properties of a working substance is its temperature. Temperature is a measure of the hotness of a substance, which is directly proportional to the kinetic energy of its individual molecules. Thermodynamic temperature is measured on the kelvin scale and is also known as *absolute temperature*. It has for its zero, the absolute zero of temperature at which all molecular movement ceases and the molecules of a substance have zero kinetic energy.

Key point

At the absolute zero of temperature all molecular movement ceases and the molecules of a substance have zero kinetic energy.

The SI unit of temperature is the kelvin, whose symbol is K. It is defined as the temperature interval between absolute zero and the triple point of water divided by 273.16. The triple point of water occurs at a very low pressure, where the boiling point has been depressed to meet the freezing point. In this condition, ice, water and steam are able to exist together in the same container, which is how the name arises. The kelvin is exactly the same temperature interval as the degree celsius (°C). The difference between the two scales lies in the points chosen for their definition. The degree celsius is defined as the temperature interval between the freezing and boiling points of water at standard atmospheric pressure of 101 325 Pa, divided by 100. The freezing point of water at this standard pressure is slightly below the triple point, being exactly 273 K (Table 6.1).

Table 6.1 *Temperature scales*

Temperature	Kelvin scale (K)	Celsius scale (°C)
Absolute zero	0	−273
Freezing point of water at standard atmospheric pressure	273	0
Boiling point of water at standard atmospheric pressure	373	100

Key point

To change the unit of pressure from bars to pascals, multiply by 10^5.

The SI unit of pressure is the pascal (Pa). A pressure of 1 Pa is exerted when a force 1 N is evenly applied at right angles to a surface area of $1\,m^2$. Another unit, which is widely used, is the 'bar'. This is almost equal to standard atmospheric pressure.

$$1\,\text{bar} = 100\,000\,\text{Pa} \quad \text{or} \quad 1 \times 10^5\,\text{Pa} \quad \text{or} \quad 100\,\text{kPa}$$

Pressure-measuring devices such as mechanical gauges and the manometers described in Chapter 5 measure the difference between the pressure inside a container and the outside atmospheric pressure. This, as you will recall, is known as *gauge pressure*. In gas calculations, however, it is often the total or *absolute pressure*, which must be used. This is obtained by adding atmospheric pressure to the recorded gauge pressure, that is,

absolute pressure = gauge pressure + atmospheric pressure

Key point

Standard temperature and pressure conditions are 0 °C and 101 325 Pa.

Normal temperature and pressure conditions are 15 °C and 101 325 Pa.

Reference has been made above to *standard atmospheric pressure*. A substance is said to be at standard temperature and pressure (STP) when its temperature is 0 °C or 273 K, and its pressure is 101 325 Pa. This definition has international acceptance.

Sometimes a substance is said to be at *normal temperature and pressure*. A substance is at normal temperature and pressure (NTP) when its temperature is 15 °C or 288 K, and its pressure is again 101 325 Pa. This definition is used in the United Kingdom and other countries with a temperate climate.

As with the fluid mechanics in Chapter 5, the density of a working substance is often required for thermodynamic calculations.

You will recall that density ρ is the mass per unit volume of a substance whose unit is $kg\,m^{-3}$. Sometimes however, and particularly in steam calculations, it is more convenient to use *specific volume*. This is the volume per unit mass of a substance whose unit is $m^3\,kg^{-1}$. Specific volume, v_s is thus the reciprocal of density, that is,

specific volume = volume occupied by 1 kg of a substance

For a substance of mass m kg and volume $V\,m^3$, its specific volume will be given by

$$\text{specific volume} = \frac{\text{volume}}{\text{mass}}$$

$$v_s = \frac{V}{m}\ (m^3\,kg^{-1})$$

or

$$v_s = \frac{1}{\rho}\ (m^3\,kg^{-1})$$

> ### Key point
>
> Like absolute temperature and absolute pressure, specific volume is a property of the working substance in a thermodynamic system.

In Chapter 5, we described the properties of an ideal fluid. We said an ideal liquid would have zero viscosity, be incompressible and would never vaporise. Similarly, an ideal gas would have zero viscosity and would never condense. We now need to extend this description to say that an ideal gas would obey the gas laws at all temperatures and pressures.

The gas laws

The gas laws which we need to consider are Boyle's law, Charles' law and Avogadro's law, which is also called Avogadro's hypothesis.

Boyle's law

This states that *the volume of a fixed mass of gas is inversely proportional to its absolute pressure provided that its temperature is constant*. When plotted on a graph of absolute pressure against volume, the process appears as shown in Figure 6.1.

Expansion and compression processes which take place at constant temperature, according to Boyle's law, are known as

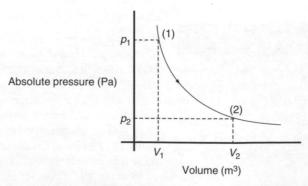

Figure 6.1 *Expansion according to Boyle's law – isothermal process*

isothermal processes. For any two points on the curve whose co-ordinates are p_1V_1 and p_2V_2, it is found that

$$\left.\begin{array}{l} p_1V_1 = p_2V_2 \\ \text{or} \\ \quad pV = \text{constant} \end{array}\right\} \tag{6.1}$$

Charles' law

This states that *the volume of a fixed mass of gas is proportional to its absolute temperature provided that its pressure is constant.* When plotted on a graph of absolute temperature against volume, the process appears as shown in Figure 6.2.

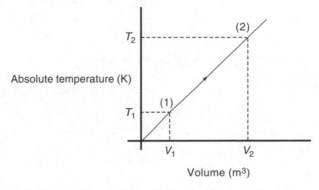

Figure 6.2 *Expansion according to Charles' law – isobaric process*

Expansion and compression processes which take place according to Charles' law are referred to as *constant pressure* or *isobaric processes.* For any two points on the curve whose co-ordinates are T_1V_1 and T_2V_2, it is found that

$$\left.\begin{array}{l} \dfrac{V_1}{T_1} = \dfrac{V_2}{T_2} \\ \text{or} \\ \quad \dfrac{V}{T} = \text{constant} \end{array}\right\} \tag{6.2}$$

You will note from Figure 6.2 that in theory, the volume of the gas should decrease uniformly until at absolute zero, which is the origin of the graph, its volume would also be zero. This is how an ideal gas would behave. Real gases obey the gas laws fairly closely at the temperatures and pressures normally encountered in power and process plant but at very low temperatures they liquefy, and may also solidify, before reaching absolute zero.

The general gas equation

This equation may be applied to any fixed mass of gas which undergoes a thermodynamic process taking it from initial conditions p_1, V_1 and T_1 to final conditions p_2, V_2 and T_2. Suppose the gas expands first according to Boyle's law to some intermediate volume V. Let it then expand further according to Charles' law to its final volume V_2.

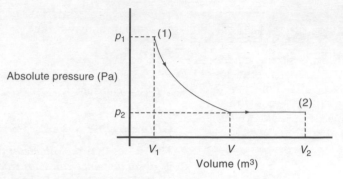

Figure 6.3 *Expansion according to Boyle's and Charles' laws*

Figure 6.3 shows the processes plotted on a graph of absolute pressure against volume. For the initial expansion according to Boyle's law,

$$p_1 V_1 = p_2 V \qquad\qquad\text{(i)}$$

For the final stage according to Charles' law,

$$\frac{V}{T_1} = \frac{V_2}{T_2} \qquad\qquad\text{(ii)}$$

Now from equation (ii),

$$V = \frac{V_2 T_1}{T_2}$$

Substituting in equation (i) for V gives

$$p_1 V_1 = \frac{p_2 V_2 T_1}{T_2}$$

$$\left.\begin{aligned} \frac{p_1 V_1}{T_1} &= \frac{p_2 V_2}{T_2} \\[2mm] \text{or}\qquad& \\[2mm] \frac{pV}{T} &= \textbf{constant} \end{aligned}\right\} \qquad (6.3)$$

This is the *general gas equation* which can be used to relate any two sets of conditions for a fixed mass of gas, irrespective of the process or processes which have caused the change. The constant in equation (6.3) is the product of two quantities. One of them is the mass m kg of the gas. The other is a constant for the particular gas known as its *characteristic gas constant*. The characteristic gas constant R has unit of joules per kilogram kelvin ($J\,kg^{-1}\,K^{-1}$) and is related to the molecular weight of the gas. Equation (6.3) can thus be written as

$$\frac{pV}{T} = mR$$

or $\qquad\qquad\qquad\qquad\qquad\qquad\qquad (6.4)$

$$pV = mRT$$

In this form it is known as the *characteristic gas equation* which is particularly useful for finding the mass of a gas whose volume, absolute pressure and absolute temperature are known.

Key point

The general gas equation can be used to relate any two sets of conditions for a fixed mass of gas, irrespective of the process or processes which have taken place.

Avogadro's hypothesis

This states that *equal volumes of different gases at the same temperature and pressure contain the same number of molecules*. It is called a hypothesis because it cannot be directly proved. It is impossible to count the very large number of molecules, even in a small volume, but there is lots of evidence to indicate that the assumption is true. Suppose we have three vessels of equal volume which contain three different gases at the same temperature and pressure, as shown in Figure 6.4.

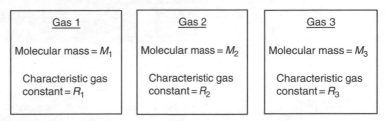

Gas 1	Gas 2	Gas 3
Molecular mass = M_1	Molecular mass = M_2	Molecular mass = M_3
Characteristic gas constant = R_1	Characteristic gas constant = R_2	Characteristic gas constant = R_3

Figure 6.4 *Equal volumes of different gases at the same temperature and pressure*

Avogadro's hypothesis states that each vessel contains the same number of molecules. Let this number be N. The mass of gas in each vessel will be different, and given by

mass of gas in vessel = number of molecules $\times$ molecular mass

$$m = N\,M \tag{i}$$

Applying the characteristic gas equation to the first vessel gives

$$pV = mR_1T$$

or

$$\frac{pV}{T} = mR_1$$

Substituting for m from equation (i)

$$\frac{pV}{T} = NM_1R_1 \tag{ii}$$

Similarly for the gases in the other two vessels, we get

$$\frac{pV}{T} = NM_2R_2 \tag{iii}$$

and

$$\frac{pV}{T} = NM_3R_3 \tag{iv}$$

Equating (i), (ii) and (iii) gives,

$$\cancel{N}M_1R_1 = \cancel{N}M_2R_2 = \cancel{N}M_3R_3 \tag{v}$$
$$M_1R_1 = M_2R_2 = M_3R_3 = \text{constant}$$

The actual mass of a molecule is very small and so in place of it, engineers and chemists use the *kilogram-molecule* or *kmol*. This is the mass of the substance, measured in kilograms, which is numerically equal to its molecular weight. The molecular weight is the weight or mass of a molecule of the gas relative to the weight or mass of a single hydrogen atom. Typical values for some common gases are shown in Table 6.2. The same symbol M is used and so for oxygen, $M = 32$ kg and for nitrogen, $M = 28$ kg, as given in Table 6.2.

Table 6.2 *Molecular weights*

Gas	Chemical symbol for molecule	Molecular weight, M
Hydrogen	H_2	2
Nitrogen	N_2	28
Oxygen	O_2	32
Carbon monoxide	CO	28
Carbon dioxide	CO_2	44

The product MR is found to have a value of 8314 J kmol^{-1} and is called the *universal gas constant*. It can be used to calculate the value of the characteristic gas constant R for a particular gas of known molecular weight.

$$MR = 8314$$

$$R = \frac{8314}{M} \tag{6.5}$$

e.g. the molecular weight of oxygen is 32 and so the value of its characteristic gas constant will be

$$R = \frac{8314}{32} = 260 \text{ J kg}^{-1}\text{K}^{-1}$$

Air is of course a mixture of oxygen and nitrogen, and so its characteristic gas constant cannot be found in this way. It can however be shown that its value is 287 J kg^{-1}K^{-1}.

Key point

A kilogram-molecule or kmol of a substance is the mass in kilograms which is numerically equal to its molecular weight.

Example 6.1

A compressor takes in carbon dioxide at a pressure of 1 bar and temperature 20 °C and delivers it through an aftercooler to a receiving vessel at a pressure of 6 bar. The initial volume in the compressor cylinder is 0.2 m^2 and the piston compresses it into a volume of 0.05 m^3 (Figure 6.5). There is no loss of pressure in the aftercooler and the molecular weight of CO_2 is 44. Determine (a) the mass of gas delivered per stroke of the piston, (b) the temperature on entering the aftercooler, (c) the volume of gas entering the receiver per stroke of the piston if it is cooled to its initial temperature in the aftercooler.

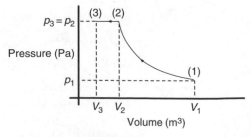

Figure 6.5 *Pressure–volume diagram*

(a) Finding characteristic gas constant for CO_2:

$$R = \frac{8314}{M} = \frac{8314}{44}$$
$$R = 189 \, \text{J} \, \text{kg}^{-1} \, \text{K}^{-1}$$

Finding mass of gas:

$$p_1 V_1 = mRT_1$$
$$m = \frac{p_1 V_1}{RT_1}$$

where $p_1 = 1 \times 10^5$ Pa and $T_1 = (20 + 273) = 293$ K

$$m = \frac{1 \times 10^5 \times 0.2}{189 \times 293}$$
$$m = 0.361 \, \text{kg}$$

(b) Finding temperature, T_2 at the end of the compression stroke:

$$\frac{p_1 V_1}{T_1} = \frac{p_2 V_2}{T_2}$$
$$T_2 = \frac{p_2 V_2 T_1}{p_1 V_1} = \frac{6 \times 10^5 \times 0.05 \times 293}{1 \times 10^5 \times 0.2}$$
$$T_2 = 440 \, \text{K} \quad \text{or} \quad 167 \, °\text{C}$$

(c) Finding volume, V_3 delivered from aftercooler to receiver per piston stroke:

$$\frac{p_1 V_1}{T_1} = \frac{p_3 V_3}{T_3}$$

But $T_3 = T_1$. These can be eliminated leaving

$$p_1 V_1 = p_3 V_3$$
$$V_3 = \frac{p_1 V_1}{p_3}$$

where $p_3 = p_2 = 6 \times 10^5$ Pa.

$$V_3 = \frac{1 \times 10^5 \times 0.2}{6 \times 10^5}$$
$$V_3 = 0.0333 \, \text{m}^3$$

There was really no need to change the pressures to pascals in parts (b) and (c). Because the pressures are in the ratios, p_2/p_1 and p_1/p_3, they could have been left in 'bars'. You might do this in future calculations but be sure to use the same units for the top and bottom lines of a ratio.

You might note that the cooling process from (2) to (3) took place at constant pressure, i.e. according to Charles' law. You might also note that because the initial and final temperatures were the same, the final condition of the CO_2 was the same as if isothermal compression had taken place from (1) to (3), i.e. according to Boyle's law.

Polytropic expansion

A fixed mass of gas can expand from an initial pressure p_1 and volume V_1 in an infinite number of ways. This is what we mean by the word *polytropic*. Figure 6.6 shows some of possible expansion processes.

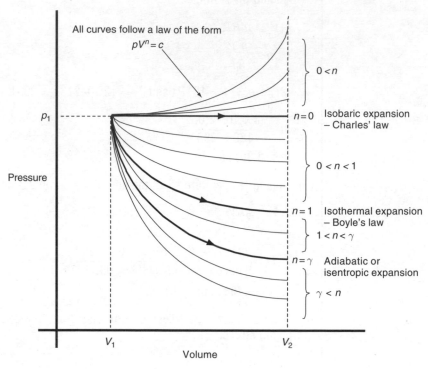

Figure 6.6 *Polytropic expansion*

The particular expansion curve followed by a gas depends on the amount and direction of the heat transfer that takes place during the expansion process. All curves have an equation of the general form

$$\left.\begin{aligned} pV^n &= \text{constant} \\ &\text{or} \\ p_1 V_1^n &= p_2 V_2^n \end{aligned}\right\} \tag{6.6}$$

For most practical expansion and compression processes, the value of the index n ranges from low negative values to positive values of around 1.5. The values of n indicated in Figure 6.5 correspond to the following conditions:

$n < 0$ Very large amounts of heat energy are being received during the expansion process. Not only does the volume increase but the temperature and pressure also increase during the process.

$n = 0$ Less heat energy being received. The temperature rises but the pressure stays constant. This is isobaric expansion which takes place according to Charles' law. Equation (6.6) becomes

$$p_1 V_1^0 = p_2 V_2^0$$

You will remember from your mathematics that any number to the power zero is 1. The equation thus reduces to $p_1 = p_2$, i.e. constant pressure.

$0 < n < 1$ Less heat energy is being received in this range. The temperature continues to rise but the pressure falls during expansion.

$n = 1$ Still less heat energy being received. The temperature stays constant and the pressure falls. This is isothermal expansion which takes place according to Boyle's law. Equation (6.6) becomes.

$$p_1 V_1 = p_2 V_2$$

As would be expected, this is exactly the same as equation (6.1).

$1 < n < \gamma$ Only small amounts of heat energy being received. Both the temperature and pressure fall during expansion.

$n = \gamma$ This is known as *adiabatic* or *isentropic* expansion where no heat transfer takes place during the expansion process. Like isobaric and isothermal expansion, it is a special case which we will discuss later in more detail. Both the temperature and pressure fall and the value of the index γ for air is around 1.4.

$\gamma < n$ In this range, heat energy is lost by the gas during expansion. Both the temperature and pressure fall by increasing amounts.

Key point

The value of the index n in a polypropic process depends on the magnitude and direction of the heat transfer that takes place.

Example 6.2

A mass of 0.1 kg of air has an initial temperature 600 °C and pressure 1 MPa. If the air is expanded according to the law $pV^{1.2}$ to a final volume of 100 l determine (a) its initial volume, (b) its final pressure, (c) its final temperature. For air, take $R = 287 \, \text{J kg}^{-1} \, \text{K}^{-1}$.

(a) Finding initial volume of air:

$$p_1 V_1 = mRT_1$$
$$V_1 = \frac{mRT}{p_1} = \frac{0.1 \times 287 \times (600 + 273)}{1 \times 10^6}$$
$$V_1 = 0.0251 \, \text{m}^3$$

(b) Finding final pressure:

$$p_1 V_1^n = p_2 V_2^n$$
$$p_2 = \frac{p_1 V_1^n}{V_2^n} = p_1 \left(\frac{V_1}{V_2}\right)^n$$

where $V_2 = 100 \, \text{l} = 0.1 \, \text{m}^3$

$$p_2 = 1 \times 10^6 \times \left(\frac{0.0251}{0.1}\right)^{1.2}$$
$$p_2 = 190 \times 10^3 \, \text{Pa} \quad \text{or} \quad 190 \, \text{kPa}$$

348   *Thermodynamics*

Test your knowledge 6.1

1. What is meant by absolute pressure and absolute temperature?
2. What are the values of STP?
3. What is specific volume?
4. What is isothermal expansion, and which of the gas laws relates to it?
5. How is the molecular weight of a substance defined?
6. State Avogadro's hypothesis and explain why it is so called.
7. What is adiabatic or isentropic expansion?

(c) Finding final temperature:

$$\frac{p_1 V_1}{T_1} = \frac{p_2 V_2}{T_2}$$

$$T_2 = \frac{p_2 V_2}{p_1 V_1} = \frac{190 \times 10^3 \times 0.1 \times 873}{1 \times 10^6 \times 0.0251}$$

$$T_2 = 661\,\text{K} \quad \text{or} \quad 388\,^\circ\text{C}$$

Alternatively, the final temperature can be found using the characteristic gas equation, that is,

$$p_2 V_2 = mRT_2$$

$$T_2 = \frac{p_2 V_2}{mR} = \frac{190 \times 10^3 \times 0.1}{0.1 \times 287}$$

$$T_2 = 661\,\text{K} \quad \text{or} \quad 388\,^\circ\text{C}$$

Activity 6.1

A petrol engine has a cylinder diameter of 90 mm and a stroke of 100 mm. The compression ratio is 8:1. At the start of the compression stroke, the cylinder is full of a petrol/air mixture at a pressure of 1 bar and temperature 25 °C. Compression takes place according to the law $pV^{1.3} = c$ and the molecular weight of the mixture can be taken to be 29. Determine (a) the mass of gas in the cylinder, (b) the pressure and temperature at the end of the compression stroke.

Problems 6.1

1. The air in a diesel engine cylinder has an initial temperature and pressure of 15 °C and 101 kPa. The initial volume is 1 l and the compression ratio is 15:1. During compression the pressure rises to 3.4 MPa. Determine (a) the mass of air in the cylinder, (b) the final temperature of the air. The characteristic gas constant for air is $287\,\text{J}\,\text{kg}^{-1}\,\text{K}^{-1}$.

[1.22×10^{-3} kg, 374 °C]

2. A vessel of volume $0.019\,\text{m}^3$ contains carbon dioxide at a pressure of 340 kPa and a temperature of 17 °C. The vessel is connected to another of the same volume which is empty and the pressure is seen to fall to 138 kPa. Determine (a) the mass of carbon dioxide present, (b) the temperature immediately after expansion into second vessel, (c) the temperature to which the connected vessels must be raised for the pressure to return to 340 kPa. The molecular weight of carbon dioxide is 44.

[0.188 kg, −37.6 °C, 307 °C]

3. A petrol engine cylinder has a diameter of 100 mm and the length of the piston stroke is 110 mm. The compression ratio

is 5.5:1 and at the start of compression the cylinder is full of a gas–air mixture at a pressure of 100 kPa and temperature 35 °C. The pressure at the end of the compression stroke is 1 MPa. (a) Calculate the mass of the cylinder contents if the equivalent molecular weight of the mixture is 28.9. (b) The temperature at the end of the compression stroke.

$$[1.23 \times 10^{-3} \text{ kg}, 287 °C]$$

4. A volume of 1 l of air at a pressure of 1 bar and temperature 30 °C is compressed into one-tenth of its original volume according to the law $pV^{1.2} = c$. Determine (a) the mass of the gas, (b) the final pressure, (c) the final temperature. Take $R = 287 \text{ J kg}^{-1} \text{ K}^{-1}$ for air.

$$[1.15 \times 10^{-3} \text{ kg}, 2 \text{ MPa}, 333 °C]$$

5. Air of mass 1.5 kg, pressure 300 kPa and volume 0.5 m³ is expanded at constant pressure to a volume of 1 m³. It is then expanded further to a volume of 1.5 m³ according to the law $pV^{1.25} = c$. Calculate its initial, intermediate and final temperatures. Take $R = 287 \text{ J kg}^{-1} \text{ K}^{-1}$.

$$[75 °C, 423 °C, 356 °C]$$

Thermodynamic systems

A thermodynamic system is defined as a quantity of matter surrounded by a real or imaginary boundary across which energy may pass in the form of heat and work. Its surroundings are considered to be all matter outside the boundary. The matter inside the system is known as the working substance, two of the most common being air and steam.

Properties of a thermodynamic system

There are six commonly used properties which are used to describe the state of the working substance in a thermodynamic system. They are:

1. Absolute pressure	(Pa)	} Intensive properties
2. Absolute temperature	(K)	
3. Specific volume	(m³)	
4. Internal energy	(J kg⁻¹)	} Extensive properties
5. Enthalpy	(J kg⁻¹)	
6. Entropy	(J kg⁻¹ K⁻¹)	

Absolute pressure and absolute temperature are called *intensive properties* because they are independent of the mass of the working substance inside a thermodynamic system. The remaining four properties are all dependant on mass, as can be seen from their units. They are called *extensive properties*. We have already described the first three properties and we will shortly be defining internal energy and enthalpy. The last of the properties, entropy, will be left for study at a higher level. It is used in advanced steam plant calculations which are beyond the range of this unit.

Key point

Temperature and pressure are independent of the mass of the working substance present in a thermodynamic system.

Two-property rule

This states that *the condition of the working substance in a thermodynamic system can be fully defined by any two of its properties provided that they are fully independent of each other*. In other words, if you know the value of any two properties such as pressure and volume or temperature and enthalpy, etc. you can use these to define the condition of a working substance and to calculate other properties.

Care must be taken when the working substance is a vapour in contact with its liquid phase, such as wet steam in a boiler. Here the temperature and pressure are not independent. The temperature at which the water is boiling to produce the steam depends on the pressure in the boiler. In such cases another property such as specific volume or enthalpy is required to fully define the condition of the steam.

Zeroth law of thermodynamics

There are three laws of thermodynamics, the zeroth, the first and the second. Presumably the zeroth was an afterthought, and should really have been the first. It defines thermal equilibrium between bodies and states that *if two bodies are each in thermal equilibrium with a third body, they must also be in thermal equilibrium with each other*.

A body is in thermal equilibrium when its temperature is constant and there is no change of state taking place, i.e. the body is not melting or evaporating. When bodies in close contact are in thermal equilibrium, there is no heat transfer between them and they are at the same temperature. A good example is a thermometer displaying a steady temperature reading. In this condition the mercury, the glass that contains it and the substance whose temperature is being measured must all be in thermal equilibrium with each other.

The first law of thermodynamics is a statement of the principle of conservation of energy applied to thermodynamic systems. It was put forward by the German physicist Rudolf Clausius (1822–1888) and states that *the total energy of a thermodynamic system and its surroundings is constant although changes may take place from one energy form to another*.

More specifically, for closed thermodynamic systems such as internal combustion engines and steam-generating plant, it can be stated that *when a closed thermodynamic system is taken through a cycle of processes, the net heat transfer is equal to the net work done*. As in other walks of life, this means that you cannot get more out of a system than you put into it.

Heat transfer

The molecules of a solid body are thought to be in a state of vibration whilst those in liquids and gases move in a random fashion. Whatever the state or *phase* of a substance, its molecules have kinetic energy. As we have already stated, temperature is an

indirect measure of molecular kinetic energy which is experienced
as hotness or coldness to the touch. The greater the mass of a
substance, the greater will be the number of molecules and the
greater will be the total amount of kinetic energy which it contains.
This is in fact its *heat energy* content but it is more usually called
internal energy. Like other energy forms, it is measured in joules
and is one of the extensive properties which can be used to define
the state of a thermodynamic system.

Because it is almost impossible to count the number of molecules in a substance or to measure their kinetic energies, it is
difficult to calculate the total internal energy of any given mass.
This does not present a problem, however, because it is the change
of internal energy, or the *heat transfer* which accompanies temperature change, which is of more importance. This can be calculated and depends upon:

● the mass of the substance
● the change of temperature which occurs
● the specific heat capacity of the substance.

Specific heat capacity

The specific heat capacity of a substance is defined as *the amount
of heat energy required to raise the temperature of a mass of 1 kg
through a temperature rise of 1 K (1 °C)*.

It is given the symbol c, and its units are $J\,kg^{-1}\,K^{-1}$. The value of
specific heat capacity for different substances is found by experiment.
Its value varies a little with the temperature at which heat transfer
takes place and those listed in Table 6.3 are accepted average values
for the temperature ranges normally encountered in process plant.

Table 6.3 *Specific heat capacities*

Substance	Specific heat capacity ($J\,kg^{-1}\,K^{-1}$)
Metals	
Aluminium	920
Carbon steels	460
Cast iron	540
Copper and brass	390
Lead	130
Non-metals	
Water	4200
Ice	700
Alcohol	230
Mercury	140

The specific heat capacity of a gas is dependant on the conditions under which heat transfer takes place. Two standard conditions are used. They are heat transfer at constant volume and heat
transfer at constant pressure. As a result, a gas has two specific
heat capacities, c_v at constant volume and c_p at constant pressure.
The values of c_v and c_p for some common gases are listed in
Table 6.4.

Table 6.4 *Specific heat capacities of gases*

Gas	Specific heat capacity ($J\,kg^{-1}\,K^{-1}$)	
	At constant volume, c_v	At constant pressure, c_p
Air	718	1005
Nitrogen	740	1004
Oxygen	660	920
Carbon monoxide	740	1040
Carbon dioxide	630	850

It will be noted that in Table 6.4, the values of specific heat capacity at constant pressure are 30–40% higher than those at constant volume. This is because heat transfer at constant pressure is accompanied by expansion of the gas. When a gas expands, it does external work in pushing back its system boundary and this is why additional heat energy is required.

Heat flow

When a substance of mass m kg and specific heat capacity $c\,J\,kg^{-1}\,K^{-1}$ undergoes a temperature change from $T_1\,°C$ to $T_2\,°C$, the heat transfer $Q\,J$ which takes place is given by the formula

$$Q = mc(T_2 - T_1) \tag{6.7}$$

If the temperature rises, heat energy is received and Q is positive. If the temperature falls, heat energy is lost and Q is negative. In the case of solid bodies the product mc is sometimes referred to as its *thermal capacity* whose units are joules per kelvin ($J\,K^{-1}$).

When water is heated in a metal container, some of the heat energy supplied goes to raise the temperature of the metal. Sometimes it is convenient to consider the mass of water to which the metal is equivalent. This is called the *water equivalent*, of the container which is defined as *the mass of water which would experience the same temperature rise when receiving the same amount of heat energy*. The water equivalent m_{wc} kg of a container whose mass is m_c kg and specific heat capacity $c_c\,J\,kg^{-1}\,K^{-1}$ is given by the formula

$$m_{wc} = \frac{c_c}{c_w}\,m_c \tag{6.8}$$

where c_w is the specific heat capacity of water. The total heat transfer to or from the container and contents is then given by

$$Q = (m_w + m_{wc})\,c_w(T_2 - T_1) \tag{6.9}$$

If the heat transfer given by equation (6.7) takes place in a time t s, the power of the heating or cooling process is given by

$$\text{power} = \frac{\text{heat transfer}}{\text{time taken}}$$

$$\left. \begin{aligned} \text{power} &= \frac{Q}{t}\,(W) \\ \text{or} \\ \text{power} &= \frac{mc(T_2 - T_1)}{t}\,(W) \end{aligned} \right\} \tag{6.10}$$

Example 6.3

A cylindrical boiler drum 5 m in length and 2 m in diameter has a mass of 4 tonnes and is lagged to prevent heat loss. The drum is made from steel of specific heat capacity 460 J kg^{-1} K^{-1} and is two-thirds full of water at a temperature of 15 °C. After supplying heat at a steady rate for 30 min the water temperature is raised to 100 °C. The density of water is 1000 kg m^{-3} and its specific heat capacity is 4200 J kg^{-1} K^{-1}. Determine (a) the water equivalent of the boiler drum, (b) the heat energy supplied, (c) the power rating of the process.

(a) Finding water equivalent of boiler drum:

$$m_{wc} = \frac{c_c}{c_w} m_c = \frac{460 \times 4 \times 10^3}{4200}$$

$m_{wc} = \textbf{438 kg}$

Finding volume of water in boiler drum:

$$V_w = \frac{2}{3}\frac{\pi d^2 l}{4} = \frac{2 \times \pi \times 2^2 \times 5}{3 \times 4}$$

$V_w = \textbf{10.5 m}^3$

Finding mass of water in drum:

$$m_w = \rho V = 1000 \times 10.5$$

$m_w = \textbf{10 500 kg}$

(b) Finding heat energy received:

$$Q = (m_w + m_{ws})c_w(T_2 - T_1)$$
$$Q = (10\,500 + 438) \times 4200 \times (100 - 15)$$

$Q = \textbf{3.90} \times \textbf{10}^9\,\textbf{J}$ or **3.90 GJ**

(c) Finding power rating of process:

$$\text{power} = \frac{Q}{t} = \frac{3.90 \times 10^9}{30 \times 60}$$

$\text{power} = \textbf{2.17} \times \textbf{10}^6\,\textbf{W}$ or **2.17 MW**

Heat transfer in mixtures

Suppose a hot body A of mass m_A, specific heat capacity c_A and temperature T_A is plunged into a cool liquid B of mass m_B, specific heat capacity c_B and temperature T_B. The substance A will loose heat energy and, assuming the container is perfectly lagged, the liquid B and its container, C, will receive the same amount. Let the mass of the container be m_c and its specific heat capacity be c_c. Eventually the mixture will achieve thermal equilibrium where heat transfer ceases and the two substances and the container are at a common final temperature T. In this condition,

heat energy lost by A = heat energy received by B and its container

$$m_A c_A (T_A - T) = m_B c_B (T - T_B) + m_c c_c (T - T_B)$$

$$m_A c_A (T_A - T) = (m_B c_B + m_c c_c)(T - T_B) \qquad (6.11)$$

This equation can be used to predict the final temperature of a mixture. Alternatively, it can be used in an experiment to find the specific heat capacity of one of the substances provided that the specific heat capacity of the other is known.

Example 6.4

An alloy specimen of mass 0.064 kg is heated to a temperature of 100 °C in boiling water. It is then quickly transferred to a lagged copper calorimeter of mass 0.097 kg containing 0.04 kg of water at 23 °C. The final temperature of the mixture is 32 °C. The specific heat capacity of copper is 390 J kg^{-1} K^{-1} and the specific heat capacity of water is 4200 J kg^{-1} K^{-1}. What is the specific heat capacity of the alloy specimen?

Finding specific heat capacity of alloy specimen. Let the alloy be material A, the water be B and the copper be material C:

heat energy lost by A = heat energy received by B and its container

$$m_A c_A (T_A - T) = m_B c_B (T - T_B) + m_c c_c (T - T_B)$$

$$m_A c_A (T_A - T) = (m_B c_B + m_c c_c)(T - T_B)$$

$$c_A = \frac{(m_B c_B + m_c c_c)(T - T_B)}{m_A (T_A - T)}$$

$$c_A = \frac{[(0.04 \times 4200) + (0.097 \times 390)](32 - 23)}{0.064(100 - 32)}$$

$$c_A = 426 \, \text{J kg}^{-1} \, \text{K}^{-1}$$

Latent heat

Matter can exist as a *solid*, a *liquid* or a *gas*. These different forms are known as phases. The substance H_2O, for example, can exist as ice, water and steam. The heat energy received by a substance that causes its temperature to rise is known as *sensible heat*. This is because its flow can be sensed by a temperature-measuring instrument, such as a thermometer or a thermocouple, and calculated using equation (6.7).

Eventually, however, a great many solids melt and liquids vaporise resulting in a change of phase. Whilst the change is taking place, the temperature remains constant, even though heat energy is still being received. It stays constant until the change is complete and then starts to rise again. The heat energy received at constant temperature to bring about the change is known as *latent heat* because its flow cannot be detected by a temperature-measuring device.

The graph of temperature against heat energy received in Figure 6.7 shows this effect. When a substance is losing heat energy and the reverse changes are taking place, the same amount of latent heat is given off and the temperature again stays constant until the changes are complete.

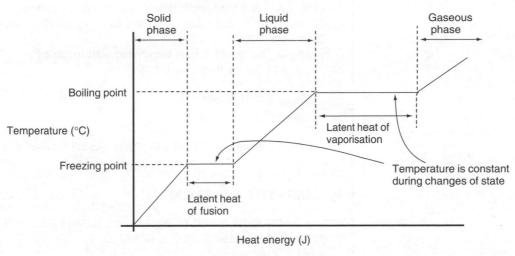

Figure 6.7 *Graph of temperature against heat energy received*

The amount of heat energy required to change 1 kg of a substance from its solid to its liquid phase at constant temperature is known as its *specific latent heat of fusion*. The working substance we shall be most concerned with, where a change of phase may occur, is water. The specific latent heat of fusion required to melt ice at standard atmospheric pressure is $335\,\text{kJ}\,\text{kg}^{-1}$. It is also known as the *specific enthalpy of fusion* and given the symbol h_{s}. You will recall that enthalpy is one of the six properties used to describe the state of a working substance and very shortly we shall be looking at its precise definition.

The latent heat required to change m kg of a substance from a solid to a liquid, or given off when a liquid changes to a solid, is given by the formula

latent heat = mass × specific latent heat of fusion

$$Q = m\,h_{\text{s}} \tag{6.12}$$

The amount of heat energy required to change 1 kg of a substance from its liquid to its gaseous phase at constant temperature is known as its *specific latent heat of vaporisation*. The specific latent heat of vaporisation required to change water to dry steam at standard atmospheric pressure is $2257\,\text{kJ}\,\text{kg}^{-1}$. It is also known as the *specific enthalpy of vaporisation* and given the symbol h_{fg}. Do not worry about the reason for using the suffix fg. It will be explained later in this chapter.

The latent heat required to change m kg of a substance from a liquid to a vapour, or given off when a vapour changes to a liquid, is given by the formula

latent heat = mass × specific latent heat of vaporisation

$$Q = m\,h_{\text{fg}} \tag{6.13}$$

Example 6.5

Calculate the heat energy required to change 20 kg of ice at $-4\,°C$ to dry steam at $100\,°C$ and atmospheric pressure. The specific heat capacity of ice is $2.1\,kJ\,kg^{-1}\,K^{-1}$ and the specific latent heat of fusion is $335\,kJ\,kg^{-1}$. The specific heat capacity of water is $4.2\,kJ\,kg^{-1}\,K^{-1}$ and the specific latent heat of vaporisation is $2257\,kJ\,kg^{-1}$. If the time taken is 25 min, calculate also the power of the heating process.

Finding Q_1, the sensible heat energy required to raise the temperature of the ice from $-4\,°C$ to $0\,°C$:

$$Q_1 = mc_{ice}(T_2 - T_1) = 20 \times 2.1 \times 10^3 \times [0 - (-4)]$$
$$Q_1 = 168 \times 10^3\,J \quad or \quad 168\,kJ$$

Finding Q_2, the latent heat required to change the ice to water at $0\,°C$:

$$Q_2 = mh_s = 20 \times 335 \times 10^3$$
$$Q_2 = 6700 \times 10^3\,J \quad or \quad 6700\,kJ$$

Finding Q_3, the sensible heat energy required to raise the temperature of the water from $0\,°C$ to $100\,°C$:

$$Q_3 = mc_w(T_3 - T_2) = 20 \times 4.2 \times 10^3 \times (100 - 0)$$
$$Q_3 = 8400 \times 10^3\,J \quad or \quad 8400\,kJ$$

Finding Q_4, the latent heat required to change the water into dry stream at $100\,°C$:

$$Q_4 = mh_{fg} = 20 \times 2257 \times 10^3$$
$$Q_4 = 45\,140 \times 10^3\,J \quad or \quad 45\,140\,kJ$$

Finding total heat energy required:

$$Q = Q_1 + Q_2 + Q_3 + Q_4$$
$$Q = 168 + 6700 + 8400 + 45\,140$$
$$Q = 60\,408\,kJ$$

Finding power of heating process:

$$power = \frac{Q}{t} = \frac{60\,408 \times 10^3}{25 \times 60}$$
$$power = 40.3 \times 10^3\,W \quad or \quad 40.3\,kW$$

Test your knowledge 6.2

1. Distinguish between intensive and extensive thermodynamic properties.
2. What is the two-property rule?
3. What is the first law of thermodynamics?
4. How are the two specific heat capacities of a gas defined?
5. What is the difference between sensible heat and latent heat?
6. What is the water equivalent of a container?

Activity 6.2

Dry steam at $100\,°C$ enters a spray condenser at the rate of $2.1\,kg\,min^{-1}$. Cooling water at a temperature of $15\,°C$ is sprayed into the steam and the condensate/cooling water mixture drains off at the rate of $2.55\,kg\,min^{-1}$. If radiation losses are neglected, what will be the final temperature of the mixture? The specific heat capacity of water is $4200\,J\,kg^{-1}\,K^{-1}$ and its specific latent heat of vaporisation is $2257\,kJ\,kg^{-1}$.

Problems 6.2

1. A boiler takes in feed water at 15 °C and supplies hot water at 100 °C. If the supply rate is 2 tonnes per hour what is the heat transfer rate to the water measured in kW. Take the specific heat capacity of water to be 4200 J kg^{-1} K^{-1}.

[198 kW]

2. An upright cylindrical tank of diameter 1.5 m and mass 50 kg contains water at a temperature of 20 °C to a depth of 1 m. An electrical heating coil raises its temperature to 90 °C in a time of 1 hour 20 min. Assuming the tank be perfectly lagged, determine (a) the heat energy received and (b) the power rating of the heater. The density and specific heat capacity of water are 1000 kg m^{-3} and 4200 J kg^{-1} K^{-1} and the specific heat capacity of the tank material is 500 J kg^{-1} K^{-1}.

[522 MJ, 109 kW]

3. A brass specimen of mass 0.032 kg is heated to 100 °C and then quickly transferred to a lagged copper calorimeter containing 0.02 kg of water at a temperature of 24 °C. The water equivalent of the calorimeter is 0.0045 kg and the final temperature inside it is 32.8 °C. Calculate the specific heat capacity of the brass.

[422 J kg^{-1} K^{-1}]

4. A block of ice of mass 8 kg and temperature −10 °C receives heat energy and is transformed into water at 60 °C. Calculate the amount of heat energy required. The specific heat capacity of ice is 2100 J kg^{-1} K^{-1} and its specific latent heat of fusion is 335 kJ kg^{-1}. The specific heat capacity of water is 4200 J kg^{-1} K^{-1}.

[4.86 MJ]

5. Ice of mass 5 kg and temperature −5 °C is transformed by heating into dry steam at 100 °C at atmospheric pressure. If heat energy is supplied at the rate of 20 kW how long does the process take? The specific heat capacity of ice is 2100 J kg^{-1} K^{-1} and the specific latent heat fusion is 335 kJ kg^{-1}. The specific heat capacity of water is 4200 J kg^{-1} K^{-1} and the specific latent heat of vaporisation is 335 kJ kg^{-1}.

[13 min]

Work transfer

When a gas expands, it does work. The hot gases in an internal combustion engine expand and push the pistons down the cylinders giving work output from the system. On the compression stroke, the opposite happens and work is put into the system. The work output from a thermodynamic system is usually regarded as positive work and the work input is taken to be negative work. This is the opposite sign convention to that used for heat transfer. You will recall that heat input to a system is positive whilst heat output, or heat loss, is negative.

Suppose that a fixed mass of gas expands in a cylinder fitted with a piston according to the general polytropic law $pV^n = c$, as shown in Figure 6.8.

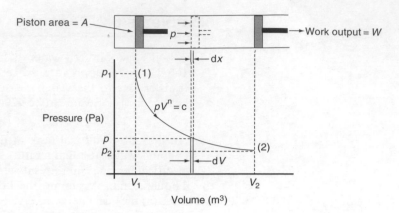

Figure 6.8 *Polytropic expansion*

At some instant during the expansion process when the pressure is p, the piston moves through a small distance dx and the volume increases by a small amount dV. During this instant a small amount of work dW is done by the gas.

work done = force on piston × distance moved by piston

$$dW = F\,dx = pA\,dx$$

But $A\,dx = dV$, the small increase in volume.

$$dW = p\,dV$$

This is the area of the elemental strip under the curve and so the total work done during expansion from (1) to (2) is the total area under the curve. This can be found by integration between the limits V_1 and V_2.

$$W = \int_{V_1}^{V_2} p\,dV \tag{i}$$

Now the equation of the curve is $pV^n = c$ and so,

$$p = \frac{c}{V} = cV^{-n}$$

Substituting for p in equation (i) gives

$$W = \int_{V_1}^{V_2} cV^{-n}\,dV$$

$$W = \left(\frac{cV^{1-n}}{1-n}\right)_{V_1}^{V_2}$$

$$W = \frac{cV_2^{1-n}}{1-n} - \frac{cV_1^{1-n}}{1-n}$$

$$W = \frac{cV_2^{1-n} - cV_1^{1-n}}{1-n} \tag{ii}$$

But $p_1V_1^n = c$ and also $p_2V_2^n = c$. Substituting for these in equation (ii) gives

$$W = \frac{p_2V_2V_2^{1-n} - p_1V_1V_1^{1-n}}{1-n}$$

$$W = \frac{p_2V_2 - p_1V_1}{1-n} \tag{iii}$$

Because the index n is usually greater than 1, it is convenient to reverse the terms in the numerator and denominator giving

$$W = \frac{p_1 V_1 - p_2 V_2}{n-1} \tag{6.14}$$

Now from the characteristic gas equation, $p_1 V_1 = mRT_1$ and $p_2 V_2 = mRT_2$. Substituting for these in equation (6.14) gives

$$W = \frac{mRT_1 - mRT_2}{n-1}$$

$$W = \frac{mR(T_1 - T_2)}{n-1} \tag{6.15}$$

When these formulae are used it will be found that the value of W is always positive for an expansion process and negative for a compression process.

The formulae can be used to find the work transfer during any polytropic process except one. That is the isothermal process which takes place according to Boyle's law. For isothermal expansion, $n = 1$ and this makes equations (6.14) and (6.15) indeterminate. We must therefore start again with the pressure v. volume graph shown in Figure 6.9.

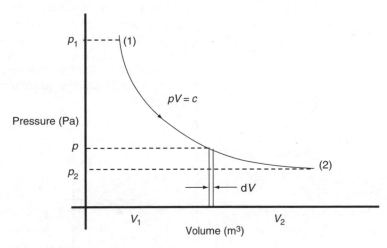

Figure 6.9 *Isothermal expansion*

The work done is again the area under the curve:

$$W = \int_{V_1}^{V_2} p\,dV \tag{i}$$

Now the equation of the curve is $pV = c$ and so,

$$p = \frac{c}{V} = cV^{-1}$$

Substituting for p in equation (i) gives

$$W = \int_{V_1}^{V_2} cV^{-1}\,dV$$

You may remember from your mathematics that this is the special case of integration which gives

$$W = (c \ln V)_{V_2}^{V_1}$$

$$W = c \ln V_2 - c \ln V_1$$

$$W = c \ln\left(\frac{V_2}{V_1}\right) \tag{ii}$$

But $pV = c$. Substituting for c in equation (ii) gives

$$W = pV \ln\left(\frac{V_2}{V_1}\right) \tag{6.16}$$

Now from the characteristic gas equation $pV = mRT$, where T is the constant temperature at which the process takes place. Substituting for pV in equation (6.16) gives

$$W = mRT \ln\left(\frac{V_2}{V_1}\right) \tag{6.17}$$

Example 6.6

Air with an initial volume of $0.1\,\text{m}^3$, pressure 1 bar and temperature $15\,^{\circ}\text{C}$ is compressed according to the law $pV^{1.3} = c$ through a 9:1 compression ratio. It is then allowed to expand isothermally back to its initial volume. Determine (a) the pressure and temperature after compression, (b) the final pressure, (c) the net work done.

The pressure–volume diagram for the processes is shown in Figure 6.10.

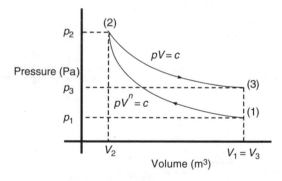

Figure 6.10 *Pressure–volume diagram*

(a) Finding pressure p_2 after compression:

$$p_1 V_1^{\,n} = p_2 V_2^{\,n}$$

$$p_2 = p_1 \frac{V_1^{\,n}}{V_2} = p_1 \left(\frac{V_1}{V_2}\right)^n$$

$$p_2 = 1 \times 10^5 \times (10)^{1.3}$$

$$\boldsymbol{p_2 = 2.0 \times 10^6\,\text{Pa}} \quad \text{or} \quad \boldsymbol{2.0\,\text{MPa}}$$

Finding temperature T_2 after compression:

$$\frac{p_1 V_1}{T_1} = \frac{p_2 V_2}{T_2}$$

$$T_2 = \frac{p_2 V_2}{p_1 V_1} T_1 = \frac{2.0 \times 10^6 \times 0.01(15 + 273)}{1 \times 10^5 \times 0.1}$$

$$T_2 = 576\,\text{K} \quad \text{or} \quad 303\,°\text{C}$$

(b) Finding final pressure, p_3:

$$p_2 V_2 = p_3 V_3$$

$$p_3 = \frac{p_2 V_2}{V_3} = \frac{2.0 \times 10^6 \times 0.01}{0.1}$$

$$p_3 = 200 \times 10^3\,\text{Pa} \quad \text{or} \quad 200\,\text{kPa}$$

(c) Finding work done, W_{1-2}, during compression process:

$$W_{1-2} = \frac{p_1 V_1 - p_2 V_2}{n - 1}$$

$$W_{1-2} = \frac{(1 \times 10^5 \times 0.1) - (2 \times 10^6 \times 0.01)}{1.3 - 1}$$

$$W_{1-2} = -33.3 \times 10^3\,\text{J} \quad \text{or} \quad -33.3\,\text{kJ}$$

(The negative sign denotes work input, or work done on the air.)

Finding work done, W_{2-3}, during the isothermal expansion process:

$$W_{2-3} = p_2 V_2 \ln\left(\frac{V_3}{V_2}\right)$$

$$W_{2-3} = 2 \times 10^6 \times 0.01 \times \ln(10)$$

$$W_{2-3} = 46.1 \times 10^3\,\text{J} \quad \text{or} \quad 46.1\,\text{kJ}$$

(The positive answer denotes work output, or work done by the air)

Finding net work done, W_{1-3}:

$$W_{1-3} = W_{1-2} + W_{2-3}$$
$$W_{1-3} = -33.3 + 46.1$$
$$W_{1-3} = 12.8\,\text{kJ}$$

(The positive answer denotes that there is a net work output.)

Test your knowledge 6.3

1. What does the area under a pressure–volume graph indicate?
2. What is the sign convention for work transfer?
3. Why does the isothermal process require a special formula for work transfer?
4. What will the general formula for work transfer reduce to for a process which obeys Charles' law?

Activity 6.3

The cylinder of petrol engine has a bore of 90 mm, a stroke of 100 mm and a compression ratio of 7.5:1. At the start of the compression process, the temperature of the cylinder contents is 30 °C and the pressure is 101 kPa. If compression takes place according to the law $pV^{1.3} = c$, determine (a) the mass of gas in the cylinder, (b) the temperature at the end of compression, (c) the work done during compression. The characteristic gas constant for air is 287 J kg^{-1} K^{-1}.

Problems 6.3

1. A cylinder contains $0.25 \, \text{m}^3$ of air at a temperature of 28 °C and pressure 1.2 bar. After compression according to the law $pV^{1.4} = c$, the pressure has risen to 6.0 bar. Calculate (a) the mass of the air, (b) the compressed volume, (c) the work input. The characteristic gas constant for air is $287 \, \text{J kg}^{-1} \text{K}^{-1}$.

   $[0.35 \, \text{kg}, \, 0.079 \, \text{m}^3, \, 44 \, \text{kJ}]$

2. A volume of 145 l of air at a pressure of 75 kPa and temperature 18 °C is compressed according to the law $pV^{1.2} = c$. If the compression ratio is 6:1, determine (a) the final pressure, (b) the final temperature, (c) the work done. The characteristic gas constant for air is $287 \, \text{J kg}^{-1} \text{K}^{-1}$.

   $[644 \, \text{kPa}, \, 143 \, °\text{C}, \, 22.9 \, \text{kJ}]$

3. Air of initial volume $1.5 \, \text{m}^3$ and pressure 100 kPa is compressed adiabatically to a pressure of 650 kPa. It is then allowed to expand isothermally back to its original volume. If $\gamma = 1.4$ for air and $R = 287 \, \text{J kg}^{-1} \text{K}^{-1}$, determine the net work done.

   $[77.3 \, \text{kJ}]$

4. Air of initial volume $0.05 \, \text{m}^3$ and temperature 1150 °C is allowed to expand according to the law $pV^{1.35} = c$ to a final pressure of 1 bar and temperature 28 °C. Determine the work done during the process. The characteristic gas constant for air is $287 \, \text{J kg}^{-1} \text{K}^{-1}$.

   $[4.5 \, \text{MJ}]$

5. Air of mass 0.5 kg and initial temperature 235 °C undergoes the following thermodynamic processes. First it is expanded isothermally from an initial pressure of 3.5 MPa to a pressure of 2.1 MPa. This is followed by cooling at constant pressure and finally it is restored to its initial state by adiabatic compression. Determine the net work done during the cycle of processes. Take $\gamma = 1.4$ for air and $R = 287 \, \text{J kg}^{-1} \text{K}^{-1}$.

   $[21.3 \, \text{kJ}]$

Closed thermodynamic system energy equation

In a closed thermodynamic system, the boundary is continuous and none of the working substance may enter or leave whilst a thermodynamic process is taking place. An example of such a system is a quantity of air enclosed in the cylinder of an engine or compressor with the valves closed. The system boundary is drawn by the cylinder head, the cylinder walls and the piston crown. Although the piston moves, it is assumed that none of the air escapes past it. Energy may however pass across the system boundary in the form of heat or work.

The working fluid in a closed system contains energy by virtue of its temperature. As we have stated, this is called *internal energy* which will change in the course a thermodynamic process if the temperature changes. Internal energy is denoted by the symbol U. Suppose that a closed system containing m kg of working substance receives a quantity of heat energy Q during an expansion process and

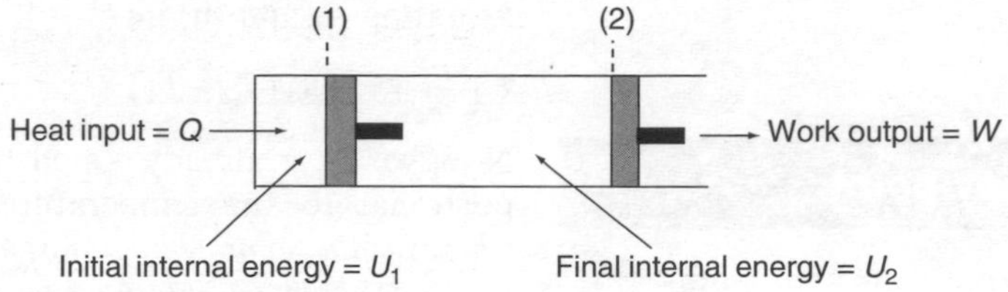

Figure 6.11 *Closed thermodynamic system*

does external work W. Let the internal energy of the substance change from U_1 to U_2 during the process as illustrated in Figure 6.11.

From the first law of thermodynamics, the total energy of the system and its surroundings is constant, that is,

$$\begin{matrix} \text{initial energy of system} \\ \text{and its surroundings} \end{matrix} = \begin{matrix} \text{final energy of system and} \\ \text{its surroundings} \end{matrix}$$

$$\left.\begin{matrix} \text{heat energy input} \\ + \\ \text{initial internal energy} \end{matrix}\right\} = \left\{\begin{matrix} \text{work output} \\ + \\ \text{final internal energy} \end{matrix}\right.$$

$$Q + U_1 = W + U_2$$
$$\boldsymbol{Q = (U_2 - U_1) + W} \qquad (6.18)$$

This is the energy equation for a closed thermodynamic system in which we have expressions (6.14) to (6.17) for calculating the work done. We also have expression (6.7) for finding the heat transfer but we can only use it for the special processes where we know the specific heat capacity of the working substance, i.e. for heat transfer at constant pressure and constant volume. In order to find the heat transfer in any polytropic process using equation (6.18), we need to be able to calculate the change in internal energy.

Consider now a closed system containing m kg of a gas which receives heat energy at constant volume, as shown in Figure 6.12.

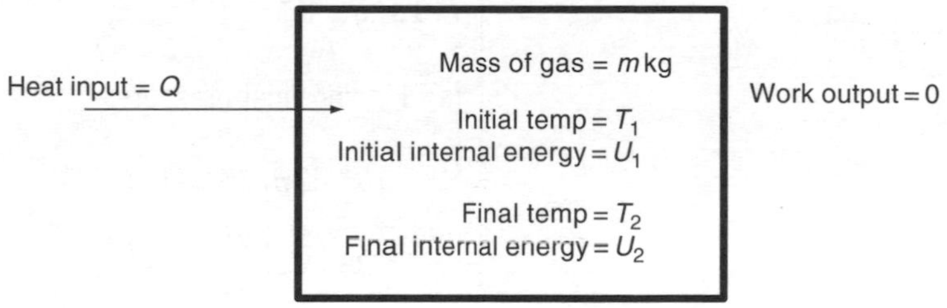

Figure 6.12 *Closed system at constant volume*

As there is no expansion, there will be no work done and all heat energy received will go into raising the internal energy of the gas.

Because no work is done, the closed system energy equation becomes

$$Q = (U_2 - U_1) + 0$$
$$\boldsymbol{Q = U_2 - U_1} \qquad \text{(i)}$$

Now if the specific heat of the gas at constant volume is c_v, the heat energy supplied will be

$$\boldsymbol{Q = mc_v(T_2 - T_1)} \qquad \text{(ii)}$$

Key point

The boundary of a closed thermodynamic system is continuous and none of the working substances may enter or leave whilst a thermodynamic is taking place.

Equating (i) and (ii) gives

$$U_2 - U_1 = mc_v(T_2 - T_1) \tag{6.19}$$

Now we have already stated that internal energy is directly proportional to the temperature of a substance. It follows that *wherever the same temperature change occurs in a given mass of gas, the same change of internal energy will take place.* This is irrespective of whether the change occurs at constant volume, constant pressure or during any other polytropic process. Equation (6.19) can thus be used as a general formula for calculating changes of internal energy for any expansion or compression process.

It may be a good idea now to summarise the sign conventions for heat transfer, work transfer and change of internal energy.

Key point

Wherever the same temperature change occurs in a given mass of gas, the same change of internal energy will take place.

Heat energy received by a system	Q **is positive** $(+)$
Heat energy rejected by a system	Q **is negative** $(-)$
Work output from a system	W **is positive** $(+)$
Work input to a system	W **is negative** $(-)$
Increase of internal energy	$(U_2 - U_1)$ **is positive** $(+)$
Decrease of internal energy	$(U_2 - U_1)$ **is negative** $(-)$

Example 6.7

0.01 kg of air is compressed in an engine cylinder into one-eighth of its original volume according to the law $pV^{1.25} = c$. The initial pressure and temperature of the air are 0.97 bar and 21 °C, respectively. Determine (a) the final pressure and temperature, (b) the work done, (c) the change in internal energy, (d) the heat transfer which takes place. The characteristic gas constant for air is 287 J kg^{-1} K^{-1} and its specific heat capacity at constant volume is 710 J kg^{-1} K^{-1}.

(a) Finding final pressure:

$$p_1 V_1^n = p_2 V_2^n$$

$$p_2 = p_1 \frac{V_1^n}{V_2^n} = p_1 \left(\frac{V_1}{V_2}\right)^n$$

$$p_2 = 0.97 \times 10^5 \times (8)^{1.25}$$

$$\boldsymbol{p_2 = 1.31 \times 10^6 \, \text{Pa} \quad \text{or} \quad 1.31 \, \text{MPa}}$$

Finding final temperature:

$$\frac{p_1 V_1}{T_1} = \frac{p_2 V_2}{T_2}$$

$$T_2 = \frac{p_2 V_2}{p_1 V_2} T_1 = \frac{1.31 \times 10^6}{0.97 \times 10^5} \times \frac{1}{8} \times (21 + 273)$$

$$\boldsymbol{T_2 = 428 \, \text{K} \quad \text{or} \quad 155 \, ^\circ\text{C}}$$

(b) Finding work done:

$$W = \frac{mR(T_1 - T_2)}{n - 1}$$

$$W = \frac{0.01 \times 287(294 - 428)}{1.25 - 1}$$

$$\mathbf{W = -1.54 \times 10^3 \, J} \quad \text{or} \quad \mathbf{-1.54 \, kJ}$$

(The negative sign denotes work input.)

(c) Finding change of internal energy:

$$U_2 - U_1 = mc_v(T_2 - T_1)$$

$$U_2 - U_1 = 0.01 \times 710(428 - 294)$$

$$\mathbf{U_2 - U_1 = 951 \, J}$$

(The positive answer denotes an increase of internal energy.)

(d) Finding heat transfer which takes place:

$$Q = (U_2 - U_1) + W$$

$$Q = 951 + (-1.54 \times 10^3)$$

$$\mathbf{Q = -589 \, J}$$

(The negative sign denotes that heat energy is lost by the air.)

Relationship between the constants R, c_p and c_v

You may have noticed that these constants for a gas have the same units, i.e. $J\,kg^{-1}\,K^{-1}$. They are in fact closely related as will be shown. Suppose a mass of m kg of a gas receives heat energy Q whilst expanding at constant pressure, i.e. whilst expanding according to Charles' law. Let the temperature rise be from T_1 to T_2 and the amount of work done be W, as shown in Figure 6.13.

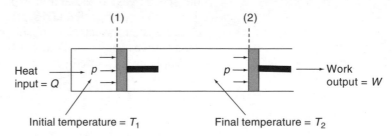

Figure 6.13 *Expansion at constant pressure*

Applying the closed system energy equation

$$Q = (U_2 - U_1) + W \tag{i}$$

The heat energy Q supplied at constant pressure can be written as

$$Q = mc_p(T_2 - T_1) \tag{ii}$$

As has been stated, the change of internal energy for any polytropic process can be written as

$$U_2 - U_1 = mc_v(T_2 - T_1) \qquad \text{(iii)}$$

The work done W can also be written in terms of the temperature change as given by equation (6.15), that is

$$W = \frac{mR(T_1 - T_2)}{n - 1}$$

You may remember that when deriving this formula, we reversed the temperatures on the top line and also the term in the denominator. This was for convenience because n is usually greater than 1. It is now convenient to change them back again giving

$$W = \frac{mR(T_2 - T_1)}{1 - n} \qquad \text{(iv)}$$

Substituting in equation (i) for these terms gives

$$\cancel{m}c_p(\cancel{T_2 - T_1}) = \cancel{m}c_v(\cancel{T_2 - T_1}) + \frac{\cancel{m}R(\cancel{T_2 - T_1})}{1 - n}$$

Cancelling all the common terms leaves

$$c_p = c_v + \frac{R}{1 - n}$$

But you may recall that for expansion at constant pressure, according to Charles' law, $n = 0$. This leaves

$$c_p = c_v + R$$

$$R = c_p - c_v \qquad \text{(6.20)}$$

It thus occurs that the characteristic gas constant is the difference between the specific heat capacities at constant pressure and constant volume.

Relationship between c_p, c_v and γ

You will remember that γ is the value of the polytripic index for the special case of adiabatic expansion and compression during which no heat transfer takes place. Suppose now that the same closed system which we considered above undergoes adiabatic expansion as shown in Figure 6.14.

Applying the closed system energy equation once again

$$Q = (U_2 - U_1) + W$$

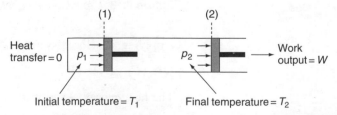

Figure 6.14 *Adiabatic expansion*

But $Q = 0$ for an adiabatic process, that is,

$$0 = (U_2 - U_1) + W$$

$$(U_2 - U_1) = -W \tag{i}$$

As always, the change of internal energy can be written as

$$U_2 - U_1 = mc_v(T_2 - T_1) \tag{ii}$$

The work done W can again be written as

$$W = \frac{mR(T_1 - T_2)}{n - 1}$$

But for adiabatic expansion $n = \gamma$ giving

$$W = \frac{mR(T_1 - T_2)}{\gamma - 1} \tag{iii}$$

Substituting in equation (i) for these terms gives

$$mc_v(T_2 - T_1) = -\frac{mR(T_1 - T_2)}{\gamma - 1}$$

Reversing the temperatures on the top right-hand side eliminates the minus sign giving

$$\cancel{mc_v(T_2 - T_1)} = \frac{\cancel{m}R(T_2 - T_1)}{\gamma - 1}$$

Cancelling the terms common to both sides leaves

$$c_v = \frac{R}{\gamma - 1}$$

$$c_v(\gamma - 1) = R$$

But we have already proved in (6.20) that $R = c_p - c_v$. Substituting gives

$$c_v(\gamma - 1) = c_p - c_v$$

$$\gamma c_v - \cancel{c_v} = c_p - \cancel{c_v}$$

$$\gamma c_v = c_p$$

$$\gamma = \frac{c_p}{c_v} \tag{6.21}$$

The adiabatic index γ is thus the ratio of the two specific heat capacities c_p and c_v.

Key point

The adiabatic index is given by the ratio of the specific heat capacities at constant pressure and constant volume.

Example 6.8

Air at a pressure of 500 kPa and temperature 450 °C occupies a volume of 0.05 m³. If the air is allowed to expand adiabatically to a final pressure of 100 kPa, determine (a) the mass of the air, (b) the final volume, (c) the final temperature, (d) the work done during expansion. For air $c_p = 1005\,\text{J}\,\text{kg}^{-1}\,\text{K}^{-1}$ and $c_v = 718\,\text{J}\,\text{kg}^{-1}\,\text{K}^{-1}$.

(a) Finding characteristic gas constant for air:

$$R = c_p - c_v = 1005 - 718$$

$$R = 287\,\text{J}\,\text{kg}^{-1}\,\text{K}^{-1}$$

Finding the mass of the air:

$$p_1 V_1 = mRT_1$$

$$m = \frac{p_1 V_1}{RT_1} = \frac{500 \times 10^3 \times 0.05}{287 \times (450 + 273)}$$

$$\mathbf{m = 0.12\,kg}$$

(b) Finding adiabatic index, γ:

$$\gamma = \frac{c_p}{c_v} = \frac{1005}{718}$$

$$\boldsymbol{\gamma = 1.4}$$

Finding final volume:

$$p_1 V_1^{\gamma} = p_2 V_2^{\gamma}$$

$$\frac{p_1}{p_2} = \frac{V_2^{\gamma}}{V_1^{\gamma}} = \left(\frac{V_2}{V_1}\right)^{\gamma}$$

Take the γth root of each side:

$$\left(\frac{p_1}{p_2}\right)^{1/\gamma} = \frac{V_2}{V_1}$$

$$V_2 = V_1 \left(\frac{p_1}{p_2}\right)^{1/\gamma} = 0.05 \left(\frac{500}{100}\right)^{1/1.4}$$

$$\boldsymbol{V_2 = 0.158\,m^3}$$

(c) Finding final temperature:

$$p_2 V_2 = mRT_2$$

$$T_2 = \frac{p_2 V_2}{mR} = \frac{100 \times 10^3 \times 0.158}{0.12 \times 287}$$

$$\boldsymbol{T_2 = 459\,K} \quad \text{or} \quad \boldsymbol{186\,°C}$$

(d) Finding change of internal energy:

$$U_2 - U_1 = mc_v(T_2 - T_1)$$

$$U_2 - U_1 = 0.12 \times 718(186 - 450)$$

$$\boldsymbol{U_2 - U_1 = -22.7 \times 10^3\,J} \quad \text{or} \quad \boldsymbol{-22.7\,kJ}$$

(The negative sign denotes a loss of internal energy.)

Finding work done:

$$Q = (U_2 - U_1) + W$$

But $Q = 0$ for adiabatic expansion

$$0 = (U_2 - U_1) + W$$

$$W = -(U_2 - U_1) = -(-22.7)$$

$$\boldsymbol{W = 22.7\,kJ}$$

(The positive answer denotes work output.)

You should note that during an adiabatic process, the work flow is equal to the change of internal energy but opposite in sign.

Test your knowledge 6.4

1. What is internal energy?
2. What is the closed system energy equation?
3. What is the sign convention for heat transfer?
4. What is the relationship between the constants R, c_p and c_v for a gas?
5. What is the relationship between the constants c_p, c_v and γ for a gas?

Activity 6.4

A marine diesel engine has a cylinder capacity of $0.25\,m^3$ and a compression ratio of 18:1. The air in the cylinder is initially at a pressure of 1 bar and a temperature of 25 °C. Compression takes place according to the law $pV^{1.35} = c$ after which expansion occurs at constant pressure until the volume is four times the compressed volume. Sketch the processes on a pressure–volume diagram and determine the net work and heat transfer which take place. For air, take $c_p = 1005\,J\,kg^{-1}\,K^{-1}$ and $c_v = 718\,J\,kg^{-1}\,K^{-1}$.

Problems 6.4

1. Air of initial volume $1.25\,m^3$, pressure 95 kPa and temperature 38 °C is compressed according to the law $pV^{1.24} = c$. If the final pressure is 650 kPa, determine (a) the mass of air present, (b) the final temperature of the air, (c) the work done during the process, (d) the heat transfer that takes place. Take $R = 287\,J\,kg^{-1}\,K^{-1}$ and $c_v = 718\,J\,kg^{-1}\,K^{-1}$.

   [1.33 kg, 178 °C, 223 kJ, 89 kJ]

2. A quantity of gas has a mass of 1 kg and occupies a volume of $0.12\,m^3$ at a pressure of 7 bar. If the gas is expanded adiabatically in a closed system to a pressure of 1.5 bar, determine (a) the final volume, (b) the work done, (c) the temperature change which occurs. Take $\gamma = 1.38$ for the gas and $c_v = 720\,J\,kg^{-1}\,K^{-1}$.

   [$0.0367\,m^2$, 72.6 kJ, 106 °C]

3. Air of mass 0.9 kg is compressed isothermally at a temperature of 37 °C in a cylinder. The initial pressure is 69 kPa and the final pressure is 565 kPa. Determine (a) the compression ratio for the process, (b) the heat transfer that takes place. Take $R = 287\,J\,kg^{-1}\,K^{-1}$.

   [8.2:1, 168 kJ]

4. A gas occupies an initial volume of $0.5\,m^3$, at a temperature 165 °C and pressure 660 kPa. If the gas expands to a pressure of 120 kPa according to the law $pV^{1.3} = c$, determine (a) the final volume and temperature, (b) the work done, (c) the change of internal energy, (d) the heat transfer that takes place. Take $R = 287\,J\,kg^{-1}\,K^{-1}$ and $c_v = 710\,J\,kg^{-1}\,K^{-1}$.

   [$1.86\,m^3$, 23 °C, 358 kJ, 265 kJ, 92.3 kJ]

5. An engine has a cylinder of diameter 350 mm and a stroke of length 405 mm. The clearance volume when the piston is at the top of its stroke is $0.0062\,m^3$ and as it moves down the cylinder the hot gases expand from a pressure of 1.92 MPa and temperature 100 °C according to the law $pV^{1.35} = c$. Determine (a) the work done on the piston, (b) the average force which acts on the piston, (c) the heat transfer which takes place.

   [17.1 kJ, 41.1 kN, 2.46 kJ]

Open thermodynamic system energy equation

In an open system, the boundary is not continuous. It has inlet and exit ports through which the working substance may enter or leave whilst work and heat transfer are taking place. Examples of such a system are steam and gas turbines through which there is a continuous and steady flow of the working substance. A closed system may consist of a number of linked open systems. A steam plant for power generation is a closed system but each of its component parts such as the boiler, superheater, turbine and condenser are open systems, linked to form a closed circuit.

Equations (6.14) and (6.15) for work done during a polytropic process must only be used for a fixed mass of gas in a closed system. They must *not* be used for open system problems. Alternative formulae are required which take into account not only the work done during expansion or compression but also the pressure-flow work which is required to maintain the flow through an open system. You may remember that this is analogous to the work done in overcoming friction in solid body dynamics and is given by equation 5.37 in Chapter 5. The net work done during a polytropic expansion or compression process in an open system is given by the shaded area in Figure 6.15.

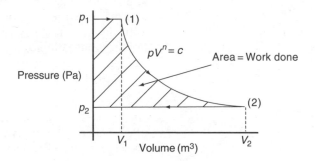

Figure 6.15 *Expansion through an open system*

$$\begin{matrix} \text{net work} \\ \text{done} \end{matrix} = \begin{matrix} \text{work done during} \\ \text{polytropic process} \end{matrix} + \begin{matrix} \text{net pressure-flow work} \\ \text{done to maintain the flow} \end{matrix}$$

$$W = \frac{p_1 V_1 - p_2 V_2}{n-1} + p_1 V_1 - p_2 V_2 \qquad (6.22)$$

The work done is given by equation (6.22). It is the area between the expansion curve and the pressure axis in Figure 6.15. Putting $n - 1$ as a common denominator gives

$$W = \frac{p_1 V_1 - p_2 V_2 + (n-1)(p_1 V_1 - p_2 V_2)}{n-1}$$

$$W = \frac{p_1 V_1 - p_2 V_2 + n p_1 V_1 - n p_2 V_2 - p_1 V_1 + p_2 V_2}{n-1}$$

Cancelling similar terms leaves

$$W = \frac{np_1V_1 - np_2V_2}{n - 1}$$

$$W = \frac{n}{n-1}(p_1V_1 - p_2V_2) \tag{6.23}$$

Since $p_1V_1 = mRT_1$ and $p_2V_2 = mRT_2$, the above equation can also be written as

$$W = \frac{n}{n-1}(mRT_1 - mRT_2)$$

$$W = \frac{nmR}{n-1}(T_1 - T_2) \tag{6.24}$$

If the expansion or compression process in an open system is isothermal, the expressions for work done are the same as for a closed system, i.e. equations (6.16) and (6.17). The reason for this is that the expansion curve $pV = c$ for an isothermal process is what is known as a *rectangular hyperbola*. With this curve, the area under the curve is the same as the area between the curve and the pressure axis and so the expressions for work done are the same.

In Chapter 5, we discussed the different forms of energy which are present in a moving fluid and derived the full steady flow energy equation. We then used it to derive Bernoulli's equation for the flow of incompressible fluids. We can now apply the same steady flow energy equation to an open thermo-dynamic system which is shown schematically in Figure 6.16.

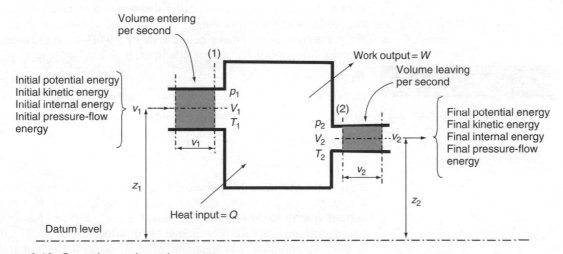

Figure 6.16 *Open thermodynamic system*

For steady flow conditions, the mass per second flowing into the system is equal to the mass per second leaving. Also, from the first law of thermodynamics, the total energy of the system and its surroundings is constant which means that the energy entering the system per second must equal the energy leaving the system per second, that is,

$$
\left.\begin{array}{c}
\text{initial potential energy} \\
+ \\
\text{initial kinetic energy} \\
+ \\
\text{initial internal energy} \\
+ \\
\text{initial pressure-flow energy} \\
+ \\
\text{heat energy input}
\end{array}\right\} = \left\{\begin{array}{c}
\text{final potential energy} \\
+ \\
\text{final kinetic energy} \\
+ \\
\text{final internal energy} \\
+ \\
\text{final pressure-flow energy} \\
+ \\
\text{work output}
\end{array}\right.
$$

In Chapter 5, we substituted expressions for each of these terms which gives

$$
\begin{aligned}
mgz_1 + \tfrac{1}{2}mv_1{}^2 &= mgz_2 + \tfrac{1}{2}m v_2^2 \\
+ U_1 + p_1 V_1 + Q &\quad + U_2 + p_2 V_2 + W
\end{aligned}
\tag{5.38}
$$

This again is the full steady flow energy equation (FSFEE) but it is unusual to apply it to open thermodynamic system problems in its full form. Some of the energy changes can be neglected, in particular changes in potential energy and kinetic energy. These tend to be very small compared to the changes of internal energy and pressure-flow energy which occur. Neglecting them leaves

$$
\begin{aligned}
U_1 + p_1 V_1 + Q &= U_2 + p_2 V_2 + W \\
Q &= U_2 - U_1 + p_2 V_2 - p_1 V_1 + W \\
Q &= U_2 + p_2 V_2 - U_1 - p_1 V_1 + W \\
Q &= (U_2 + p_2 V_2) - (U_1 + p_1 V_1) + W
\end{aligned}
\tag{6.26}
$$

The bracketed terms are the *sum of the initial internal and pressure-flow energy* and the *sum of the final internal and pressure-flow energy*. It is convenient to consider these as a single energy quantity which is known as *enthalpy*. This, you will remember is one of the properties used to describe the state of a working substance. It is given the symbol H, so that in equation (6.26),

$$
H_1 = U_1 + p_1 V_1
\tag{6.27}
$$

and

$$
H_2 = U_2 + p_2 V_2
\tag{6.28}
$$

Equation (6.26) can thus be written as

$$
Q = (H_2 - H_1) + W
\tag{6.29}
$$

You will note that this shortened form of the energy equation (6.29), for open systems, bears a close resemblance to the closed system energy equation. The difference is that change of internal energy has been replaced by change of enthalpy. Consider now the change in enthalpy which is obtained by subtracting equation (6.27) from equation (6.28):

$$
\begin{aligned}
H_2 - H_1 &= U_2 + p_2 V_2 - U_1 - p_1 V_1 \\
H_2 - H_1 &= (U_2 - U_1) + (p_2 V_2 - p_1 V_1)
\end{aligned}
\tag{6.30}
$$

Now the change of internal energy can be written as $U_2 - U_1 = m c_v (T_2 - T_1)$ and from the characteristic gas

Key point

The enthalpy of a working substance is the sum of its internal energy and its pressure-flow energy.

equation, $p_2 V_2 = mRT_2$ and $p_1 V_1 = mRT_1$ Substituting in equation (6.30), we get

$$H_2 - H_1 = m\, c_v(T_2 - T_1) + (mRT_2 - mRT_1)$$
$$H_2 - H_1 = m\, c_v(T_2 - T_1) + m\, R(T_2 - T_1)$$
$$H_2 - H_1 = m(T_2 - T_1)(c_v + R)$$

We have shown that $R = c_p - c_v$ and so

$$H_2 - H_1 = m(T_2 - T_1)(\cancel{c_v} + c_p - \cancel{c_v})$$
$$\boldsymbol{H_2 - H_1 = m\, c_p(T_2 - T_1)} \tag{6.31}$$

You will note that this expression bears a marked resemblance to equation (6.19) for change of internal energy. The difference is that c_p replaces c_v.

Example 6.9

A centrifugal compressor takes in 420 m³ of air per minute at a temperature of 20 °C and pressure 100 kPa and delivers it at a pressure of 225 kPa. Compression takes place according to the law $pV^{1.3} = c$ and the input power is 630 kW. Determine (a) the mass flow rate of the air, (b) the delivery temperature, (c) the heat transfer rate.

Take $c_p = 1005\ \text{J kg}^{-1}\,\text{K}^{-1}$ and $c_v = 718\ \text{J kg}^{-1}\,\text{K}^{-1}$.

(a) Finding characteristic gas constant:

$$R = c_p - c_v = 1005 - 718$$
$$\boldsymbol{R = 287\ \text{J kg}^{-1}\,\text{K}^{-1}}$$

Finding mass flow rate:

$$p_1 V_1 = mRT_1 \quad (\text{where } V_1 = 420/60 = 7.0\ \text{m}^3\,\text{s}^{-1})$$
$$m = \frac{p_1 V_1}{R\, T_1} = \frac{100 \times 10^3 \times 7.0}{287(20 + 273)}$$
$$\boldsymbol{m = 8.32\ \text{kg s}^{-1}}$$

(b) Finding volume delivered per second:

$$p_1 V_1^n = p_2 V_2^n$$
$$\frac{p_1}{p_2} = \frac{V_2^n}{V_1^n} = \left(\frac{V_2}{V_1}\right)^n$$

Taking the nth root of each side, we obtain

$$\left(\frac{p_1}{p_2}\right)^{1/n} = \frac{V_2}{V_1}$$
$$V_2 = V_1\left(\frac{p_1}{p_2}\right)^{1/n} = 7.0\left(\frac{100}{225}\right)^{1/1.3}$$
$$\boldsymbol{V_2 = 3.75\ \text{m}^3\,\text{s}^{-1}}$$

Finding delivery temperature:

$$p_2 V_2 = mRT_2$$

$$T_2 = \frac{p_2 V_2}{mR} = \frac{225 \times 10^3 \times 3.75}{8.32 \times 287}$$

$$T_2 = 353\,\text{K} \quad \text{or} \quad 80\,°\text{C}$$

(c) Finding change of enthalpy:

$$H_2 - H_1 = m\,c_p(T_2 - T_1)$$

$$H_2 - H_1 = 8.32 \times 1005(80 - 20)$$

$$H_2 - H_1 = 502 \times 10^3\,\text{W} \quad \text{or} \quad 502\,\text{kW}$$

Note that because m is the mass flow rate, the change of enthalpy has unit of joules per second or Watts.

Finding heat transfer rate:

$$Q = (H_2 - H_1) + W \quad \text{(where } W = -630\,\text{kW, the input power)}$$

$$Q = 502 + (-630)$$

$$Q = -128\,\text{kW}$$

(The negative sign denotes heat lost during the compression process.)

Test your knowledge 6.5

1. What is the essential difference between a closed and an open thermodynamic system?
2. What are the two energy forms which together make up the property enthalpy?
3. What is the shortened version of the steady flow energy equation which can be applied to open thermodynamic systems?
4. What is the expression used for calculating changes of enthalpy?

Activity 6.5

A steady flow cooler takes in oxygen at a pressure of 2 MPa and temperature 150 °C and delivers it at a pressure of 100 kPa. Expansion in the cooler takes place according to the law $pV^{1.45} = c$ and the mass flow rate is 5 kg s^{-1}. Ignoring any potential or kinetic energy changes, determine the heat transfer rate. Take $c_p = 917\,\text{J kg}^{-1}\,\text{K}^{-1}$ and $c_v = 657\,\text{J kg}^{-1}\,\text{K}^{-1}$.

Problems 6.5

1. A gas enters an open system at the rate of $3\,\text{m}^3$ per minute at a pressure of 14 bar and leaves at a pressure of 5 bar after expanding according to the law $pV^{1.3} = c$. Assuming steady flow conditions, what will be the change of pressure-flow energy per second?

   [14.8 kW]

2. The internal energy of a gas flowing steadily through an open system decreases by 92 kJ kg^{-1}. At the same time there is a heat transfer of 7 kJ kg^{-1} out of the system and the pressure-flow energy increases by 44 kJ kg^{-1}. If there is no appreciable change in potential and kinetic energy, determine (a) the change of enthalpy per kilogram, (b) the work output for a flow rate of 50 kg min^{-1}.

   [48 kJ kg^{-1}, 34.2 kW]

3. Oxygen at an initial temperature of 200 °C and pressure 90 bar flows steadily through an open system at a rate of 600 kg min^{-1}. During its passage it expands adiabatically and emerges at a pressure of 30 bar. If kinetic and potential energy changes are negligible, determine (a) the final temperature, (b) the work done per minute. Take $c_v = 657$ J kg^{-1} K^{-1} and $c_p = 917$ J kg^{-1} K^{-1}.

[73.5 °C, 70 MJ min^{-1}]

4. Carbon dioxide at a temperature of 15 °C and pressure 100 kPa enters a rotary compressor at the rate of 7.5 kg s^{-1} and leaves at a pressure of 500 kPa. If compression takes place according to the law $pV^{1.25} = c$, determine (a) the work done per second on the gas, (b) the change in enthalpy per second, (c) the rate of heat transfer. Neglect changes in potential and kinetic energy and take $R = 220$ J kg^{-1} K^{-1} and $c_p = 630$ J kg^{-1} K^{-1}.

[924 kW, 529 kW, 395 kW]

5. Air at a temperature of 20 °C and pressure 1 bar is taken in by a rotary compressor at a steady rate of 1.5 kg s^{-1}. Compression follows the law $pV^{1.2} = c$ and the final pressure is 6 bar. If kinetic and potential energy changes are neglected, determine (a) the final temperature, (b) the input power, (c) the heat transfer rate. Take $R = 287$ J kg^{-1} K^{-1} and $c_p = 1005$ J kg^{-1} K^{-1}.

[122 °C, 263 kW, 109 kW]

Combustion of fuels

A fuel (other than nuclear) is a substance whose main constituents are carbon and hydrogen. During combustion, these elements combine chemically with oxygen from the atmosphere, giving off appreciable amounts of heat energy during the process. Sulphur is also present in some fuels and this too combines with oxygen to give off heat energy. Its heating value is however outweighed by the production of sulphur dioxide. This combines with the moisture, from the combustion of hydrogen, to form sulphurous acid that has a corrosive effect on flues and exhaust pipes. *The calorific value of a fuel is the heat energy liberated when 1 kg is completely burnt.* The separate calorific values of carbon, hydrogen and sulphur are as follows:

calorific value of carbon = 33.7 MJ kg^{-1}

calorific value of hydrogen = 144 MJ kg^{-1}

calorific value of sulphur = 9.1 MJ kg^{-1}

Key point

Calorific value is the heat energy liberated from the complete combustion of 1 kg of fuel.

Solid fuels

The main solid fuels that are used industrially throughout the developed world are lignite or brown coal, bituminous coal and anthracite. These are derived from compressed vegetable matter that was laid down in pre-historic times.

Lignite This is an inferior type of coal found in eastern Europe. It has a higher moisture content than true coal and needs to be dried before it will burn satisfactorily. Lignite has a calorific value of around 23 MJ kg^{-1}.

Bituminous coal This is the type of coal which is still mined in United Kingdom and which is imported for power generation and domestic use. The calorific value of good quality bituminous coal is around $32\,MJ\,kg^{-1}$.

Anthracite This is a high-quality coal which is hard and brittle. It has a higher carbon content than bituminous coal and was formerly mined in South Wales. It is difficult to ignite but burns with little smoke and leaves comparatively little ash. Anthracite has a higher carbon content than lignite and bituminous coal and a calorific value of around $35\,MJ\,kg^{-1}$.

Liquid fuels

Crude petroleum is the main source of liquid fuels. It is heated to temperatures in excess of $400\,°C$ and the vapours given off are condensed during a process known as fractional distillation. The main fuels obtained by this means are petrol, diesel oil and kerosene. Coal can also yield liquid fuels by the distillation of coal tar. Benzene and toluene are obtained in this way. Vegetable matter can be fermented to produce alcohol and this too can be used as a fuel.

The terms *flash point* and *ignition temperature* are often quoted for liquid fuels, and the two are sometimes confused. Flash point is the temperature at which inflammable vapour starts to be given off by a fuel. It is important to know this and provide adequate ventilation when considering storage. The ignition temperature is the temperature at which the fuel will ignite spontaneously, without the application of a flame or spark.

Fuel oils Fuel oils include diesel oil and heating oil. These are condensed during the fractional distillation from crude petroleum vapours at temperatures between $425\,°C$ and $345\,°C$. Their calorific value is $44\,MJ\,kg^{-1}$.

Kerosene Kerosene or paraffin is condensed out as the vapour cool down from $345\,°C$ to $220\,°C$. Its calorific value is $44\,MJ\,kg^{-1}$.

Petrol Petrol or gasoline is condensed out as the vapour cool down from $220\,°C$ to $65\,°C$. Its calorific value is $46.5\,MJ\,kg^{-1}$.

Benzole This is a mixture of benzene, toluene and xylene which are obtained by distillation from coal tar. Its availability and use has declined with the replacement of coal gas with natural gas and the subsequent decline of the coal industry. Its calorific value is $40\,MJ\,kg^{-1}$.

Alcohol Alcohol, C_2H_5OH, is formed by the fermentation of vegetable matter. It is not used so much as a fuel but as an additive to petrol. Its effect is to enable higher compression ratios to be used and higher thermal efficiencies to be achieved. It is also capable of absorbing traces of moisture in the fuel. When used alone, its calorific value is $30\,MJ\,kg^{-1}$.

Gaseous fuels

The most commonly used gaseous fuels for domestic and industrial use are natural gas, butane and propane. Coal gas or town gas is still used in some parts of the world and was the main gaseous fuel in this country prior to the discovery of natural gas in the 1960s. The calorific value of gaseous fuels is usually quoted in $MJ\,m^{-3}$ at normal temperature and pressure conditions.

Natural gas   Natural gas contains 80–90% methane whose chemical formula is CH_4. The remainder is made up of other gases which include small amounts of butane and propane. The calorific value of the natural gas supplied in the United Kingdom is $37.8\,MJ\,m^{-3}$ at NTP.

Butane   Butane has the chemical formula C_4H_{10} and is obtained during the distillation of crude petroleum. It is liquefied under pressure for storage and has a calorific value of $120\,MJ\,m^{-3}$ at NTP. Its main drawback is its relatively high freezing point which is around $-5\,°C$.

Propane   Propane has the chemical formula C_3H_8 and is also obtained during the distillation of crude petroleum and liquefied. It is better suited to outdoor use and storage than butane on account of its lower freezing point which is around $-40\,°C$. The calorific value of propane is $93\,MJ\,m^{-3}$ at NTP.

Key point

The calorific value of gaseous fuels is measured in $MJ\,m^{-3}$ at NTP.

The bomb calorimeter

The bomb calorimeter is used to determine the calorific value of solid fuels and some of the less volatile liquid fuels such as fuel oil. It is also used by food technologists to determine the calorific value of foods. As with the fuels which we use, the foods which we eat are chemical combinations of carbon and hydrogen, and when the moisture content has been removed, they can be burned like a fuel to release their stored energy.

There are several different designs of bomb calorimeter but basically the apparatus consists of a screw-topped pressure vessel, the bomb, surrounded by water in a lagged copper calorimeter. The screw top of the bomb contains insulated electrical connections and a pressurising valve. This enables it to be charged with oxygen to ensure that there is sufficient for complete combustion of the fuel. A small quantity of water, about 10 ml, is also placed inside the bomb. Its purpose is to ensure condensation of the steam formed during combustion and absorb any acid products which are formed.

Leads from the electrical connections extend inside the bomb and one of them supports a small porcelain crucible containing a measured sample of fuel m_f. For solid fuels the fuse were connected between the leads is positioned in contact with the fuel sample. With liquid fuels a cotton thread is hung from the fuse wire into the fuel sample. The water equivalent of the bomb and copper calorimeter m_{wc} is supplied by the makers of the apparatus. The arrangement is shown in Figure 6.17.

The assembled and pressurised bomb is placed inside the insulated copper calorimeter and the electrical leads are connected to a low-voltage power supply. A measured quantity of water m_w is added to the calorimeter to above the level of the bomb but slightly below the level of the electrical connections. The stirrer, insulated lid and sensitive thermometer are then placed in position.

The apparatus is allowed to stand for a period of time to allow the temperature inside the calorimeter to stabilise. The thermometer

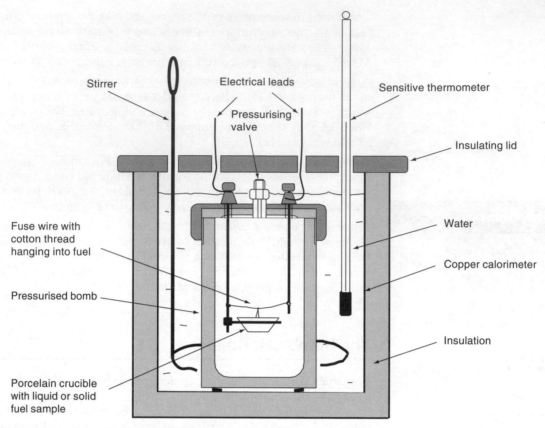

Figure 6.17 *General arrangement of bomb calorimeter*

reading is then taken and the bomb is fired by switching on the power supply. Thermometer readings are then taken at 1-min intervals as the water temperature rises whilst all the time operating the stirrer. Eventually, after rising by one or two degrees, the temperature is observed to level-off and may then begin to fall as heat energy is lost to the atmosphere. A correction to the recorded temperature rise must be made for heat loss during the experiment. This is done by plotting a graph of temperature against time, as shown in Figure 6.18.

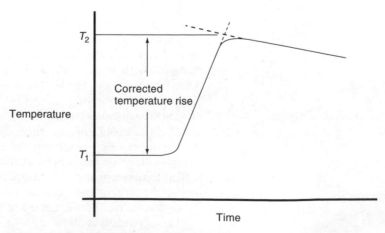

Figure 6.18 *Graph of temperature v. time*

The calorific value of the fuel can now be found by equating the heat released by the fuel to the heat received by the water, bomb and the copper calorimeter, that is,

$$\begin{array}{c}\text{heat supplied} \\ \text{by fuel}\end{array} = \begin{array}{c}\text{heat received by the water,} \\ \text{bomb and calorimeter}\end{array}$$

$$\begin{array}{c}\text{mass of} \\ \text{fuel}\end{array} \times \begin{array}{c}\text{calorific} \\ \text{value}\end{array} = \left(\begin{array}{c}\text{mass of} \\ \text{water}\end{array} + \begin{array}{c}\text{water equivalent} \\ \text{of apparatus}\end{array}\right)$$

$$\times \begin{array}{c}\text{specific heat} \\ \text{capacity}\end{array} \times \begin{array}{c}\text{corrected} \\ \text{temperature rise}\end{array}$$

$$m_\text{f} \times \text{CV} = (m_\text{w} + m_\text{wc})\, c_\text{w}(T_2 - T_1)$$

$$\mathbf{CV} = \frac{(\boldsymbol{m_\text{w}} + \boldsymbol{m_\text{wc}})\, \boldsymbol{c_\text{w}}(\boldsymbol{T_2} - \boldsymbol{T_1})}{\boldsymbol{m_\text{f}}} \qquad (6.32)$$

Combustion of the hydrogen content of the fuel produces H_2O in the form of steam. This condenses inside the bomb, giving up its latent heat, which becomes a part of the total heat received by the apparatus. Because this is included, the value of the calorific value obtained from the bomb calorimeter is known as the *higher* or *gross calorific value* of the fuel.

When a fuel is used in practical situations such as in an internal combustion engine, the steam from combustion leaves with the exhaust gases and condenses in the atmosphere or in the exhaust pipe. As a result its latent heat is not available for conversion into useful work and is subtracted from the higher calorific value of the fuel. This is then known as the *lower* or *net calorific value*. Fuel technologists are able to calculate its value from a knowledge of the hydrogen content and it is the lower calorific value which is usually quoted by oil and gas suppliers.

Key point

The higher calorific value of a fuel includes the latent heat which is given off by steam from the combustion of hydrogen as it condenses.

Example 6.10

The following readings were obtained from a test on a sample of fuel oil using a bomb calorimeter.

$$\text{mass of oil sample} = 0.68\,\text{g}$$
$$\text{total mass of water in calorimeter} = 2.750\,\text{kg}$$
$$\text{water equivalent of apparatus} = 1.100\,\text{kg}$$
$$\text{corrected temperature rise} = 1.75\,°\text{C}$$

Taking the specific heat capacity of water to be $4187\,\text{J}\,\text{kg}^{-1}\,\text{K}^{-1}$, determine the gross calorific value of the fuel.

$$\begin{array}{c}\text{heat supplied} \\ \text{by fuel}\end{array} = \begin{array}{c}\text{heat received by the water,} \\ \text{bomb and calorimeter}\end{array}$$

$$\begin{array}{c}\text{mass of} \\ \text{fuel}\end{array} \times \begin{array}{c}\text{calorific} \\ \text{value}\end{array} = \left(\begin{array}{c}\text{mass of} \\ \text{water}\end{array} + \begin{array}{c}\text{water equivalent} \\ \text{of apparatus}\end{array}\right)$$

$$\times \begin{array}{c}\text{specific heat} \\ \text{capacity}\end{array} \times \begin{array}{c}\text{corrected} \\ \text{temperature rise}\end{array}$$

$$m_f \times CV = (m_w + m_{wc})c_w(T_2 - T_1)$$

$$CV = \frac{(m_w + m_{wc})c_w(T_2 - T_1)}{m_f}$$

$$CV = \frac{(2.75 + 1.10) \times 4187 \times 1.75}{0.68 \times 10^{-3}}$$

$$CV = 41.5 \times 10^6 \, J \, kg^{-1} \quad \text{or} \quad 41.5 \, MJ \, kg^{-1}$$

Boy's gas calorimeter

Boy's gas calorimeter is used to determine the calorific value of gaseous fuels. It contains a burner from which the hot gases pass over a set of cooling tubes. A controlled flow of water through the tubes cools the hot gases down to the gas supply temperature and thus extracts all the heat which has been released during combustion. The burner receives a metered flow of gas at a steady supply pressure and a drain at the base of the calorimeter enables the condensed steam, formed from the combustion of hydrogen, to be collected. The arrangement is shown in Figure 6.19.

Having adjusted the cooling water flow rate so that the exhaust gases emerge at approximately the ambient temperature, the cooling water inlet and outlet temperatures T_i and T_o are recorded. The gas supply pressure h_g measured in metres of water on the U-tube manometer is also recorded together with the gas supply and exhaust temperature T_s. A gas meter reading is then taken and simultaneously the measuring jars are placed in position to collect the cooling water and condensate. After a period of around 5 min the gas meter reading is again taken and the volume of gas V_s which has been supplied is calculated. The mass of cooling water m_w and the mass of condensate m_c which have been collected are also recorded.

Key point

The cooling water flow rate through Boy's gas calorimeter must be adjusted so that the exhaust gases emerge approximately at the ambient temperature.

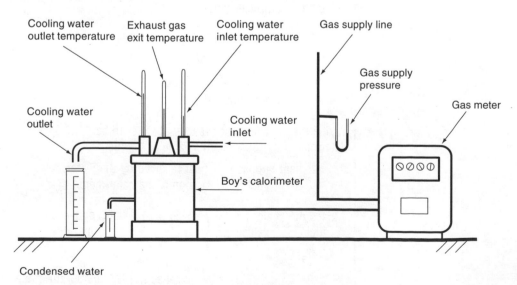

Cooling water outlet temperature

Exhaust gas exit temperature

Cooling water inlet temperature

Gas supply line

Gas supply pressure

Gas meter

Cooling water outlet

Cooling water inlet

Boy's calorimeter

Condensed water

Figure 6.19 *General arrangement of Boy's gas calorimeter*

The calorific value of a gas is measured in $MJ\,m^{-3}$ at (NTP) and some preliminary calculations are required before its value can be determined. First, the gas supply pressure p_s must be converted from metres of water into pascals, that is,

$$p_g = h_g \rho g$$

This of course is the gauge pressure of the gas and it must now be converted to absolute supply pressure p_s by adding to it the prevailing value of atmospheric pressure p_a obtained from a Fortin or precision aneroid barometer, that is,

$$p_s = p_g + p_a$$

The next task is to convert the volume of gas which has been used to the volume which it would occupy at NTP. You will recall that normal temperature, $T_n = 15\,°C$, and normal pressure, $p_n = 101.325\,kPa$. This is done using the general gas equation:

$$\frac{p_n V_n}{T_n} = \frac{p_s V_s}{T_s}$$

$$V_n = \frac{p_s V_s T_n}{p_n T_s} \qquad (6.33)$$

The higher or gross calorific value of the gas can now be calculated by equating the heat released by the gas to the heat received by the cooling water:

heat released by gas = heat received by cooling water

$$\begin{array}{c} \text{volume of} \\ \text{gas at NTP} \end{array} \times \begin{array}{c} \text{gross calorific} \\ \text{value} \end{array} = \begin{array}{c} \text{mass of cooling} \\ \text{water} \end{array} \times \begin{array}{c} \text{specific heat} \\ \text{capacity} \end{array}$$

$$\times \begin{array}{c} \text{temperature} \\ \text{rise} \end{array}$$

$$V_n \times \text{gross CV} = m_w c_w (T_o - T_i)$$

$$\textbf{gross CV} = \frac{m_w c_w (T_o - T_i)}{V_n} \qquad (6.34)$$

The lower or net calorific value of the gas can be found by subtracting the latent heat given up by the condensate from the heat received by the cooling water:

$$V_n \times \text{net CV} = m_w c_w (T_o - T_i) - m_c h_{fg}$$

$$\textbf{net CV} = \frac{m_w c_w (T_o - T_i) - m_c h_{fg}}{V_n} \qquad (6.35)$$

The value of the specific latent heat of vaporisation at NTP can be taken to be $h_{fg} = 2453\,kJ\,Kg^{-1}$.

Example 6.11

The following data were obtained during a test on a gaseous fuel using a Boy's gas calorimeter:

$$\text{volume of gas consumed} = 10\,l$$
$$\text{gas pressure} = 20\,mmH_2O$$
$$\text{gas temperature} = 20\,^\circ C$$
$$\text{atmospheric pressure} = 104.1\,kPa$$
$$\text{cooling water collected} = 2.5\,kg$$
$$\text{cooling water inlet temperature} = 14.5\,^\circ C$$
$$\text{cooling water outlet temperature} = 38.3\,^\circ C$$
$$\text{condensate collected} = 5.13\,g$$

Determine the gross and net calorific values of the gas at NTP, i.e. at 15 °C and 101.325 kPa. The specific latent heat of the condensate is 2453 kJ kg^{-1} at NTP.

Finding gauge pressure of gas:

$$p_g = h_g \rho g = 20 \times 10^{-3} \times 1000 \times 9.81$$
$$\boldsymbol{p_g = 196.2\,Pa}$$

Finding absolute supply pressure of gas:

$$p_s = p_g + p_a = 196.2 + (104.1 \times 10^3)$$
$$\boldsymbol{p_s = 104.296 \times 10^3\,Pa} \quad \text{or} \quad \boldsymbol{104.296\,kPa}$$

Finding volume occupied by gas at NTP:

$$\frac{p_n V_n}{T_n} = \frac{p_s V_s}{T_s}$$
$$V_n = \frac{p_s V_s}{p_n T_s} T_n = \frac{104.296 \times 10^3 \times 10 \times 10^{-3}(20+273)}{101.325 \times 10^3(15+273)}$$
$$\boldsymbol{V_n = 0.0105\,m^3}$$

Finding gross calorific value of gas:

$$\text{gross CV} = \frac{m_{wc} c_w (T_o - T_i)}{V_n}$$
$$\text{gross CV} = \frac{2.5 \times 4187(38.3 - 14.5)}{0.0105}$$
$$\boldsymbol{\text{gross CV} = 23.7 \times 10^6\,J\,m^3} \quad \text{or} \quad \boldsymbol{23.7\,MJ\,m^3}$$

Finding net calorific value of gas:

$$\text{net CV} = \frac{m_w c_w (T_o - T_i) - m_c h_{fg}}{V_n}$$
$$\text{net CV} = \frac{[2.5 \times 4187(38.3-14.5)] - (5.13 \times 10^{-3} \times 2453 \times 10^3)}{0.0105}$$
$$\boldsymbol{\text{net CV} = 22.5 \times 10^6\,J\,m^3} \quad \text{or} \quad \boldsymbol{22.5\,MJ\,m^{-3}}$$

Test your knowledge 6.6

1. What are the three main combustible constituents of fuels?
2. How is the calorific value of a gaseous fuel defined?
3. Distinguish between the flash point and the ignition temperature of a liquid fuel.
4. Why does the bomb calorimeter need to be charged with oxygen before firing?
5. Distinguish between the gross and net calorific values of a fuel.

Activity 6.6

The following results were obtained during a test on an oil sample using a bomb calorimeter:

mass of oil sample $= 0.70\,g$

water equivalent of apparatus $= 0.374\,kg$

mass of water in bomb and calorimeter $= 2.51\,kg$

corrected temperature rise $= 2.65\,°C$

estimated mass of steam condensate $= 0.945\,g$

Determine the gross and net calorific values of the fuel given that the specific heat capacity of water is $4187\,J\,kg^{-1}\,K^{-1}$ and specific latent heat of vaporisation of water is $2453\,kJ\,kg^{-1}$.

Combustion processes

Fuel is expensive, and inefficient combustion processes are wasteful. Furthermore, inefficient combustion results in atmospheric pollution which can lead to prosecution. To understand what happens during combustion processes you need a little knowledge of chemistry. The main combustible elements which combine with oxygen are hydrogen, carbon and sulphur. Their atomic and molecular weights, together with that of nitrogen are listed in Table 6.5.

Table 6.5 *Atomic and molecular weights*

Element	Chemical symbol	Atomic weight	Molecular weight, M
Hydrogen	H_2	1	$(2 \times 1) = 2$
Carbon	C	12	12
Sulphur	S	32	32
Oxygen	O_2	16	$(2 \times 16) = 32$
Nitrogen	N_2	14	$(2 \times 14) = 28$

Key point

The atomic weight of an element is a comparison of the weight of its atoms to those of hydrogen.

You will recall that the atomic weight of an element tells you how much heavier its atoms are than an atom of hydrogen. The atoms of the gases hydrogen, oxygen and nitrogen combine together naturally in pairs to form molecules which is why they are listed in Table 6.5 as H_2, O_2 and N_2. Their molecular weights are thus twice their atomic weights.

During combustion, hydrogen in a fuel combines with oxygen in the air to form H_2O which eventually condenses into water. When completely burnt, carbon in the fuel combines with oxygen to form CO_2, which is carbon dioxide. If there is insufficient oxygen, however, some of the carbon combines to form CO, which is carbon monoxide. This is a dangerous gas whose toxic effects have been responsible for many deaths due to badly ventilated heating systems. The presence of carbon monoxide in exhaust gas

Key point

The incomplete combustion of carbon results in the production of carbon monoxide.

Key point

By mass, air contains 77% nitrogen and 23% oxygen.

always indicates inefficient combustion. Sometimes however this is unavoidable. It is always present in motor vehicle exhaust gases but in much smaller quantities than previously, due to advances in engine management and exhaust systems.

Sulphur in fuel combines with oxygen to form SO_2, which is sulphur dioxide. This can also have undesirable effects. It combines with the water produced from the combustion of hydrogen, to form sulphurous acid which is a corrosive pollutant. Oxygen for combustion comes from the atmosphere which is made up of about 23% oxygen and 77% nitrogen by mass. There are very small amounts of other gases present but these can be neglected. Nitrogen in the air plays no part in the combustion process. It passes through the system and leaves with the exhaust gases. Its is an unwelcome passenger, since it carries heat energy away with it to the atmosphere but its presence is unavoidable. There might also be unburned oxygen present in the exhaust gases, which has the same effect.

The molecular weights of the products of combustion, H_2O, CO, CO_2 and SO_2, can be found by adding together the atomic weights of their component elements as shown in Table 6.6. Oxygen and nitrogen are also included because of their likely presence in exhaust gases.

Table 6.6 *Molecular weights of exhaust gas components*

Compound	Chemical symbol	Molecular weight, *M*
Water	H_2O	$(2 \times 1) + 16 = 18$
Carbon monoxide	CO	$12 + 16 = 28$
Carbon dioxide	CO_2	$12 + (2 \times 16) = 44$
Sulphur dioxide	SO_2	$32 + (2 \times 16) = 64$
Oxygen	O_2	$(2 \times 16) = 32$
Nitrogen	N_2	$(2 \times 14) = 28$

Stoichiometric air to fuel ratio

The word *stoichiometric* means 'chemically correct' and the *stoichiometric air to fuel ratio* is that which will in theory provide a sufficient mass of oxygen for the complete combustion of a given mass of fuel. In order to find the mass of oxygen required per kilogram of fuel, we need to know how much is needed to burn the hydrogen content, how much is needed to burn the carbon content and how much is needed to burn the sulphur content. These can then be added together to find the total amount of oxygen required per kilogram of fuel. Because we know that air is 23% oxygen and 77% nitrogen by mass, we can then calculate the total amount of air required. We now need to take a closer look at the chemical equations for the combustion of hydrogen, carbon and sulphur.

Complete combustion of hydrogen

The chemical equation for this process which results in the formation of steam is

$$2H_2 + O_2 = 2H_2O$$

This indicates that two molecules of hydrogen combine with one molecule of oxygen to form two molecules of water. The molecular weights on both sides of the equation must be equal, that is,

$$(2 \times 2) + 32 = (2 \times 18)$$
$$4 + 32 = 36$$

Dividing both sides by 4 gives

$$\mathbf{1 + 8 = 9} \tag{6.36}$$

This tells us that for the complete combustion of 1 kg of hydrogen, it requires 8 kg of oxygen, and produces 9 kg of steam.

Complete combustion of carbon

The chemical equation for this process which results in the formation of carbon dioxide is

$$C + O_2 = CO_2$$

This indicates that one atom of carbon combines with one molecule of oxygen to form one molecule of carbon dioxide.

Molecular weights: $12 + 32 = 44$

Dividing both sides by 12 gives

$$1 + \frac{32}{12} = \frac{44}{12}$$

This cancels down to

$$\mathbf{1 + \frac{8}{3} = \frac{11}{3}} \tag{6.37}$$

This tells us that for the complete combustion of 1 kg of carbon, it requires $2\frac{2}{3}$ kg of oxygen, and produces $3\frac{2}{3}$ kg of carbon dioxide.

Incomplete combustion of carbon

If there is insufficient oxygen for complete combustion, carbon monoxide will be formed. The chemical equation for this process which results in the formation of carbon monoxide is

$$2C + O_2 = 2CO$$

This indicates that two atoms of carbon combines with one molecule of oxygen to form two molecules of carbon monoxide.

Molecular weights: $(2 \times 12) + 32 = (2 \times 28)$
$$24 + 32 = 56$$

Dividing both sides by 24 gives

$$1 + \frac{32}{24} = \frac{56}{24}$$

This cancels down to

$$\mathbf{1 + \frac{4}{3} = \frac{7}{3}} \tag{6.38}$$

This tells us that 1 kg of carbon will combine with $1\frac{1}{3}$ kg of oxygen to produce $2\frac{1}{3}$ kg of carbon monoxide.

Complete combustion of sulphur

The chemical equation for this process which results in the formation of sulphur dioxide is

$$S + O_2 = SO_2$$

This indicates that one atom of sulphur combines with one molecule of oxygen to form one molecule of sulphur dioxide.

Molecular weights: $32 + 32 = 64$

Dividing both sides by 32 gives

$$1 + 1 = 2 \qquad (6.39)$$

This tells us that for the complete combustion of 1 kg of sulphur, it requires 1 kg of oxygen, and produces 2 kg of sulphur dioxide.

Suppose now that 1 kg of a fuel is made up of H kg of hydrogen, C kg of carbon and S kg of sulphur. Using the above values, the total oxygen required to completely burn the kilogram of fuel will be

$$\left(8H + \frac{8}{3}C + S\right) \text{kg of O}_2 \text{ per kg fuel} \qquad (6.40)$$

There may be oxygen of some form already present in the fuel. Let this be O kg per kilogram of fuel. We can assume that this will be used for combustion so that it requires a little less oxygen from the air. The above formula can then be written as

$$\left(8H + \frac{8}{3}C + S - O\right) \text{kg of O}_2 \text{ per kg fuel} \qquad (6.41)$$

Now, you will recall that air contains 23% of oxygen by mass and so the mass of air required per kilogram of fuel, which is the stoichiometric air to fuel ratio, will be given by

$$\frac{100}{23}\left(8H + \frac{8}{3}C + S - O\right) \text{ kg of air per kg fuel} \qquad (6.42)$$

Sometimes, and particularly with coal-fired furnaces, it is found necessary to supply more air than is theoretically required. This is because of the difficulty in mixing the fuel and air together, even when the coal has been pulverised. The percentage of excess air required varies with the type of boiler or furnace. Too much is a disadvantage as it carries heat away to the atmosphere.

Key point

The complete combustion of a fuel generally requires more air than is theoretically required.

Example 6.12

The chemical analysis of a solid fuel shows it to contain 81.5% carbon, 4.5% hydrogen, 0.8% sulphur and 2.6% oxygen. The remainder consists of incombustible solids. Determine (a) the stoichiometric air to fuel ratio, (b) the actual mass of air supplied per kilogram of fuel when 20% excess air is required for complete combustion, (c) the mass of the products of combustion per kilogram of fuel. Assume that air contains 23% oxygen and 77% nitrogen by mass.

(a) Finding stoichiometric air to fuel ratio, i.e. theoretical mass of air to burn 1 kg of fuel:

$$\text{theoretical air supply} = \frac{100}{23}\left(8H + \frac{8}{3}C + S - O\right)$$

In 1 kg of the fuel:

mass of hydrogen, $H = 0.045\,\text{kg}$

mass of carbon, $C = 0.815\,\text{kg}$

mass of sulphur, $S = 0.008\,\text{kg}$

mass of oxygen, $O = 0.026\,\text{kg}$

$$\text{theoretical air supply} = \frac{100}{23}\left[(8 \times 0.045) + \left(\frac{8}{3} \times 0.815\right)\right.$$
$$\left. + 0.008 - 0.026\right]$$

theoretical air supply = 10.6 kg of air per kg

the stoichiometric air to fuel ratio is thus 10.6 : 1.

(b) Finding actual mass of air required per kg of fuel with 20% excess:

actual mass of air required $= \dfrac{120}{100} \times 10.6 = $ **12.72 kg per kg fuel**

(c) Finding mass of steam produced per kg of fuel:
From equation (6.36), 1 kg of hydrogen produces 9 kg of H_2O

H_2O produced per kg of fuel $= 9H = 9 \times 0.045 = $ **0.405 kg**

Finding mass of carbon dioxide produced per kg of fuel:
From equation (6.37), 1 kg of carbon produces 11/3 kg of CO_2

CO_2 produced per kg of fuel $= \dfrac{11}{3}C = \dfrac{11}{3} \times 0.815 = $ **2.988 kg**

Finding mass of sulphur dioxide produced per kg of fuel:
From equation (6.39), 1 kg of sulphur produces 1 kg of SO_2

SO_2 produced per kg of fuel $= S = $ **0.008 kg**

Finding mass of unused oxygen in exhaust per kg of fuel, i.e. 23% of excess air:

$$O_2 \text{ in exhaust per kg of fuel} = \frac{23}{100} \times \text{mass of excess air}$$

O_2 in exhaust per kg of fuel $= \dfrac{23}{100} \times (12.72 - 10.6) = $ **0.488 kg**

Finding mass of nitrogen in exhaust per kg of fuel, i.e. 77% of actual air supply:

$$N_2 \text{ in exhaust per kg of fuel} = \frac{77}{100} \times \text{actual mass of air}$$

N_2 in exhaust per kg of fuel $= \dfrac{77}{100} \times 12.72 = $ **9.794 kg**

A complete analysis of the exhaust gases per kg of fuel is as follows:

Test your knowledge 6.7

1. What is the molecular weight of oxygen?
2. What is the composition of air by mass?
3. What is meant by the stoichiometric air to fuel ratio?
4. How many kilograms of CO_2 are produced when 1 kg of carbon is completely burnt?
5. Why is excess air required in some combustion processes?

Exhaust product	Mass per kg of fuel (kg)	Percentage composition (%)
H_2O	0.405	2.96
CO_2	2.988	21.84
SO_2	0.008	0.05
O_2	0.488	3.57
N_2	9.794	71.58
Total	13.683	100

Activity 6.7

The fuel oil supplied to a boiler contains 86% carbon and 14% hydrogen by mass. The boiler house temperature is 20 °C and the temperature of the exhaust gases is 350 °C. Calculate (a) the stoichiometric air to fuel ratio for the oil, (b) the air supply required if 24% excess is needed for complete combustion, (c) the mass of the products of combustion per kilogram of fuel burnt, (d) the heat energy per kilogram of fuel burnt which is carried away by the excess oxygen and the nitrogen components.

The composition of air by mass is 23% oxygen and 77% nitrogen. The specific heat capacities of oxygen and nitrogen at constant pressure are 910 J kg^{-1} K^{-1} and 1040 J kg^{-1} K^{-1}, respectively.

Problems 6.6

1. The following observations were taken during a test to determine the calorific value of a liquid fuel using a bomb calorimeter:

   mass of oil sample 515×10^{-6} kg
   mass of water in calorimeter 2.42 kg
   water equivalent of calorimeter 0.39 kg
   initial water temperature 15 °C
   final water temperature 16.9 °C

   Determine the higher or 'gross' calorific value of the fuel.
   [42.9 MJ kg^{-1}]

2. The following observations were taken during a test to determine the calorific value of a gaseous fuel using a Boy's calorimeter:

   mass of water collected 2.5 kg
   inlet water temperature 16.5 °C
   outlet water temperature 30.8 °C
   volume of gas burnt 10 l
   gas supply pressure 89 mmH$_2$O
   gas supply temperature 17 °C
   barometric pressure 780 mmHg

   Calculate the higher calorific value of the fuel per cubic metre at NTP.
   [16.6 MJ m^{-3}]

3. The analysis of a fuel by mass is 91% carbon, 7% hydrogen and 1% sulphur. The remainder being composed of incombustible solids. Determine the minimum mass of air that is theoretically required for complete combustion per kilogram of fuel.

[13.04 kg]

4. The analysis of a fuel oil revealed the following composition by mass. Carbon 84%, hydrogen 14%, sulphur 2%. Calculate (a) the minimum mass of air required for complete combustion of 1 kg of the fuel, (b) the mass of nitrogen present in the exhaust gases per kilogram of fuel burnt.

[14.7 kg, 11.3 kg]

5. A coal sample has the following analysis per kg. Carbon 0.6 kg, hydrogen 0.15 kg, sulphur 0.03 kg, oxygen 0.15 kg. The remainder is composed of incombustible solids. Determine the minimum mass of air required for combustion per kilogram of coal.

[11.7 kg]

6. A sample of fuel oil has the following analysis by mass. Carbon 84%, hydrogen 14%, oxygen 1.5% and nitrogen 0.5%. Determine (a) the minimum mass of air required per kilogram of fuel, (b) the mass of each constituent in the flue gases per kilogram of fuel burnt if 30% excess air is supplied.

[14.55 kg, CO_2 3.08 kg, H_2O 1.26 kg, O_2 1.004 kg, N_2 14.56 kg]

Properties and use of steam

In both the open and closed thermodynamic systems the transfer and conversion of energy needs a working substance. In power plant such as internal combustion engines and steam turbines its purpose is to receive heat energy from the fuel and then release it in the form of external work. Steam is an excellent working substance. It can carry large amounts of heat energy. It is produced from water which is plentiful and it is environmentally friendly. In most of our larger power stations, the electrical generators are driven by steam turbines. Steam is also widely used as a heat source in industrial processes and in hospitals for central heating and the sterilisation of equipment.

Phases of a substance

As you very well know, the substance H_2O can exist in three different states. It can exist as a solid in the form of ice, as a liquid which is water and as a gas, which is of course steam. These different states are known as *phases*. When a substance is of the same nature throughout its mass, it is said to be of a *single phase*. If two or more phases can exist together, the substance is then said to be *two-phase mixture*. In a single phase the substance is said to be *homogenous* and in a two phase mixture it is said to be *heterogeneous*.

Vapours

When a substance is in its gaseous phase it may or may not be classed as a *vapour*. A vapour is a gas which can be changed back into a liquid by compressing it. This can only happen if the gas is below a certain temperature known as its *critical temperature*. In the case of steam, the critical temperature is 374.15 °C. The substance H_2O cannot exist as water above this temperature. It can only exist as a gas which is known as *supercritical steam*. Now a word of warning. Vapours cannot be trusted to obey the gas laws and you should not use the general or polytropic gas equations when dealing with the expansion and compression of steam.

Saturation temperature (t_s°C)

When water receives heat energy, its temperature rises. This, you may recall, is known as *sensible heat* because its flow can be sensed by a temperature measuring device and that the specific heat capacity of water c_w is $4187 \, J \, kg^{-1} \, K^{-1}$. Eventually a condition is reached where the water cannot absorb any more heat energy without undergoing a change of phase. It is then said to be *saturated* with sensible heat and is known as *saturated water*. This might sound a little strange but it is one of the many terms concerned with steam generation that you will soon get used to. The temperature at which this occurs, which you know as the boiling point, is also called the *saturation temperature*, t_s °C.

The saturation temperature, or boiling point increases with pressure. At a normal pressure of 101.325 kPa or 1.01325 bar, it is exactly 100 °C. At a pressure of 10 bar, it rises to just under 180 °C and at 100 bar it is 311 °C. The highest possible boiling point of water is 374.15 °C which occurs when the pressure is 221.2 bar. This is the critical temperature above which water cannot exist as a liquid. You can find the saturation temperature at any given pressure, and a great deal more useful information, in steam property tables. In order to solve energy transfer problems where steam is the working substance, you will need to obtain a copy of these. If they are not provided by your school or college, you will unfortunately, need to beg, borrow or even purchase a set.

Specific enthalpy (*h*)

When we were investigating open thermodynamic systems, we defined enthalpy as being the sum of the internal energy and pressure-flow energy of the working substance. Steam boilers, turbines and condensers are all examples of open systems through which there is a steady flow of steam and water. From now on we shall refer to the energy of the water in a boiler, or the steam produced, as being its enthalpy.

The enthalpy per kilogram or *specific enthalpy* values, for water and steam at given temperatures and pressures are given in steam property tables. For convenience it is assumed that water at 0 °C has zero enthalpy and the listed values are thus not the total enthalpy but its value above this base line. This is perfectly acceptable since it is enthalpy changes which are of importance in thermodynamic steam calculations, and these are the same wherever the base line is taken.

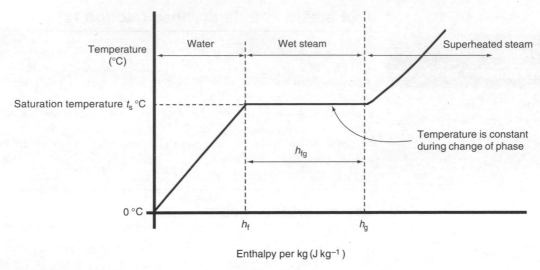

Figure 6.20 *Graph of temperature against specific enthalpy*

Figure 6.20 shows how the enthalpy of water and steam at a given pressure increases with temperature. The following subscripts indicate the condition of the working substance:

h_w = specific enthalpy of water below its saturation temperature

h_f = specific enthalpy of saturated water, i.e. water at $t_s°C$

h_{fg} = specific enthalpy of vaporisation, i.e. the specific latent heat

h_g = specific enthalpy of dry saturated steam, i.e. dry steam at $t_s°C$

h_{sup} = specific enthalpy of superheated steam

Specific enthalpy of vaporisation (h_{fg})

If saturated water continues to receive heat energy, a change of phase starts to take place. The water begins to evaporate and the temperature stays constant at the boiling point, or saturation temperature, $t_s°C$ whilst the change is taking place. The water is then receiving its latent heat of vaporisation. You have probably remembered that it is called latent heat because its flow cannot be detected by a thermometer or other temperature measuring device.

The amount of heat energy required to change 1 kg of saturated water completely into steam is of course its *specific latent heat of vaporisation*. It is also called *the specific enthalpy of vaporisation* which is given the symbol h_{fg}. The value of the specific enthalpy of vaporisation depends on the pressure at which the steam is being generated. At a normal pressure of 101.325 kPa or 1.01325 bar where the saturation temperature is 100 °C, its value is 2256.7 kJ kg^{-1}. As the pressure and saturation temperature increase, the specific enthalpy of vaporisation h_{fg} becomes less, falling to zero at the critical pressure and temperature. Its value at any pressure may be obtained from the column headed h_{fg} in steam property tables.

Wet steam and its dryness fraction (x)

As the steam bubbles rise out of the water in a boiler, they carry with them small droplets of water. This is known as *wet steam*. Wet steam has not received all of the latent heat required to change it completely to dry steam. The amount of latent heat which it has received is given by its *dryness fraction*, x. For example, if wet steam has a dryness fraction of $x = 0.9$, this means that it has received 90% of its latent heat and that one tenth of its mass will be made up of water droplets.

Dry saturated steam

This is another term which sounds rather strange. *Dry saturated steam* is steam which has just received all of its latent heat, h_{fg} so that it is dry, but still at the saturation temperature, $t_s °C$. The enthalpy per kilogram, h_g and specific volume, v_g of dry saturated steam at any given pressure is given in steam property tables. The units of specific volume are $m^3 kg^{-1}$, i.e. cubic metres per kilogram.

Superheated steam

If dry saturated steam continues to receive heat energy, its temperature will start to rise again. It is then known as *superheated steam*, which is of course a vapour until its temperature exceeds 374.15 °C, the critical temperature. Thereafter it becomes supercritical steam, as we have described. The number of degrees by which the temperature of superheated steam exceeds its saturation temperature, $t_s °C$, is known as its *degrees of superheat*. The enthalpy per kilogram, h_{sup} and specific volume, v_{sup} of superheated steam at given temperatures and pressure are given in steam property tables.

The temperature v. enthalpy diagram, Figure 6.21, shows how the various values of specific enthalpy values of saturated water, h_f, dry saturated steam, h_g, and vaporisation h_{fg}, vary with saturation temperature and pressure.

Use of steam property tables

Steam property tables are set out in particular way to assist the solution of problems concerned with steam plant. The first page contains information on notation and units. The second page contains the properties of saturated water and steam up to a saturation temperature of 100 °C. As you will find, this is particularly useful for finding the enthalpy of boiler feed water. The second page contains almost the same information but tabulated in pressure increments up to 1 bar. This is particularly useful for finding the enthalpy of condensate. Successive pages list the properties of saturated water and steam up to the critical temperature and also the properties of superheated steam.

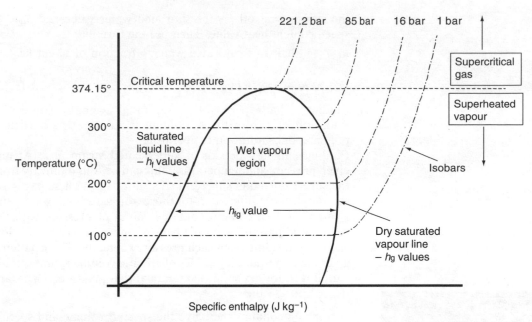

Figure 6.21 *Temperature–enthalpy diagram*

In the saturated water and steam section, the columns which you will use are the first two which list pressure, p (bar) and corresponding saturation temperature, t_s (°C), the third which is headed v_g (m^3 kg^{-1}), the corresponding specific volume of dry saturated steam, and the specific enthalpy columns which are headed h_f, h_{fg} and h_g (kJ kg^{-1}). Do not use the columns headed s_f, s_{fg} and s_g, they are for more advanced work on steam plant. Also, do not use the columns headed u_f and u_g which are internal energy values. They are useful for some closed system industrial process but should not be used for open systems.

1. Finding enthalpy per kg of boiler feed water, h_w: This can be done in two ways. The first is to calculate the energy required to heat 1 kg of the water from 0 °C up to the temperature, t °C at which it enters the boiler. The specific heat capacity of water is 4178 J kg^{-1}K^{-1} or 4.178 kJ kg^{-1} K^{-1}, that is,

   h_w = mass × specific heat capacity × temperature change

   $h_w = 1 \times c_w \times (t - 0)$

   $$h_w = c_w t = 4.187\,t \; (\text{kJ kg}^{-1}) \tag{6.43}$$

   The second method, which is quicker and easier, is to say that the enthalpy of the feed water at any particular temperature, t °C is the same as it would be if that were its boiling point. All you then need to do is look in tables for the h_f value at that particular temperature. That is,

   $$h_w = h_f \text{ value for } t_s \text{ equal to } t \, (\text{kJ kg}^{-1}) \tag{6.44}$$

2. Finding enthalpy per kg of saturated water, h_f: This can be found directly from the h_f column in tables at its particular pressure and saturation temperature.

3. Finding enthalpy per kg of wet steam, h_{wet} whose dryness fraction is x: At the particular temperature and pressure of

the steam, read off h_f, the saturated water value and h_{fg}, the specific latent heat value. Then use the formula

h_{wet} = enthalpy of saturated water + fraction of latent heat received

$$h_{wet} = h_f + x\,h_{fg}\,(kJ\,kg^{-1}) \tag{6.45}$$

4. Finding enthalpy per kg of dry saturated steam, h_g: This can be found directly from the h_g column in tables at its particular pressure and saturation temperature.
5. Finding enthalpy per kg of superheated steam h_{sup}: At any given pressure and temperature, this can be found directly from the superheated and supercritical steam pages. These pages are set out rather differently to those for saturated water and steam. The first column contains values of pressure with the corresponding saturation temperature given in brackets below each one. Leading from each pressure value, the second column gives the corresponding values of specific volume v_g and specific enthalpy h_g for dry saturated steam at that pressure. (Disregard the u_g and s_g values.)

Succeeding columns then give the specific volume and specific enthalpy values at temperatures rising in $50\,°C$ and later in $100\,°C$ increments. If you are confronted with superheated steam whose temperature is somewhere between these values, then you will have to do a little bit of interpolation to find its specific volume and enthalpy.

Key point

Interpolation involves calculating intermediate values as accurately as possible. It is not guesswork.

Calculation of energy transfer

As we have stated, there is a continuous flow of water and steam through the various components of a steam plant. The boiler, superheater, turbine and condenser in a power generation plant are all examples of open thermodynamic systems and you may recall that the shortened form of the energy equation for an open system was given by equation (6.29). That is,

heat transfer = change of enthalpy + work done
$$Q = (H_2 - H_1) + W$$

This describes the energy changes in a mass of m kg of a working substance. In steam plant calculations it is usual to consider the energy changes to a mass 1 kg as it passes through a system. Lower case letters are used to indicate that a specific mass of 1 kg is being considered. That is,

$$q = (h_2 - h_1) + w \tag{6.46}$$

When this equation is applied to a boiler, superheater or condenser there is no work input or output from the system and so the term work, w is zero. For these components, it can thus be written that the heat transfer is equal to the change of enthalpy. That is,

$$q = (h_2 - h_1) \tag{6.47}$$

In the case of a steam turbine, it is assumed that adiabatic expansion takes place within it so that there is no heat transfer.

If this is the case then the term heat transfer term, q is zero which gives

$$0 = (h_2 - h_1) + w$$
$$-w = (h_2 - h_1)$$

or

$$w = (h_1 - h_2) \tag{6.48}$$

Example 6.13

A boiler takes in feed water at a temperature of 75 °C and produces wet steam of dryness fraction 0.93 at a pressure of 15 bar. The wet steam then passes through a superheater where it receives heat at constant pressure and emerges at a temperature of 400 °C. Determine (a) the temperature of the steam as it enters the superheater, (b) the heat energy per kilogram received in the boiler, (c) the heat energy per kilogram received in the superheater, (d) the total heat energy received per kilogram of steam.

(a) Finding temperature of wet steam:

temperature of wet steam = saturation temperature at
a pressure of 15 bar

From tables, t_s at 15 bar = 198.3 °C

(b) Finding enthalpy per kg of boiler feed water, h_1:
     By the calculation method,

$$h_1 = 4.187 \times t = 4.187 \times 75$$
$$\mathbf{h_1 = 314 \, kJ \, kg^{-1}}$$

Alternative method from tables,

$$h_1 = h_f \text{ at } 75 \,°C$$
$$\mathbf{h_1 = 314 \, kJ \, kg^{-1}}$$

Finding enthalpy per kg of wet steam leaving boiler, h_2:

$$h_2 = h_f + xh_{fg} \text{ at 15 bar pressure and 0.93 dry}$$
$$h_2 = 845 + (0.93 \times 1947)$$
$$\mathbf{h_2 = 2656 \, kJ \, kg^{-1}}$$

Finding heat energy per kg received in boiler, q_{1-2}:

$$q_{1-2} = h_2 - h_1 = 2656 - 314$$
$$\mathbf{q_{1-2} = 2342 \, kJ \, kg^{-1}}$$

(c) Finding enthalpy per kg of superheated steam, h_3:

$$h_3 = h_{sup} \text{ at 15 bar and } 400 \,°C$$
$$\mathbf{h_3 = 3256 \, kJ \, kg^{-1}}$$

Finding heat energy per kg received in superheater, q_{2-3}:

$$q_{2-3} = h_s - h_2 = 3256 - 2342$$

$$\boldsymbol{q_{2-3} = 914\,kJ\,kg^{-1}}$$

(d) Finding total heat energy received by steam, q_{1-3}:

$$q_{1-3} = q_{1-2} + q_{2-3} = 3256 + 914$$

$$\boldsymbol{q_{1-3} = 4179\,kJ\,kg^{-1}}$$

Example 6.14

The high pressure stage of a turbine receives superheated steam at a pressure of 50 bar and temperature 450 °C. The steam leaves at a pressure of 20 bar and temperature 250 °C at which it enters the low pressure stage and exhausts as wet steam of dryness fraction 0.85 into a condenser at a pressure of 0.2 bar. Determine (a) the work done per kilogram of steam in the high pressure stage, (b) the work done per kilogram of steam in the low pressure stage, (c) the total work done per kilogram of steam.

(a) Finding enthalpy per kg of steam entering high pressure stage, h_1:

$$h_1 = h_{sup} \text{ at 50 bar and 450 °C}$$

$$\boldsymbol{h_1 = 3316\,kJ\,kg^{-1}}$$

Finding enthalpy per kg of steam leaving high pressure stage, h_2:

$$h_2 = h_{sup} \text{ at 20 bar and 250 °C}$$

$$\boldsymbol{h_2 = 2904\,kJ\,kg^{-1}}$$

Finding work done per kg of steam in high pressure stage, w_{1-2}:

$$w_{1-2} = (h_1 - h_2) = 3316 - 2904$$

$$\boldsymbol{w_{1-2} = 412\,kJ\,kg^{-1}}$$

(b) Finding enthalpy per kg of wet steam exhausting to condenser, h_3:

$$h_3 = h_f + x h_{fg} \text{ at 0.2 bar and 0.85 dry}$$

$$h_3 = 251 + (0.85 \times 2358)$$

$$\boldsymbol{h_3 = 2255\,kJ\,kg^{-1}}$$

Finding work done per kg of steam in low pressure stage, w_{2-3}:

$$w_{2-3} = (h_2 - h_3) = 2904 - 2255$$

$$\boldsymbol{w_{2-3} = 649\,kJ\,kg^{-1}}$$

Finding total work done per kg of steam, w_{1-3}:

$$w_{1-3} = w_{1-2} + w_{2-3} = 412 + 649$$

$$\boldsymbol{w_{1-3} = 1061\,kJ\,kg^{-1}}$$

Throttling

Throttling occurs when steam at high pressure is forced through a small orifice or throttling valve to emerge at a lower pressure as in Figure 6.22. The steam also emerges at a lower temperature, which may be above the saturation temperature at this lower pressure. If the fluid up-stream of the orifice is wet steam of dryness fraction x, the effect of throttling is to cause the water droplets to evaporate. This can give superheated steam or wet steam with a higher dryness fraction, depending on the pressure drop. Throttling can thus be used as a way of drying out or superheating the steam. Because the process takes place very quickly there is no loss of energy to the surroundings and the enthalpy of the steam before and after throttling is the same.

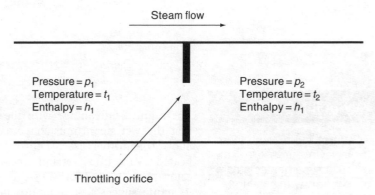

Figure 6.22 *Throttling process*

Throttling calorimeter

The throttling calorimeter is a device used to determine the dryness fraction of wet steam by throttling it down to atmospheric pressure as shown diagrammatically in Figure 6.23.

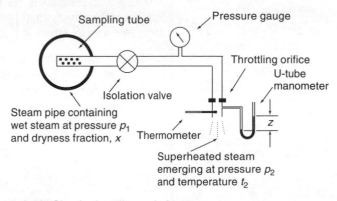

Figure 6.23 *Simple throttling calorimeter*

The isolation valve is opened allowing wet steam to flow through the throttling orifice to the atmosphere. When the flow is steady, the pressure of the wet steam is recorded from the pressure gauge and the pressure and temperature of the emerging steam are obtained from the manometer and thermometer. The enthalpy values, h_1 and h_2, before and after throttling are given by

$$h_1 = h_f + x\,h_{fg} \text{ at pressure } p_1$$

and

$$h_2 = h_{sup} \text{ at pressure } p_2 \text{ and temperature } t_2$$

The values of h_f and h_{fg} before throttling and the value of h_{sup} after throttling can all be obtained from steam property tables. Assuming that there are no energy losses during throttling, it can be written that,

$$\text{enthalpy per kg before throttling} = \text{enthalpy per kg after throttling}$$

That is,

$$h_1 = h_2$$
$$h_f + x\,h_{fg} = h_{sup}$$
$$x = \frac{h_{sup} - h_f}{h_{fg}}$$

$$(6.49)$$

Key point

The throttling of wet steam to a lower pressure can produce superheated steam or steam with a higher dryness fraction.

The main disadvantage with the simple throttling calorimeter is that the wet steam must have a high dryness fraction, typically above 0.95, otherwise the throttled steam will not emerge superheated. A modified version, known as a throttling and separating calorimeter, is used for wet steam with a high moisture content. As its name suggests, it separates out some of the moisture before throttling so that the steam will emerge in superheated condition.

Example 6.15

Wet steam in a pipeline has a pressure of 9 bar and is sampled by a simple throttling calorimeter. Its temperature after throttling is 110 °C and its pressure is 100 kPa. Determine (a) the dryness fraction of the wet steam, (b) the least dryness fraction which can be measured for wet steam at this initial pressure.

(a) Finding the enthalpy, h_{sup} of the emerging superheated steam by interpolating between the values at 100 °C and 150 °C at a pressure of 1 bar and temperature 110 °C:

$$h_{sup} \text{ at } 110\,°C = \frac{\text{value at}}{100\,°C} + \frac{1/5 \text{ of difference between}}{\text{values at 150 and 100}\,°C}$$

$$h_{sup} = 2676 + \frac{10}{50}(2777 - 2676)$$

$$\mathbf{h_{sup} = 2696\,J\,kg^{-1}}$$

Finding dryness fraction of wet steam sample:

$$\text{enthalpy per kg before throttling} = \text{enthalpy per kg after throttling}$$

That is,

$$h_1 = h_2$$
$$h_f + xh_{fg} \text{ at } 9\,\text{bar} = h_{sup} \text{ at } 1\,\text{bar and } 110\,°C$$
$$x = \frac{h_{sup} - h_f}{h_{fg}} = \frac{2696 - 743}{2031}$$
$$x = 0.962$$

(b) Finding least dryness fraction which can be measured for wet steam at this initial pressure. This will occur when the throttled steam emerges in dry saturated condition with enthalpy, h_g at a pressure of 1 bar.

enthalpy per kg before throttling = enthalpy per kg after throttling

i.e.

$$h_1 = h_2$$
$$h_f + xh_{fg} \text{ at } 9\,\text{bar} = h_g \text{ at } 1\,\text{bar}$$
$$x = \frac{h_g - h_f}{h_{fg}} = \frac{2675 - 743}{2031}$$
$$x = 0.951$$

Activity 6.8

A boiler receives feed water at a temperature of 85 °C and produces wet steam at a pressure of 20 bar. The steam is sampled by a simple throttling calorimeter and emerges with a temperature of 112 °C at a pressure of 1 bar. The wet steam passes through a superheater where its temperature rises to 500 °C at constant pressure and is then used to drive to a turbine. The steam finally exhausts to a condenser at a pressure of 0.1 bar with a dryness fraction of 0.85. Determine (a) the dryness fraction of the wet steam leaving the boiler, (b) the total heat energy received per kilogram of steam, (c) the heat energy received per kilogram of steam in the superheater given as a percentage of the total heat received, (d) the work done per kilogram of steam in the turbine.

Problems 6.7

1. A heat exchanger takes in feed water at 30 °C and generates dry saturated steam at a pressure of 4 bar. Determine the heat energy received per kilogram of steam produced.

$$[2613\,\text{kJ kg}^{-1}]$$

2. A boiler takes in feed water at a temperature of 70 °C generates wet steam at a pressure of 6 bar and dryness fraction of 0.9. Determine (a) the temperature of the steam, (b) the heat energy received per kilogram.

$$[158.8\,°C,\ 2254\,kJ\,kg^{-1}]$$

3. A superheater takes in wet steam at a pressure of 8 bar and 0.92 dryness fraction. The steam emerges at the same pressure but with 300 °C of superheat. What is the final temperature of the steam and the amount of heat energy received per kilogram?

$$[470.4\,°C,\ 812\,kJ\,kg^{-1}]$$

4. A boiler and superheater generate steam at a pressure of 10 bar and 300 °C. The boiler takes in feed water at 50 °C and the steam enters the superheater with a dryness fraction of 0.9. Calculate the heat supplied in the superheater as a percentage of the total heat supplied.
(Note: The pressure is the same in the boiler and superheater.)

$$[15.3\%]$$

5. Steam enters a turbine at a pressure of 10 bar and with 350 °C of superheat. It exhausts at a pressure of 5 bar with 98 °C of superheat. Determine (a) the work done per kilogram of steam, (b) the temperature change that occurs.

$$[582\,kJ\,kg^{-1},\ 280\,°C]$$

6. Wet steam enters a simple throttling calorimeter at a pressure of 10 bar and emerges at a pressure of 1 bar and temperature 135 °C. Determine the dryness fraction of the wet steam.

$$[0.984]$$

Power generating plant

By far the greater part of our electricity is produced by power stations in which the generators are powered by steam turbines. An approximate breakdown of the generating capacity in the United Kingdom is 37% coal fired, 31% gas fired, 25% nuclear, 2% oil fired and 5% from renewable sources such as hydro-electric and wind power. Whatever the heat source, steam generating plants have the same major components. A typical arrangement is shown in Figure 6.24.

Boilers and superheaters

In coal, gas and oil fired systems, the fuel and air enter the boiler where the hot gases from combustion heat the feed water to produce wet steam. There are two basic kinds of boiler, the *fire tube* type and the *water tube* type. In fire tube boilers the hot gases from combustion pass through a system of tubes around which water is circulating. These are usually to be found in small installations where low pressure steam is required for industrial processes and space heating. In water tube boilers, the hot gases from combustion circulate around a system of tubes containing water. This is the type used in power stations for producing large quantities of high pressure steam.

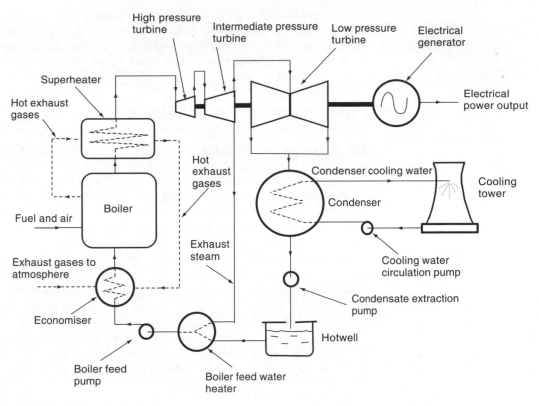

Figure 6.24 *Steam plant circuit*

The wet steam passes through a system of tubes in the *super-heater* where additional heat energy is supplied from the combustion gases to produce superheated steam. Every possible unit of heat energy is extracted from the exhaust gases and before escaping to the atmosphere, they are used to heat up the boiler feed water in the *economiser*. They are also used to pre-heat the incoming air, but this is not shown in Figure 6.24.

In nuclear installations, the heat source is enriched uranium. This is bombarded with neutrons in the reactor vessel causing some of the uranium atoms to split and release heat energy. The process is called *nuclear fission*. There are many different designs of reactor in operation throughout the world. In Britain pressurised carbon dioxide is used to transfer heat energy from the reactor to the boilers, superheaters and economisers whilst in American designs, pressurised water is preferred.

Turbines

The superheated steam passes to the *high pressure turbine* where it expands and does work on the rotor blades. It then passes to the *intermediate pressure turbine* where the blades have a larger diameter. Here it continues to expand and do work. You will note that some of the exhaust steam from the intermediate pressure turbine is fed to the boiler feed water heater where it is injected into the feed water from the hotwell. The remainder of the steam from the intermediate pressure turbine passes to the *low pressure turbine*. Here the blades are of a still larger diameter and arranged so that the steam enters centrally, as shown in Figure 6.24, and expands outwards through the two sets of low pressure blades. All three turbines are connected by a common shaft which drives the electrical generator.

Condensers and feed heaters

The exhaust steam from the low pressure turbine passes to the condenser as low pressure wet steam. There are two basic types of condenser. In the *spray type*, cooling water is injected into the steam causing it to condense. In the *surface type*, the steam condenses on the surface of a system of pipes through which there is a flow of cooling water. The surface type is used in all large power stations. Sea water and river water are used for cooling wherever practical and a great many of our nuclear and gas fired power stations are sited on the coast. Coal fired power stations tend to be sited inland near the remaining coal fields, to reduce transportation costs. Here the cooling water for the condensers is generally re-circulated through cooling towers as shown in Figure 6.24.

When the steam condenses it occupies a much smaller volume and as a result, the pressure in the condenser is well below atmospheric pressure. This is beneficial because it creates as large a pressure drop as is possible across the low pressure turbine, allowing steam to expand freely and do the maximum possible amount of work. The condensed steam must however be extracted from the condenser by the *condensate extraction pump*.

The condensate passes to a reservoir called the *hotwell* where make-up water is added for evaporation losses. The feed water from the hotwell is heated first in the *feed water heater* by exhaust steam, and then in the economiser by the exhaust gases from the boiler. *The boiler feed pump* delivers the feed water through the economiser to the boiler. The objective of the feed water heaters is to raise the temperature of the water to as close to its saturation pressure as possible before it enters the boiler.

Power rating

The output power of a boiler is the heat energy received per second by the feed water as it is changed into steam.

$$\frac{\text{power}}{\text{rating}} = \frac{\text{heat energy received}}{\text{per kg of steam (J kg}^{-1})} \times \frac{\text{steam flow}}{\text{rate (kg s}^{-1})}$$

If the initial and final enthalpy values of the feed water and steam are h_1 and h_2, this can be written as

$$\frac{\text{boiler power}}{\text{rating}} = \frac{\text{change of enthalpy}}{\text{per kg of steam (J kg}^{-1})} \times \frac{\text{steam flow}}{\text{rate (kg s}^{-1})}$$

boiler power rating $= (h_2 - h_1)m_s \ (W)$ (6.50)

The output power of a turbine is its shaft work output per second. This is the product of its output torque, T (N m) and angular velocity, ω (rad s^{-1}).

$$\text{shaft output power} = \text{output torque (N m)}$$
$$\times \text{angular velocity (rad s}^{-1})$$

$$\textbf{shaft output power} = \textbf{\textit{T}}\boldsymbol{\omega}\textbf{ (W)} \tag{6.51}$$

Thermal and mechanical efficiency (η)

The thermal efficiency of a boiler gives a comparison of the heat energy received per second by the water and steam and the heat energy available per second in the fuel.

$$\text{thermal efficiency of boiler} = \frac{\begin{array}{c}\text{heat energy received}\\ \text{per second by water and steam}\end{array}}{\begin{array}{c}\text{heat energy available}\\ \text{per second in fuel}\end{array}}$$

If the fuel consumption rate is m_f kilograms per second and its calorific value is CV then,

$$\textbf{Thermal }\boldsymbol{\eta}\textbf{ of boiler} = \frac{(h_2 - h_1)m_s}{m_f\ \text{CV}} \tag{6.52}$$

The mechanical efficiency of a turbine gives a comparison of the work done per second by the expanding steam and the actual shaft output power after losses in the blades and friction losses.

$$\text{mechanical efficiency of turbine} = \frac{\text{shaft output power}}{\begin{array}{c}\text{work done per second}\\ \text{by steam}\end{array}}$$

If the initial and final enthalpy values of the steam as it passes through the turbine are h_1 and h_2, this can be written as

$$\textbf{mechanical }\boldsymbol{\eta}\textbf{ of turbine} = \frac{T\omega}{(h_1 - h_2)m_s} \tag{6.53}$$

The overall thermal efficiency of a steam plant gives a comparison of the heat energy available per second in the fuel and the actual shaft output power.

$$\text{overall thermal efficiency of plant} = \frac{\begin{array}{c}\text{shaft output power}\\ \text{from turbine}\end{array}}{\begin{array}{c}\text{heat energy available}\\ \text{per second in fuel}\end{array}}$$

$$\textbf{overall thermal }\boldsymbol{\eta}\textbf{ of plant} = \frac{T\omega}{m_f\ \text{CV}} \tag{6.54}$$

When using these formulae you should be careful to use the correct units. Steam flow rates and fuel consumption are often given in tonnes per hour which need to be converted to kg s^{-1}.

Calorific values are often given in MJ kg^{-1} and the enthalpy values from steam property tables are given in kJ kg^{-1}. Both need to be converted to J kg^{-1}.

Example 6.16

A coal fired steam plant takes in feed water at a temperature of 80 °C and produces 12.5 tonnes of steam per hour at a pressure of 60 bar and temperature 400 °C. The fuel consumption rate is 1.25 tonnes per hour and the calorific value of the fuel is 36 MJ kg^{-1}. Determine the power rating of the boiler and its thermal efficiency.

Finding enthalpy of feed water, h_1 and output steam, h_2:

$h_1 = h_f$ value at 80 °C

$h_1 = 335\,\text{kJ kg}^{-1}$

$h_2 = h_{sup}$ at 60 bar and 400 °C

$h_2 = 3177\,\text{kJ kg}^{-1}$

Finding steam flow rate in kg s^{-1}:

$$m_s = \frac{12.5 \times 10^3}{60^2}$$

$m_s = 3.47\,\text{kg s}^{-1}$

Finding power rating of boiler:

boiler power rating $= (h_2 - h_1)m_s$

boiler power rating $= (3177 - 335) \times 10^3 \times 3.47$

boiler power rating $= 9.86 \times 16^6$ W   or   9.86 MW

Finding fuel consumption in kg s^{-1}:

$$m_f = \frac{1.25 \times 10^3}{60^2}$$

$m_f = 0.347\,\text{kg s}^{-1}$

Finding thermal efficiency of boiler:

$$\text{thermal } \eta \text{ of boiler} = \frac{(h_2 - h_1)m_s}{m_f\,\text{CV}} = \frac{9.86 \times 10^6}{0.347 \times 36 \times 10^6}$$

thermal η of boiler $= 0.789$ or   78.9%

Example 6.17

A steam turbine set receives steam at a pressure of 80 bar and temperature 425 °C and exhausts to a condenser at a pressure of 0.09 bar and 0.89 dry. The steam flow rate is 11.5 tonnes per hour and the shaft output power is 2.0 MW. Determine (a) the thermal efficiency of the turbine, (b) the overall thermal efficiency of the plant if its fuel consumption is 950 kg of coal per hour with a calorific value of 35 MJ kg^{-1}.

(a) Finding initial enthalpy of steam, h_1:

$h_1 = h_{sup}$ at 80 bar and 425 °C

$h_1 = 3207\,\text{kJ kg}^{-1}$

Finding enthalpy of exhaust steam, h_2:

$h_2 = h_f + xh_{fg}$ at a pressure of 0.09 bar and 0.89 dry
$h_2 = 183 + (0.89 \times 2397)$

$\mathbf{h_2 = 2316\ kJ\ kg^{-1}}$

Finding steam flow rate in $kg\ s^{-1}$:

$$m_s = \frac{11.5 \times 10^3}{60^2}$$

$\mathbf{m_s = 3.19\ kg\ s^{-1}}$

Finding mechanical efficiency of turbine:

thermal η of turbine $= \dfrac{T\omega}{(h_1 - h_2)\,m_s}$

where $T\omega = 2.0\ MW$, the given output power of the turbine.

mechanical η of turbine $= \dfrac{2.0 \times 10^6}{(3207 - 2316) \times 10^3 \times 3.19}$

mechanical η of turbine $= 0.704$ or   70.4%

(b) Finding fuel consumption in $kg\ s^{-1}$:

$$m_f = \frac{950}{60^2}$$

$\mathbf{m_f = 0.264\ kg\ s^{-1}}$

Finding overall thermal efficiency of plant:

overall thermal η of plant $= \dfrac{T\omega}{m_f\ CV}$

overall thermal η of plant $= \dfrac{2.0 \times 10^6}{0.264 \times 35 \times 10^6}$

overall thermal η of plant $= 0.216$ or   21.6%

Test your knowledge 6.9

1. What are the two basic types of boiler used in steam plant?
2. How does the feed water heater receive a supply of heat energy?
3. What are the two basic types of condenser used in steam plant?
4. What is the function of the economiser in a steam plant and from where does it receive a supply of heat energy?
5. What is the purpose of cooling towers in steam generating plant?

Activity 6.9

An oil-fired steam generating plant raises 15 tonnes of steam per hour from feed water at a temperature of 75 °C. The fuel consumption rate is 1.5 tonnes per hour with a calorific value of 42 MJ kg^{-1}. The steam enters the high pressure turbine stage at 70 bar and temperature 500 °C and exhausts to the condenser at a pressure of 0.08 bar and 0.89 dry. The mechanical efficiency of the turbine is 72%. Determine (a) the power output from the boiler, (b) the boiler efficiency, (c) the power output from the turbine, (d) the overall thermal efficiency of the plant.

Problems 6.8

1. A boiler takes in feed water at 80 °C and generates steam at a pressure of 7 bar and dryness fraction 0.95. If the steam generation rate is 2 tonnes per hour determine the power rating of the boiler.

    [1.29 MW]

2. A boiler takes in feed water at 75 °C and generates steam at a pressure of 20 bar and temperature 275 °C. The calorific value of the fuel used is $30 \, MJ \, kg^{-1}$ and the fuel consumption rate is 1.18 tonnes per hour. If the steam generation rate is 10 tonnes per hour determine the boiler efficiency.

    [75%]

3. Steam enters a turbine at a pressure of 20 bar and temperature 300 °C and exhausts at a pressure of 1 bar and dryness fraction 0.95. The steam consumption rate is 18.3 tonnes per hour and the shaft output power is 2 MW. Determine the turbine efficiency.

    [85%]

4. Superheated steam enters the high pressure stage of a turbine set at a pressure of 70 bar and temperature 500 °C. It leaves for the low pressure stage at a pressure of 15 bar from which it exhausts at a pressure of 1.4 bar and 0.98 dryness fraction. Determine (a) the total work done per kilogram of steam, (b) the steam temperature at the entry to the low pressure stage given that the two stages develop the same output power.

    [765 KJ kg^{-1}, 295 °C]

5. The boiler of an oil fired steam generating plant takes in feed water at 70 °C and produces steam at a pressure of 80 bar and temperature 400 °C. The steam flow rate is 12 tonnes per hour and the fuel consumption rate is 1.1 tonnes per hour. The calorific value of the fuel oil is $43 \, MJ \, kg^{-1}$. The steam exhausts from the turbine, which is 85% efficient, at a pressure of 0.075 bar and dryness fraction 0.8. Determine (a) the power rating of the boiler, (b) the boiler efficiency, (c) the power output from the turbine, (d) the overall thermal efficiency of the plant.

    [9.49 MW, 72.2%, 2.96 MW, 22.5%]

Index